4th International Conference on Renewable Energy Technologies (ICRET 2018)

IOP Conference Series: Earth and Environmental Science
Volume 168

Kuala Lumpur, Malaysia
16 – 18 January 2018

ISBN: 978-1-5108-6584-6
ISSN: 1755-1307

Printed from e-media with permission by:

Curran Associates, Inc.
57 Morehouse Lane
Red Hook, NY 12571

Some format issues inherent in the e-media version may also appear in this print version.

Printed by Curran Associates, Inc. (2018)

Institute of Physics
Dirac House, Temple Back
Bristol BS1 6BE UK

Phone: 44 1 17 929 7481
Fax: 44 1 17 920 0979

techtracking@iop.org

Additional copies of this publication are available from:

Curran Associates, Inc.
57 Morehouse Lane
Red Hook, NY 12571 USA
Phone: 845-758-0400
Fax: 845-758-2633
Email: curran@proceedings.com
Web: www.proceedings.com

Table of contents

Volume 168

2018 the 4th International Conference on Renewable Energy Technologies (ICRET 2018)

16–18 January 2018, Kuala Lumpur, Malaysia

Accepted papers received: 15 June 2018
Published online: 2 July 2018

Preface

Papers

Power System and Energy Technology

Power Machinery and Electromechanical Systems

Renewable Energy and Energy Engineering

Energy Planning and Management

PREFACE

2018 4th International Conference on Renewable Energy Technologies (ICRET218) held in University of Malaya, Kuala Lumpur, Malaysia during January 16-18, 2018. It is dedicated to discuss and resolve issues related to Renewable Energy Technologies. The major goal of the conference is to bring academic scientists, engineers, industry researchers together to exchange and share their experiences and research results, and discuss the practical challenges encountered and the solutions adopted. This conference is sponsored by University of Malaya, Widyatama University, Indonesia and ASEPTEC SDN. BHD Company.

The proceedings present a selection from papers submitted to the conference from universities, research institutes and industries. 48 papers were accepted from submissions of 15 countries including India, Kuwait, Oman, Switzerland, Kingdom of Bahrain, Indonesia, Australia, China, Saudi Arabia, South Korea, Malaysia, Singapore, Russia, Thailand, Algeria, provide researches in Renewable energy and new energy development and Power System and Energy Technology. All of the papers were subjected to peer-review by conference committee members and international reviewers. The papers selected depended on their quality and their relevancy to the conference. The selection of papers was based on the quality and relevance to the conference. Here we would like to thank all the technical program committee members who made great efforts on paper reviewing.

This conference program is highlighted by our invited speakers to share with us their researches, and they are Prof. Chan Siew Hwa from Nanyang Technology University, Singapore; Prof. Saad Mekhilef from University of Malaya, Malaysia; Prof. Chul H. Jo from Inha University, South Korea; Prof. Ir. Dr. Masjuki Bin Haji Hassan from University of Malaya, Malaysia; Prof. Jerry B. Superales from JH Cerilles State College, Philippines; Prof. Dr. Hazlie Mokhlis from University of Malaya, Malaysia. In order to provide more opportunities for the delegates to communicate with each other face to face, the conference organizing group arranged Lab visiting on January 16, 2018 in University of Malaya. The participants visited 3 labs. Here we sincerely express our gratitude to the staff in labs.

Last but not least, our appreciation also goes to the chairs, technical program committee members, the invited speakers, sponsors, and conference secretary, who made great effort to make the conference run smoothly.

Next year, the conference will be held it in Seoul, Korea. And we look forward to seeing you in Seoul!

Program Chair
Prof. JB Superales,
JH Cerilles State College, Philippines
March 13, 2018

IOP Conf. Series: Earth and Environmental Science **168** (2018) 011002 doi:10.1088/1755-1315/168/1/011002

COMMITTEES

Conference Chairs

Prof. Dr. Saad Mekhilef, University of Malaya, Malaysia
Prof. Chan Siew Hwa, Nanyang Technology University, Singapore

Program Chairs

Prof. Hazlie Mokhlis, University of Malaya, Malaysia
Prof.Chul H. Jo, Inha University, South Korea
Prof. JB Superales, JH Cerilles State College, Philippines

Local Organizing Committee

Mr. Leong Wen Chek, University of Malaya, Malaysia
Mr. Nasser Sabeur, University of Malaya, Malaysia

Technical Committee

Dr. Yang Xi, Diodes Inc., United States
Dr. Mehrdad Ahmadi Kamarposhti, Islamic Azad University, Iran
Prof. Madhusudhanarao Gurrala, TKR College of Engineering and Technology, India
Prof. Iyiola Olusola Oluwaleye, Ekiti State University, Nigeria
Prof. Madhusudhanarao Gurrala, TKR College of Engineering and Technology, India
Assoc. Prof. Ghaleb Ibrahim, American University in Dubai,United Arab Emirates
Lecturer Sheikh Abujar, Daffodil International University, Bangladesh
Prof. Dr. Husham Ahmed, AMA International University, Bahrain
Dr. Long Shi, RMIT University, Australia
Dr.Jin Weixin, Environmental and Water Technology Centre of Innovation, Ngee Ann Polytechnic, Singapore
Assoc. Prof. Dr. U. K. SINHA, National Institute of Technology Jamshedpur, India
Dr. S. R. Jufar, Universiti Teknologi PETRONAS, Malaysia
Dr. Sajid Ali Ansari, Dongguk University, Korea
Dr.Vibha Vyas, college of Engineering Pune, India
Dr. Agus Didit Haryanto,Universitas Padjadjaran (UNPAD), Indonesia
Dr. Azman Ismail, Universiti Kuala Lumpur, Malaysia

Peer review statement

All papers published in this volume of *IOP Conference Series: Earth and Environmental Science* have been peer reviewed through processes administered by the proceedings Editors. Reviews were conducted by expert referees to the professional and scientific standards expected of a proceedings journal published by IOP Publishing.

IOP Conf. Series: Earth and Environmental Science **168** (2018) 012001 doi:10.1088/1755-1315/168/1/012001

A Nature Inspired Algorithm for location and Performance analysis of UPFC

G MadhusudhanaRao[1] and Srinivas Singirikonda[2]

[1]IEEE Member, Professor, TKR College of Engineering and Technology(autonomous), Hyderabad, India
[2]IEEE Member, Professor, TKR College of Engineering and Technology(autonomous), Hyderabad, India

Email:gurralamadhu@gmail.com , s.srinivas241@gmail.com

Abstract: The paper presents a PSO based approach for optimal location of UPFC in transmission system. Two objectives reduction of power loss and improving the bus voltages are considered in this paper. In this regard the power flow injection model of UPFC is derived and injected into load flow studies. In this work the effect of placement of UPFC at optimal location on the system parameter is studied. The effectiveness of the method is tested on IEEE 30 bus system. The results obtained shoe that the proposed method is capable of finding the optimal location and improving the system stability.

1. Introduction

There is a vast increase in the power flow transactions due to power system restructuring. High cost and environmental are major hurdles for expansion of power transmission network, which provokes the need for study of unused potential of the transmission system capacity. FACT devices can increase power transfer capability reduce system losses, and stability. There are many advanced approaches proposed in the literature for optimizing location FACT devices and their parameter settings.

Unified Power Flow Controller (UPFC), Phase Shifting Transformer (PST) and Optimal UPFC can be utilized to control the power flow in the lines by changing their parameters to achieve various objectives. FACTS devices can control steady state power flow as well as system parameters in dynamic state [1-4]. Without changing the generation schedule and topology of power system network, the power flow can be controlled by placing the FACTS devices in appropriate locations [5]. There is an increased interest in FACTS due to the development in modern power electronic devices [6] combined with deregulation of power industry. The power flow control is a cost-effective means of dispatching specified power transactions. FACTS devices can relieve the system from congestion and help in utilizing the maximum capacity of the transmission network without threatening the stability and network security. There are several methods to find optimal locations of specified type of FACT device. However, there is no generalized approach for placement of any type of FACT device. This paper presents a generalized method to determine ideal location for placement of any FACT device with a fixed parameter set. According to the proposed method, the objective function is differentiated with respect to the parameter of the corresponding FACT device to be optimized. The basic concept of the generalized method is initially identifying the control parameters of the respective FACT devices and then to determine sensitivity index with FACT device located in each line. Sensitivity index is

IOP Conf. Series: Earth and Environmental Science **168** (2018) 012001 doi:10.1088/1755-1315/168/1/012001

obtained by differentiating the objective function with respect to device parameters. The parameter that influences power flow in a line is the angle of injected voltage of the series converter. The generalized approach cannot be applied to the systems where analytical model of the FACTS device is not available.

The proposed method is tested on a 5 Bus and IEEE14 Bus systems for placement of three devices, viz., UPFC and IPFC respectively. The rest of the paper is organized as follows. Section II contains the static model of UPFC and STATCOM, section III contains optimal location of FACTS devices. Section IV gives the result analysis and Section V gives the conclusions.

2. Modelling of UPFC

UPFC is designed by combining the series compensator (SSSC) and shunt compensator (STATCOM) coupled with a common DC capacitor. It provides the ability to simultaneously control all the transmission parameters of power systems, i.e. voltage, impedance and phase angle. It consists of two converters – one connected in series with the transmission line through a series inserted transformer and the other one connected in shunt with the transmission line through a shunt transformer. The DC terminal of the two converters is connected together with a DC capacitor. The series converter control to inject voltage 7 magnitude and phase angle in series with the line to control the active and reactive power flows on the transmission line. Hence the series converter will exchange active and reactive power with the line.

The UPFC can perform the function of STATCOM and SSSC and phase angle regulator. Besides that, the UPFC also provides an additional flexibility by combining some of the function above. UPFC has also a unique capability to control real and reactive power flow simultaneously on a transmission system as well as to regulate the voltage at the bus where it's connected [7].

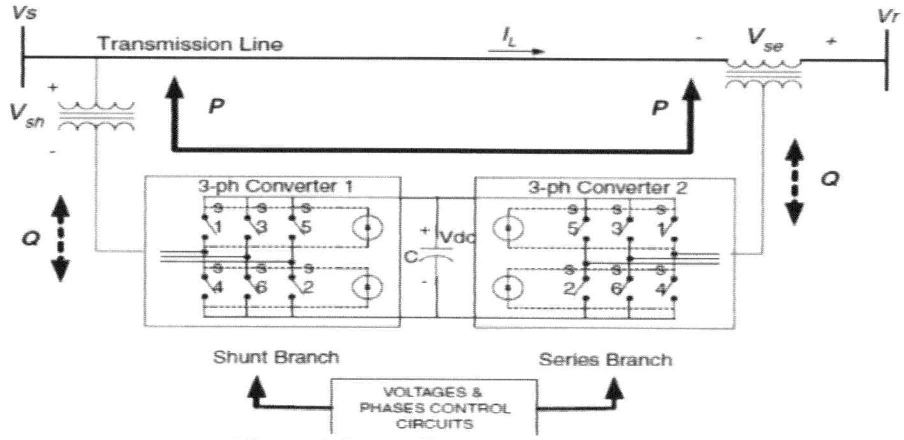

Figure 1. Power flow model of UPFC

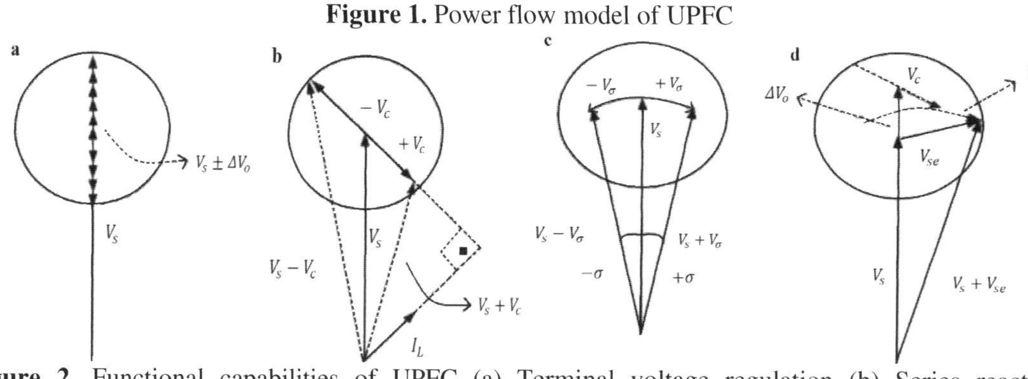

Figure 2. Functional capabilities of UPFC (a) Terminal voltage regulation (b) Series reactive compensation (c) Phase angle shifting (d) multi-function control mode

The steady-state UPFC mathematical model is developed by replacing voltage source V_s by a current source I_s parallel with the transmission line [8], where

$$b_s = \frac{1}{x_s} \quad (1)$$

$$I_s = -jb_sV_s \quad (2)$$

The detailed model of UPFC is given in [9]

The complete steady state model of UPFC is given by the following equations

$$P_{i,upfc} = rb_sV^2{}_i\sin\gamma - rb_sV_iV_j\sin(\theta_i - \theta_j + \gamma) \quad (3)$$

$$P_{j,upfc} = rb_sV_iV_j\sin(\theta_i - \theta_j + \gamma) \quad (4)$$

$$Q_{i,upfc} = -rb_sV^2{}_i\cos\gamma \quad (5)$$

$$Q_{j,upfc} = rb_sV_iV_j\cos(\theta_i - \theta_j + \gamma) \quad (6)$$

General nodal power flow equations and linearized power system model can be expressed in rectangular form by the following equations:

$$\begin{bmatrix} \Delta P \\ \Delta Q \end{bmatrix}^n = \begin{bmatrix} H & N \\ J & L \end{bmatrix}^n \begin{bmatrix} \Delta\theta \\ \Delta V/V \end{bmatrix}^n \quad (7)$$

where P and Q are vectors of real and reactive nodal power injections, which are function of nodal voltages, (V\h), and network conductance and susceptance, (G and B), respectively. (ΛP = Pspe _ Pcal) is the real power mismatch vector and (ΛQ = Qspe _ Qcal) is the reactive power mismatch vector. (ΛV and Λh) are vectors of incremental changes in nodal voltages. H, N, J, and L denote the basic elements in the Jacobian matrix. Derived injected power model can be incorporated into a general NR power flow algorithm by modifying the related elements in the normal Jacobian matrix and the corresponding power mismatch equations as well [10].

3. Optimal location of UPQC

PSO is an evolutionary computation technique developed by Eberhart and Kennedy in 1995 and was inspired by the social behavior of bird flocking and fish schooling. PSO has its roots in artificial life and social psychology as well as in engineering and computer science. It utilizes a population of individuals, called particles, which fly through the problem hyperspace with some given initial velocities. In each iteration, the velocities of the particles are stochastically adjusted considering the historical best position of the particles and their neighborhood best position; where these positions are determined according to some predefined fitness function. Then, the movement of each particle naturally evolves to an optimal or near-optimal solution [11-14]. The name of "swarm" comes from the irregular movements of the particles in the problem space, more similar to a swarm of mosquitoes rather than flock of birds or school of fish.

The following equation represents the fundamental PSO

$$V_i^{k+1} = w\, V_i^k + c_1\, rand_1(\ldots) \times (p\, best_i - s_i^k) + c_2\, rand_2(\ldots) \times (gbest - s_i^k)$$

Where v_i^k: velocity of agent i at iteration k, w: weighting function, c_j: weighting factor, rand: uniformly distributed random number between 0 and 1, s_i^k : current position of agent i at iteration, $pbest_i$: pbest of agent i, gbest: gbest of the group.

Algorithm for optimal location is given as below

A. Because UPFC are expensive, therefore the minimum device installed is searched for economic efficiency reasons.

B. Generator buses where voltages are regulated by the generator do not need UPFC installation.

C. Each bus is limited to the installation of one device. Installing more does not represent a significant effect.

D. If the bus voltage is above 0. 95p.u then UPFC is not installed.

The algorithms applied for IEEE 30 bus system and results are obtained as follows

4. Results and analysis

A 30-Bus test system is used for this paper. The test system consists of 5 generators and 24 PQ bus (or load bus). The problem to be addressed consists of finding the optimal location and power rating of UPFC with power flow injection model (PFI). In this case the PSO is able to find different options for capacity of the UPFC with the PFI model [15].

TABLE I. Optimal location of UPFC

UPFC Bus	V (pu)	Theta (Degree)	P (pu)	Q (pu)
18	1.0000	-16.4708	0.9343	0.9206
27	1.0000	-15.5960	0.9343	0.9206

TABLE II. Power flow with UPFC

Bus No.	V (pu)	Angle degree	P Flow (pu)	Q Flow (pu)
1	1.0600	0.0000	0.9451	0.9314
2	1.0430	-5.3585	0.9421	0.9283
3	1.0189	-7.5267	0.9377	0.9239
4	1.0094	-9.2798	0.9360	0.9223
5	1.0100	-14.1846	0.9361	0.9224
6	1.0086	-11.0506	0.9359	0.9221
7	1.0014	-12.8653	0.9346	0.9208
8	1.0100	-11.8277	0.9361	0.9224
9	1.0366	-14.0831	0.9409	0.9271
10	1.0166	-15.7094	0.9373	0.9235
11	1.0820	-14.0831	0.9491	0.9353
12	1.0466	-15.1839	0.9427	0.9289
13	1.0710	-15.1839	0.9471	0.9333
14	1.0276	-16.0515	0.9393	0.9255
15	1.0194	-16.0117	0.9378	0.9240
16	1.0266	-15.6766	0.9391	0.9253
17	1.0142	-15.9128	0.9369	0.9231
18	1.0000	-16.4708	0.9343	0.9206
19	0.9972	-16.6961	0.9338	0.9200
20	1.0012	-16.5100	0.9346	0.9208
21	1.0028	-16.2514	0.9348	0.9210
22	1.0060	-16.0166	0.9354	0.9216
23	1.0029	-16.2595	0.9349	0.9211
24	0.9917	-16.3227	0.9329	0.9191
25	0.9914	-16.0447	0.9328	0.9190
26	0.9732	-16.4871	0.9295	0.9157
27	1.0000	-15.5960	0.9343	0.9206
28	1.0056	-11.6790	0.9354	0.9216
29	0.9796	-16.8850	0.9307	0.9169

The results in the above table show that placing the UPFC at bus 18 or 27 the voltage profile of the system is improved and the voltages at all buses is more than 1.0 p.u which is in the stable region. The results in table III show that the power loss also greatly reduced from 0.2p.u to 0.002116 p.u there by increasing the power flow.

TABLE III. Power loss with and without UPFC

Total power loss	Without UPFC (p.u)	With UPFC (p.u)
	0.2 p.u	0.002116

Analysis of Results of results is divides between voltage profile and power loss

A. Voltage profile

The results show that wit out UPFC majority of the buses are facing under voltage problem i.e below 0.95 p.u. When UPFC is injected at optimal location using PSO all the buses are in stable voltage region and is above 1.0 p.u value. The active and reactive power flow at each bus is also improve and can be seen from table 1. The results also show that the placements of UPFC with proposed algorithm has improved the system performance.

B. Power loss

The optimal location of UPFC with the proposed algorithm has reduced the total power loss in the system. The power loss obtained without UPFC is 0.2 p.u. When the UPFC is placed in optimal location the power loss is drastically reduced to 0.002116 p.u as shown in table III. Hence the placement of UPFC not only increased the power flow but also reduced the power loss to an acceptable limit.

5. CONCLUSIONS

In this paper a PSO algorithm is proposed for the optimal location of UPFC.A power flow injection model of UPFC is also derived to incorporate this model into load flow studies. The results show that the optimal location of UPFC increases the voltage profile and reduces the power loss. The power flow capability of the system is also improved and is observed from the results obtained. The proposed algorithm can also be extended for advanced FACTS devices like DPFC in the presence of contingencies and its security.

6. REFERENCES

[1] Verma KS, Singh SN, Gupta HO. Location of unified power flow controller for congestion management. Electric Power Syst Res 2001;58(2):89–96.
[2] Singh SN, Erlich I. Locating unified power flow controller for enhancing power system loadability. In: International conference on future power system. Amsterdam, Netherlands; November 2005, p. 162–66.
[3] Hingorani NG, Gyugyi L. Understanding FACTS: concepts and technology of flexible AC transmission systems. New York: Wiley IEEE Press; 2000.
[4] Gyugyi L. A unified power flow control concept for flexible AC transmission systems. IEE Proc Part C 1992;139(4):323–31.
[5] Galiana GD et al. Assessment and control of the impact of FACTS devices on power system performance. IEEE Trans Power Syst 1996;11(4):1931–6.
[6] Larsen E, Miller N, Nilsson S, Lindgren S. Benefits of GTO Based compensation systems for electric utility applications. IEEE Trans Power Del 1992;7(4):2056–64.
[7] Gyugyi L, Shauder CD, Williams SL, Rietman TR, Torgerson DR, Edris A. The unified power flow controller: a new approach to power transmission control. IEEE Trans Power Del 1995; 10(2):1085–97.
[8] Nabavi-Niaki A, Iravani MR. Steady-state and dynamic models of unified power flow controller (UPFC) for power system studies. IEEE Trans Power System 1996; 11(4):1937–43.

[9] Ambriz-Perez H, Acha E, Fuerte-Esquivel CR, De la Torre A. Incorporation of a UPFC model in an optimal power flow using Newton's method. IEE Proc, Gener, Trans Distrib 1998; 145(3):336–44.
[10]Fuerte-Esquivel CR, Acha E. Unified power flow controller: a critical comparison of Newton-Raphson UPFC algorithms in power flow studies. IEE Proc, Gener, Trans, Distrib, 1997; 144(5):437–44.

[11] Lashkar Ara A, Kazemi A, Gahramani S, Behshad M. Optimal reactive power flow using multi-objective mathematical programming. Sci Iranica Trans D: Comput Sci Eng Electr Eng 2012; 19(6):1829–36.
[12] R. Z. Minano, "Optimal power flow with stability constraints," Ph.D. dissertation, Universidad de Castilla - La Mancha, 2010.
[13] V. Ajjarapu, Computational Techniques for Voltage Stability Assessment and Control. Springer, 2006.
[14] P. H. E. Azadani, S. Hosseinian, "Optimal placement of multiple statcom for voltage stability margin enhancement using particle swarm optimization," Springer Berlin Electrical Engineering (Archiv fur Elektrotechnik), pp. 503–510, 2008.
[15] Y. del Valle, J. C. Hernandez, G. K. Venayaga moorthy, and R. G. Harley, "Multiple statcom allocation and sizing using particle swarm optimization," Power Systems Conference and Exposition, pp. 1884–1891, 2006.

IOP Conf. Series: Earth and Environmental Science **168** (2018) 012002 doi:10.1088/1755-1315/168/1/012002

Development and validation of an intelligent algorithm for synchronizing a low-environmental-impact electricity supply with a building's electricity consumption

T Schafer[1], E-L Niederhäuser[1], G Magnin[1] and D Vuarnoz[2]

[1]Energy institute, University of Applied Science of Western Switzerland, HES-SO, Fribourg, Switzerland

[2]Building 2050 Research Group, Swiss Federal Institute of Technology, Lausanne, Switzerland

Abstract. Standard algorithm of building's energy strategy often use electricity and its tariff as the sole criterion of choice. This paper introduced an algorithmic regulation using global warming potential (GWP) of energy flux, to select which installation will satisfy the building energy demand (BED).

In the frame of the Correlation Carbon project conducted by the Smart Living Lab (SLL), a research center dedicated to the building of the future, this paper presents the algorithm behind the design, the selection and the operation of appropriate energy installations for low-carbon buildings. The control strategy governing the building's energy flow takes into account demand, supply and storage. The latter being mainly linked to economic or energy benefits, which standard algorithms often use as the only criterion of choice, thus increasing the environmental impact of the energy flowing through the building. In contrast to those, the algorithm introduced in this paper controls the energy fluxes of the electrical grid, photovoltaic (PV) panels, static and vehicle batteries according to their GWP. A simulation over one year at an hourly time step – enabling one to take into account small but significant variations of the GWP of the electricity from grid – is presented for different scenarios, in order to draw an annual balance for a given combination of energy systems. The algorithm chooses the "cleanest" instantaneous energy source combination to meet the BED. Simulations show that a building using this algorithm can satisfy its BED while reducing its GWP by as much as 40% compared to an energy supply using the grid only.

1. Introduction

Most electricity consumers in the electrical grid support a certain type of "clean" energy production. However, they feel less concerned by the provenance or by the waste generated by the energy, they consume every day. Different research have recently been done for assessing hourly GWP of national mixes (e.g. Switzerland [1], France [2], Belgium [3] and Sweden [4]). The electricity available on the grid, thanks also to international energy markets, comes from multiple power plants, whose quantities of pollutants produced are different and which do not always operate continuously and simultaneously. Thus, each country has his own energy mix with its own GWP ($CO_{2\text{-eq}}$ per kWh) fluctuating during the day and throughout the year (See figure 1).

For this reason, an environment-friendly control algorithm has been developed on LabVIEW [5] in the Correlation Carbon project with the aim of reducing grid dependency when it has an unfavorable carbon footprint [6]. This regulation is different of mostly building energy management based on cost or consumption optimization [7] - [10].

A hypothetical building is designed to define energy requirements and to test the control strategy. The complete technical equipment of the building is composed of building integrated photovoltaic panels (BiPVs) installed on the east and west facades as well as on the roof, fixed batteries, and the batteries of electric vehicles connected to the building. Furthermore, in order to reduce the users' energy requirements during critical phases and to compensate them during more appropriate periods with higher needs (by load shift), the developed algorithm considers possibilities to influence the users' behavior [11].

The algorithm also includes the possibility to use only parts or none of the available equipment, and to interfere or not on the users' behavior.

As algorithm was conceived for a given context building, the structure of the current paper is:

- Definition of building case study and his context,
- Review of possible energy supply,
- Shape of a decision tree that is the algorithm's main part
- Validation process of the algorithm,
- Results and conclusions.

2. Context

A feasibility study of the smart living building, planned to host the research activities of the SLL by 2021 in Fribourg, Switzerland, is considered as a case study for the simulations of the control algorithm. This structure will be a giant laboratory dedicated to studying the building of the future; it will host an experimental hall, offices, common areas, conference rooms, restaurants and even some apartments. Electricity needs of the case study [6] have been defined at the hourly time step for a typical year for Switzerland on the base of national standards [12].

3. Carbon footprint

The reference parameter to investigate is the Global Warming Potential (GWP), representing the amount of CO_2 per kWh of energy produced. The fluctuations of the electrical grid are thus represented by this parameter, same as the energy coming from the photovoltaic panels and the batteries. Each part of the installation have his own GWP index, calculated on the basis of the lifetime's production prediction and the cradle-to-cradle environmental impacts [13].

3.1 Electrical grid

In the context of the case study, Switzerland, which is partly bordered by Italy, France, Germany and Austria, is considered. Several electricity purchase and sale transactions are realized: imports from France, Germany and Austria as well as exports to France and Italy. The electrical energy consumed in Switzerland depends therefore mainly on the production means of the importing countries. In addition, each of these nations has its own energy supply and production policy. While Germany has abandoned nuclear power and has returned to older and polluting techniques based on coal, Austria has never used atomic power for production and France continues to give priority to nuclear power. The GWP index of the Swiss electricity grid is therefore very closely related to the transactions with these three countries and to its own energy strategy. For this reason, an hourly GWP of electric grid is needed. Overall, the grid represents an "unclean" energy source for the building, and despite the fluctuations of its GWP, it competes with solar energy from photovoltaic panel with most appropriate orientation only during less than ten hours over the whole year. GWP index of the Swiss grid fluctuates between 0.41 and 0.03 kg $CO_{2\text{-eq.}}$/kWh during a whole year [1].

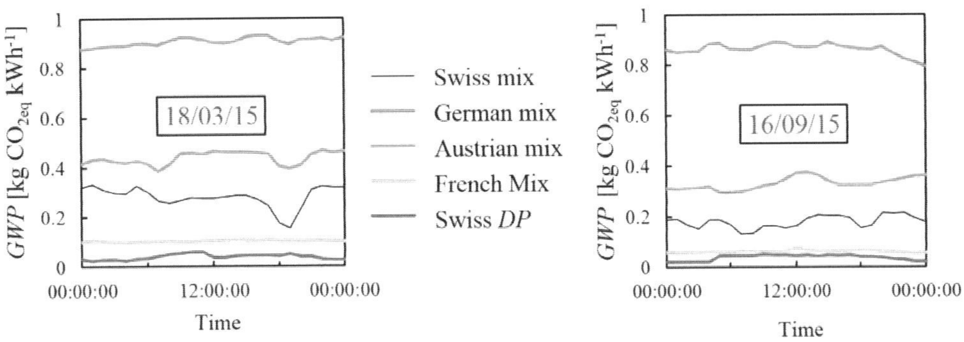

Figure 1. *GWP* of the contributors to the Swiss mix (DP: domestic production) [1]

3.2 *Building integrated photovoltaic panels (BiPV)*

Depending on the PV technology and the resulting characteristics – lifetime, efficiency, etc. – the GWP index for solar production varies. Photovoltaic electricity production is ensured by three clusters of panels installed on the east/west facades and roof of the building, each cluster having its own GWP index due to the total lifetime's production. For example, in the context of the case study, GWP of PV on the roof (0.041 kg $CO_{2\text{-eq.}}$/kWh) is less than index for PV on the east side of the building (0.094 kg $CO_{2\text{-eq.}}$/kWh), because of roof mounted PV will produce more energy. These values are considered constant throughout the simulated year. The algorithm therefore has the possibility to favor the direct use for the building of the energy from the "cleanest" cluster and, according to the energy strategy, to deliver to the electrical grid the energy from the most polluting cluster. Regardless of the PV technology and the side of the building on which the panels are placed, their index is lower than the yearly average index of the grid, varying between 40-80% of its value.

3.3 *Batteries*

In the frame of the case study, two types of Li-Ion batteries are considered. The first are fixed storage batteries and the second are those of the various electric vehicles associated to the building mobility. The GWP index of the energy delivered from the batteries is calculated from a weighted average according to the index and the amount of energy stored. An initial GWP index is defined for each battery. Through the control period, the GWP index of each battery can therefore vary from the best value of the solar panels to the maximum value of the grid. In addition, the battery index of vehicles varies when these are used, and they are only available when people are at their workplace.

At time *i*, the GWP calculation of the batteries (b) is described by Eq.1:

$$GWP_{b,i} = \frac{E_{b,i-1} \times GWP_{b,i-1} + E_{charge}\big|_{i-1}^{i} \times GWP_{charge} - E_{discharge}\big|_{i-1}^{i} \times GWP_{discharge}}{E_{b,i-1} + E_{charge}\big|_{i-1}^{i} - E_{discharge}\big|_{i-1}^{i}} \quad (1)$$

The letter E represents the amount of energy transiting during a time interval $(i\text{-}1) \rightarrow i$, which is multiplied by the GWP index of the source to obtain the Greenhouse Gases (GHG) emissions value.

4. Decision tree

Thanks to the GWP index, the different sources of energy available to supply the building could be compared. On this basis, a decision tree is created, comprising all technical equipment – thus simulating scenario no. 5 – in order to define the best energy management strategy. Other scenarios were also tested by limiting the algorithm to the essential branches of the decision tree. Based on figure 3, the green parts of output cases represent the sources of electricity and those in red the electric consumers.

The first case of the decision tree (vertical branch on the far left) visible on figure 3 is the ideal one: when the production of BiPVs exceeds the building's needs, the remaining energy is used to charge the batteries and the rest is sent to the grid. In contrast, case number 12, the most critical, is also represented: when there is no energy produced by the BiPVs and the batteries are empty, the only source of energy is the grid. Cases 7, 10 and 13 are used when the supply conditions of the building are not optimal and a load shift is allowed but not necessarily possible. In these cases, the algorithm restarts with a new consumption value, if it has been possible to reduce the building's demand.

The decision tree shown in figure 3 is a simplified version. Indeed, in each case, the algorithm must define what it is the part of energy that is managed. Moreover, solar production is ensured by three faces of the building with different GWP indices and storage possible in two different types of batteries, creating the necessity for a second level of decision tree that set which type of battery or PV from which side is first to be used.

5. Process validation

In order to validate the accuracy of the algorithm, two types of tests were carried out: the first one to confirm the quality of the decision tree. Artificial values were entered in the decision tree subprogram. Output cases were compared to the expected ones from quality point of view.

The second test was performed to demonstrate the consistency of energy values for different parts of the installation by energy conservation balance.

6. Results

Several scenarios, shown in figure2, were tested in order to assess the impact of each element on the carbon footprint of the building.

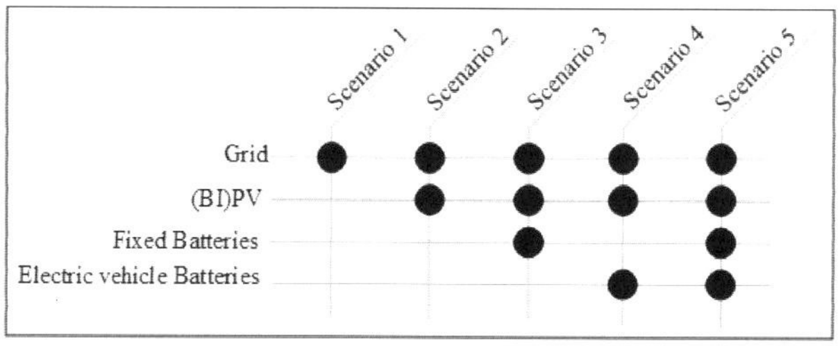

Figure 2. Tested scenarios in the carbon-based energy strategy for building [6]

Some scenarios are tested several times by varying the technology of an equipment or the assumptions of operation, for example with several types of PV panels or batteries, various possibilities or quantities of needs distributed differently, etc. The control algorithm is tested over one year using fixed meteorological values.

According to the simulation performed over a whole year, the results show that the algorithm allows improving the GWP of the consumed electricity. Figure 4 shows different combinations of mix supply and their respective GWP. A building using only the grid obtains an average annual GWP supply of 0.206 kg $CO_{2\text{-eq.}}$/kWh. A user equipped with solar panels, fixed batteries and electric vehicle can use the algorithm to lower its time average annual GWP supply to 0.122 kg $CO_{2\text{-eq.}}$/kWh. This value represents a gain of 40% on the GWP supplied by the grid. Figure 4 shows also that using fixed batteries alone and/or using fixed batteries coupled with vehicle batteries allows obtaining similar curves. Coupling these two types of batteries allows a gain of only 0.3% compared to the use of static battery alone. The recurring actions chosen by the algorithm are numbers 8 and 1 (27% each). There is therefore a frequent transit between solar panels and batteries. Case 3 (battery and PV consumption) appears in 15% of cases. These three cases (1, 3 and 8), representing 69% of appearance, do not involve the grid, meaning that only one-third of the cases require energy network back-up.

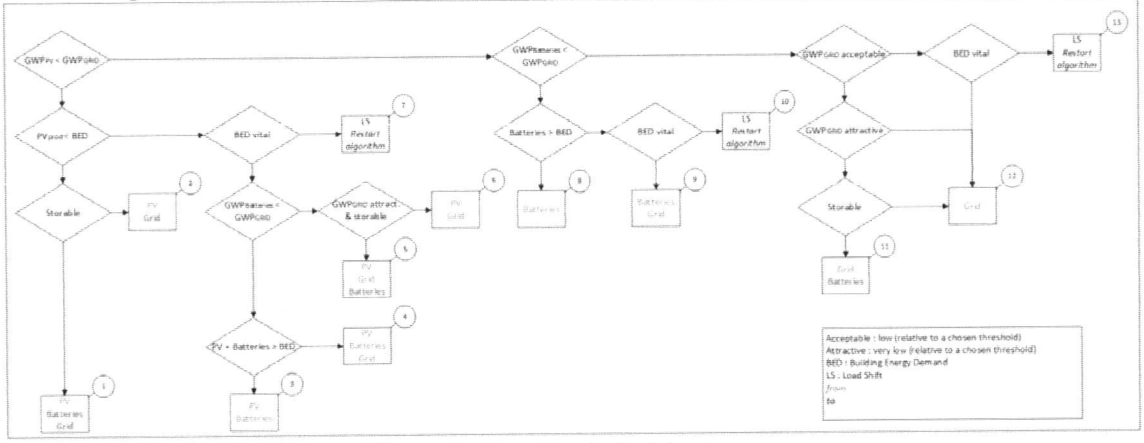

Figure 3. Algorithm's decision tree

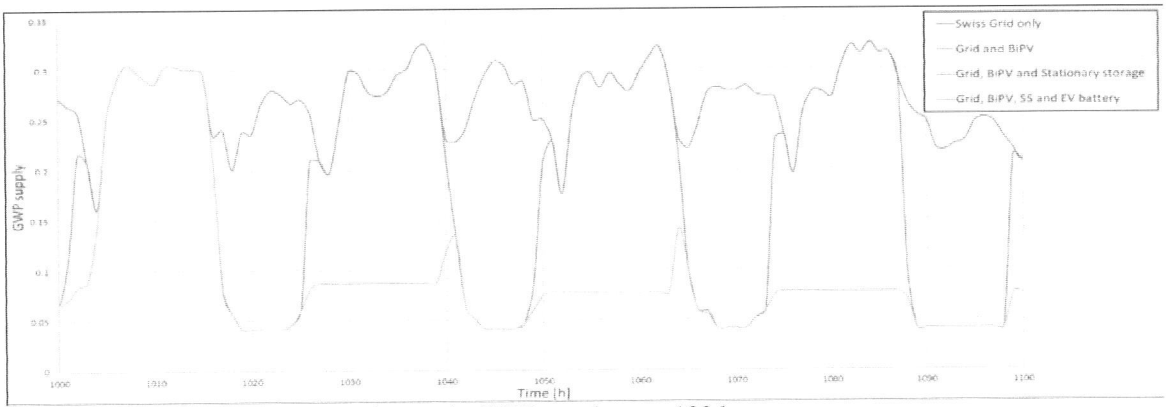

Figure 4. Hourly GWP supply on a 100 hour range

7. Conclusions

The algorithm presented in this paper allows improving the global warming potential of electricity supply by a smart control of the energy flow necessary to satisfy the building electricity demand. Simulations were carried out in order to validate the methodology developed. Moreover, the tests of different technologies integrated in algorithm allowed to dimension and select the best option during the design phase of a building.

This method can be used to identify the impact of different technical installations on a future or on an existing building. Using an hourly time step for simulations with iterative values, always adapted to the last actions, the results give an indication about the rate of building's autonomy.

In the frame of the case study in Fribourg, simulations demonstrated that the building's autonomy is not directly linked to the batteries capacity whereas this last affect environmental building's impact.

A project is currently underway to apply the regulation to an existing building. This real application will allow further improvements and optimization of the algorithm.

8. References

[1] Vuarnoz D and Jusselme T 2017 *Temporal varitation in the primary energy use and greenhouse gas emissions of the electricity provided by Swiss grid.* (submitted)

[2] Roux C, Schalbart P and Peuportier B 2016 *Accounting for temporal variation of electricity production and consumption in the LCA of an energy-efficient house*. J. Clean. Prod. **113**, 532-540.
[3] Messagie M, Mertens J, Oliveira L, Rangaraju S, Sanfelix J, Coosemans T, Van Mielo J and Macharis C 2014. *The hourly life cycle carbon footprint of electricity generation in Belgium, bringing a temporal resolution in life cycle assessment*, Applied Energy **134**, 469-476.
[4] Kristinsdóttir A R, Stoll P, Nilsson A and Brandt N 2013. *Description of climate impact calculation methods of the CO2e signal for the Active house project*. KTH Royal Institute of Technology, 1-23.
[5] Labview software. http://www.ni.com/labview/
[6] Vuarnoz D, Cozza S, Jusselme T, Magnin G, Schafer T, Couty P and Niederhäuser E-L 2017 *Carbon-based energy strategy for building* (submitted)
[7] Neurobat porducts NOL and NIQ: http://neurobat.net/fr/produits/
[8] Loxone – smart home product Miniserver: http://shop.loxone.com/dech/miniserver.html
[9] Siemens – Synoco product: http://www.buildingtechnologies.siemens.com/bt/global/en/buildingautomation-hvac/building-automation/building-automation-for-small-applications-synco/pages/hvac-building-automation.aspx
[10] Gridsense. https://www.gridsense.ch/en/home.html
[11] Vuarnoz D, Jusselme T, Cozza S, Rey E and Andersen M 2016 *Studying the dynamic relationship between energy supply carbon content and building energy demand*, PLEA conf. Los Angeles
[12] SIA 2024, 2015. Données d'utilisation des locaux pour l'énergie et les installations du bâtiment, **1-152**.
[13] KBOB, eco-bau et IPB (2014b) Recommandation KBOB 2009/1:2014: Données des écobilans dans la construction, état d'avril 2014.Coordination des services fédéraux de la construction et de l'immobilier p.a. Office fédéral des constructions et de la logistique, retrieved from: www.bbl.admin.ch/kbob/00493/00495/index.html?lang=fr

Numerical approach of Al_2O_3-water nanofluid in photovoltaic cooling system using mixture multiphase model

B Sutanto* and Y S Indartono

Faculty of Mechanical and Aerospace Engineering, Institut Teknologi Bandung, Indonesia

Email: *bayusutanto@students.itb.ac.id

Abstract. This article aims to numerically analyze the cooling system of photovoltaic device using mixture multiphase model. Aluminum oxide (Al_2O_3)-water nanofluids at various concentration of 0, 0.5, 1 and 2 vol.% were used as cooling fluid for photovoltaic module. The constant and uniform heat flux of 800 W/m^2 was applied on photovoltaic (PV) module and the various cooling fluids flowed under PV module at Reynold number of 200 and 500. The temperature distribution and local heat transfer coefficient of cooling fluid were observed. The results show that increasing concentration of Al_2O_3 nanoparticle enhances the local heat transfer coefficient and decreases the temperature of PV cell. The significant enhancement of convection heat transfer of Al_2O_3-water nanofluid occur at entrance region. The operating temperature of PV cell can be decreased till 303 K for 2 vol.% Al_2O_3-water nanofluid at Reynold number of 500.

1. Introduction

Solar energy is one of the most abundant renewable energy resources. The electrical devices which convert solar energy into electrical energy via photovoltaic mechanism such as silicon have been developed to gain maximum performance. One of the main factor that can decrease the efficiency of photovoltaic cells is temperature [1]. Moreover, the effect of temperature in photovoltaic devices also decrease the life time and stability of PV system. The cooling mechanism is one of the solution which is applied by researcher to decrease the operating temperature of PV panels. The various types of fluids which can be used as PV cooling system are air, water, ethylene glycol and nanofluid.

Nanofluid is a fluid which dispersed nanomaterials with size typically of 1 to 100 nm. The use of nanomaterials have attracted the interest for many applications such as semiconductor, solar cells, nanofluid, sensor, and etc [2-5]. The dispersion process of nanoparticle will enhance thermal conductivity and specific heat of fluid [6]. The addition of Al_2O_3 nanoparticles into water have been reported to increase the performance of convection heat transfer [6]. The heat transfer characteristic of nanofluid had been successfully investigated by single-phase and multi-phase model [7, 8]. However, the multi-phase model produce heat transfer prediction which is more accurate than the single-phase model, because the disorder movement of nanoparticle and its interaction with base fluid also considered. Movareji, et al [8] showed that Al_2O_3-water nanofluid under heat flux constant in the discrete phase model had enhancement in heat transfer coefficient and a good agreement with experimental results. Thus, this work will investigate the numerical investigation of Al_2O_3-water nanofluid as cooling fluid for photovoltaic cells by using two-phase mixture model approach. The laminar flow condition was chosen to show the influence of nanoparticle contribution particularly at low velocity which is easier for engineering application.

2. Methods

2.1. Problem Description

Generally, the intensity of solar radiation which is accepted by PV module is 1000 W/m^2. The efficiency of PV cell at the market is range from 12 to 17% and the portion of 3 to 10% of solar radiation is reflected by glass [9, 10]. Thus for this study assume that the total of 20% of solar energy is converted into electrical energy and back to surrounding. Meanwhile the portion of solar radiation which is not converted will be changed as heat input to the PV module that is 800 W/m^2. There are 6 main layers components in the PV module as shown in table 1. The layers are arranged like a sandwich, and the cooling fluid steadily flows under the PVF layer. At this work, the cooling fluids steadily flow in 5 mm channel along 1 m horizontal PV panels. The 0 (pure water), 0.5, 1 and 2 vol.% of Al_2O_3-water nanofluid were chosen due to its high thermal conductivity and minor agglomeration tendency [6, 11].

Table 1. Properties of PV panel's layers [9, 12].

Layer	Thickness (m)	Thermal conductivity (W/m.K)	Density (kg/m^3)	Specific heat capacity (J/kg.K)
Glass	3 x 10^{-3}	1.8	3000	500
ARC	1 x 10^{-7}	32	2400	691
PV cells	2.25 x 10^{-4}	148	2330	677
EVA	5 x 10^{-4}	0.35	960	2090
Rear contact	1 x 10^{-5}	237	2700	900
PVF	1 x 10^{-4}	0.2	1200	1250

Table 2. Properties of Al_2O_3 nanoparticle [13-15]

Property	Value	Unit
Diameter size	50	nm
Density	3970	kg/m^3
Thermal conductivity	42	W/m.K
Specific heat	765	J/kg.K

2.2. Nanofluid's properties

The properties of Al_2O_3 nanoparticle which is used for this study were demonstrated in table 2. The Al_2O_3 nanoparticles were assumed that had perfect ball shape with uniform and constant properties. The density (ρ_{nf}) and specific heat ($c_{p,nf}$) of nanofluid in any various volume fraction (ϕ) can be calculated by equation (1) and (2). The subscript *nf, np* and *bf* mean nanofluid, nanoparticle and base fluid, respectively. The viscosity of nanofluid (μ_{nf}) can be evaluated by Bachelor equation as shown in equation (3). The experiment results from Ghanbarpour M, et al [16] also showed that the thermal conductivity of Al_2O_3-water nanofluid (k_{nf}) can be predicted by equation (4).

$$\rho_{nf} = (1-\phi)\rho_{bf} + \phi\rho_{np} \tag{1}$$

$$c_{p,nf} = \frac{(1-\phi)(\rho.c_p)_{bf} + \phi(\rho.c_p)_{np}}{(1-\phi)\rho_{bf} + \phi\rho_{np}} \tag{2}$$

$$\mu_{nf} = (1+2.5\phi+6.2\phi^2)\mu_{bf} \tag{3}$$

$$k_{nf} = (1+3.5\phi+2.5\phi^2)k_{bf} \tag{4}$$

2.3. Governing Equation

The interaction between Al_2O_3 nanoparticle and water can be modelled in mixture multiphase by Euler-Euler approach using ANSYS-Fluent. Mixture model can be used to analyze multiphase flow with very strong coupling, and phases moving at the same or different velocity. Equation (5), (6) and (7) show the continuity, momentum and energy equation for mixture multiphase flow, respectively.

$$\frac{\partial}{\partial t}(\rho_m) + \nabla.(\rho_m.\vec{V}_m) = 0 \tag{5}$$

$$\frac{\partial}{\partial t}(\rho_m.\vec{V}_m) + \nabla.(\rho_m.\vec{V}_m.\vec{V}_m) = -\nabla p + \nabla.\left[\mu_m\left(\nabla\vec{V}_m + \nabla\vec{V}_m^T\right)\right] + \rho_m.\vec{g} + \vec{F} + \nabla.\left(\sum_{k=1}^{n}\phi_k\rho_k\vec{V}_{dr,k}.\vec{V}_{dr,k}\right) \tag{6}$$

$$\frac{\partial}{\partial t}\sum_{k=1}^{n}(\phi_k\rho_k E_k) + \nabla.\sum_{k=1}^{n}\left(\phi_k\vec{V}_k(\rho_k E_k + p)\right) = \nabla.(k_{eff}\nabla T) + S_E \text{ with } E_k = h_k - \frac{p}{\rho_m} + \frac{V_k^2}{2} \tag{7}$$

3. Results and Discussion

3.1. Verification and Validation

The computational fluid dynamic (CFD) tool, ANSYS Fluent 14.5 was used to analysis this studied models. The Semi-Implicit Method for Pressure-Linked Equation (SIMPLE) procedure was used to solve the velocity-pressure coupling algebraic equations. The validation was conducted by analyzing the heat transfer coefficient (h) of Al_2O_3-water nanofluid which had been experimentally analyzed by Wen and Ding [11]. Figure 1 shows the experiment result that conducted by Wen and Ding (2004) and the numerical simulation by mixture multiphase model at Re = 1050±50. The simulation has good fit prediction with experimental results with errors less than 15%. The work by Wen and Ding had mentioned before that the deviation under their conditions of experiment setup was 6.2% due to heat losses and errors of apparatus.

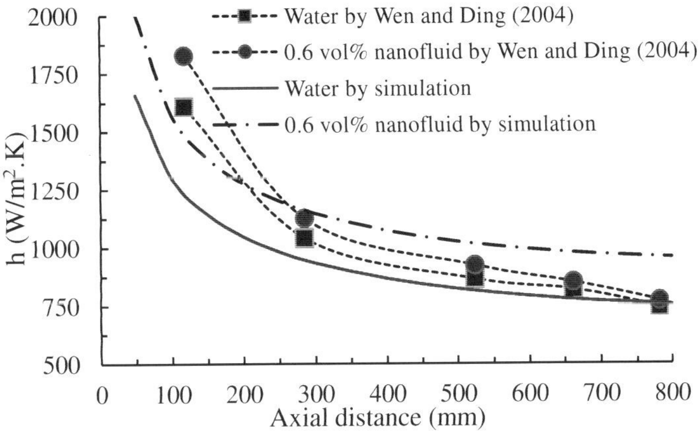

Figure 1. Local heat transfer coefficient analysed by Wen and Ding (2004) and mixture multiphase model for water and 0.6 vol.% Al_2O_3-water nanofluid at Re = 1050±50.

3.2. Thermal Distribution

Figure 2 shows the thermal distribution of PV module and cooling fluid under Reynold number of 200 and 500. It reveals that increasing of Reynold number will decrease the temperature of PV module. The fluid temperature increases and develops from entrance region till outlet section. At the entrance region, the PV cell has the lowest temperature and then increase along axial direction. The distribution of PV cell temperature is highly affected by the amount of heat removal from it. The 2 vol.% Al_2O_3-water nanofluid at Reynold number of 500 can decrease the maximum operating temperature of PV cell till 303 K. This result has important role to efficiency of PV module [1]. The local heat transfer coefficient ($h(x)$) was defined by equation (8). The q'', T_{wall} and T_{mean} are the heat flux, the wall temperature and the mean temperature, respectively.

$$h(x) = \frac{q''}{T_{wall}(x) - T_{mean}(x)} \text{ with } T_{mean}(x) = \frac{\int u.T.y.dy}{\int u.y.dy} \tag{8}$$

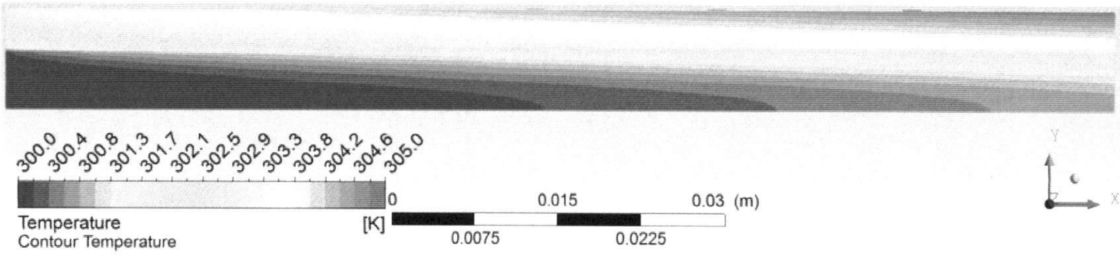

(a)

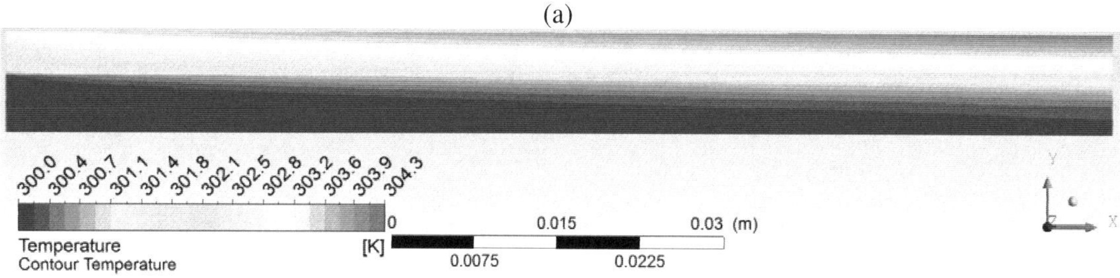

(b)

Figure 2. Thermal distribution in PV panels at Reynold number of (a) 200, and (b) 500 of 2 vol% Al_2O_3-water nanofluid

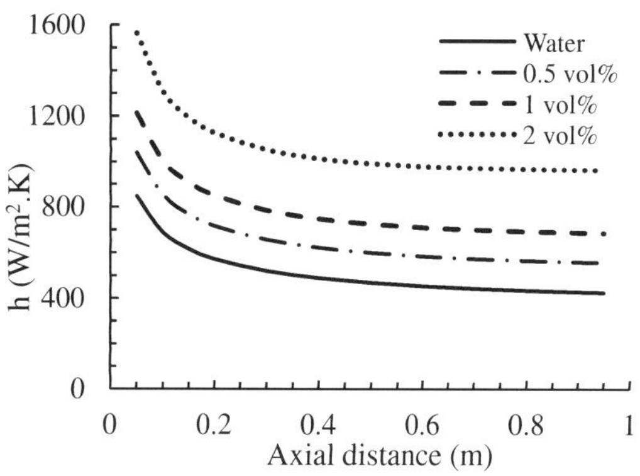

(a)

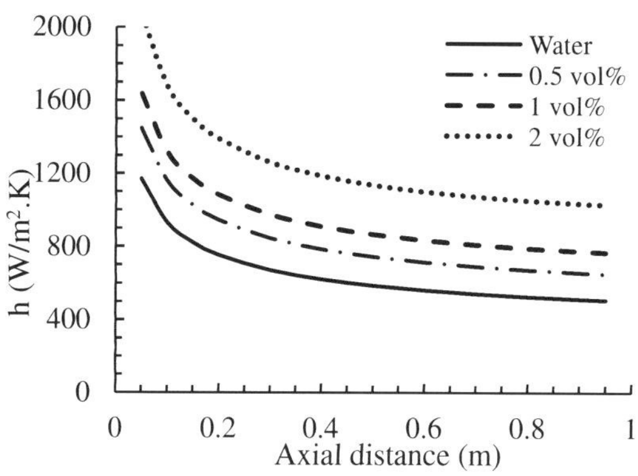

(b)

Figure 3. Local heat transfer coefficient at Reynold number of (a) 200, and (b) 500 of nanofluid

Table 3. Pressure drop of various cooling fluid

Cooling fluid	Pressure drop (Pa)	
	Re = 200	Re = 500
Water	18.94	50.65
0.5 vol.% Al_2O_3-water	21.50	50.82
1 vol.% Al_2O_3-water	22.00	52.10
2 vol.% Al_2O_3-water	22.05	55.44

Figure 3 shows the local heat transfer coefficient along axial direction of fluid. At the entrance region, the heat transfer coefficient has high value, then decreases along axial direction. This phenomena is strongly influenced by the temperature gradient at the PVF surface. The heat transfer coefficient of 2 vol.% Al_2O_3-water is higher than other cooling fluid types. That is caused by the thermal conductivity and specific heat of nanofluid which is proportional with volume fraction of nanoparticles in base fluid [7, 11]. At the Reynold number of 500 has higher local heat transfer coefficient for all variation of volume fraction in nanofluid than that of Reynold number of 200. That is related with mass flow rate of cooling fluid. Table 3 shows the pressure drop in various cooling fluid of photovoltaic system. The increasing volume fraction of nanoparticles and Reynold number have linear effect on the pressure drop. However, the increasing of pressure drop due to addition of nanoparticles into base fluid has minor consideration if compared with enhancement of heat transfer coefficient.

4. Conclusions

The numerical approaches of PV cooling system using Al_2O_3-water nanofluid at laminar flow under constant heat flux have been carried out by mixture multiphase model. The various volume fraction of Al_2O_3-water nanofluid at Reynold numbers of 200 and 500 have been numerically investigated to observe their influence on heat transfer coefficient and temperature of PV cell. The results show that nanofluids have significant enhancement of heat transfer coefficient and decreasing of the temperature of PV cells.

References

[1] Dubey S, Sarvaiya J N and Seshadri B 2013 *Energy Procedia* **33** 311-321.
[2] Li J and Zhang J Z 2009 *Coordination Chemistry Reviews* **253** 3015-3041.
[3] Sutanto B, Arifin Z, Suyitno, Hadi S, Pranoto L M and Agustia Y V 2016 *Int. Conf. on Sustainable Energy and Advanced Materials* vol 1717, ed A T Wijayanta and A Rivai (Surakarta: AIP Conference Proceedings) p 040006.
[4] Kumar S, Nehra M, Deep A, Kedia D, Dilbaghi N and Kim K-H 2017 *Renewable and Sustainable Energy Reviews* **73** 821-839.
[5] Agustia Y V, Suyitno, Arifin Z and Sutanto B 2016 *Int. Conf. on Sustainable Energy and Advanced Materials* vol 1717, ed A T Wijayanta and A Rivai (Surakarta: AIP Conference Proceedings) p 040005.
[6] Sridhara V and Satapathy L N 2011 *Nanoscale Research Letters* **6** 456.
[7] He Y, Men Y, Zhao Y, Lu H and Ding Y 2009 *Applied Thermal Engineering* **29** 1965–1972.
[8] Moraveji M K and Esmaeili E 2012 *International Communications in Heat and Mass Transfer* **39** 1297–1302.
[9] Popovici C G, Hudişteanu S V, Mateescu T D and Cherecheş N-C 2016 *Energy Procedia* **85** 425-432.
[10] Sahouane N and Zerga A 2014 *Energy Procedia* **44** 118-125.
[11] Wen D and Ding Y 2004 *International Journal of Heat and Mass Transfer* **47** 5181-5188.
[12] Armstrong S and Hurley W G 2010 *Applied Thermal Engineering* **30** 1488-1495.
[13] Darzi A A R, Farhadi M and Sedighi K 2013 *International Communications in Heat and Mass Transfer* **47** 105-112.
[14] Sheikhzadeh G A, Fakhari M M and Khorasanizadeh H 2017 *Journal ofApplied Fluid Mechanics* **10** 209-219.
[15] Zakaria I, Azmi W H, Mamat A M I, Mamat R, Saidur R, Talib S F A and Mohamed W A N W 2016 *International Journal of Hydrogen Energy* **41** 5096-5112.
[16] Ghanbarpour M, Bitaraf E and Khodabandeh R 2014 *Experimental Thermal and Fluid Science* **53** 227-235.

Acknowledgments

The first author thanks to the LPDP (Indonesia Endowment Fund for Education) for the scholarship program No.PRJ-6854/LPDP.3/2016. The authors also thank to Penelitian Unggulan dan Strategis Nasional (PUSNAS) project 2015-2017 which support this research.

IOP Conf. Series: Earth and Environmental Science **168** (2018) 012004 doi:10.1088/1755-1315/168/1/012004

Study of Applications of Artificial Intelligence Algorithm for Analysis and Investigation of Hybrid Energy Systems to give Optimum Power Generation

Ashwin Vyas[1] **, Vibha Vyas***[2]

[1]Vishwakarma Institute of Technology Pune, India
[*2]College of Engineering Pune, India

Email: Ashwin.vyas17@vit.edu; vsv.extc@coep.ac.in

Abstract. Philosophy, psychology and Computer Science together contribute to Artificial Intelligence (AI). AI helps and play prime role for prediction, modelling and analysing of performance of renewable energy. It also plays vital role for study and control of various parameters for the said system. Complex rules which are the outcome of very strong mathematical equations and routines are rigid methods or approaches. On the other hand, AI based approaches are the flexible adaptive ways to learn key information patterns within information domain which is heaving multiple objective optimization algorithmic requirement. Further, hybrid energy systems are investigated which is formed by creating constructive superposition of advantages of the solar, wind and hydro energy systems. Study of simulation results is supporting the application of AI based techniques for hybrid energy resources based power generations.

1. Introduction

Artificial intelligent based techniques cover various algorithms from machine intelligence to expert systems. Basic fundamental thing is, machine is empowered with the ability to think. Researchers are proceeding in the area to make machine think intelligently in an adaptive way. Earlier methods with fixed coefficient values and rigid parameters are giving very much limited output. To excel in the output optimisations so that multiple objectives which are have tread off can be optimized and maximum output can be achieved. To compensate earlier problems, AI based techniques are emerging as very strong candidate. AI based approaches find applications in prediction and modelling of radiations from solar system and performance control and designing of hybrid system based photo voltaic systems.

The term hybrid energy systems basically points towards multiple energy conversion devices combined to aid to the energy supply system. They are best alternative to more conventional systems. The renewable energy resources like wind solar, water, and even ocean waves are combined either all together or minimum two of them finds huge contribution toward energy and power generation and can improve the performance tremendously[1][2][3][4].

Hybrid systems are having number of advantages as compared to the system based on single sources like

- Reliability is very high
- Life cycle cost is minimized
- Efficiency is enhanced

- Electricity generation is achieved for maximum hours.

The Paper is organised as follows: section 2 gives the brief discussions on AI based techniques. Section three discusses about the energy sources. Sections four discusses the conclusions and study some simulation results. Last section, fifth, lists the references.

2. AI based Algorithms

There are various AI based techniques like Genetic algorithms and advancement like real coded genetic algorithm, artificial Neural network (ANN) and advancement likes convolutional Neural Networks, R-CNN and Faster R-CNN.

2.1.Artificial Neural Networks

Methods proposed earlier were based on handcrafted features and a classifier combination. These which were basically dependent on human ingenuity for feature design. Old methods required manual analysis of real data and suitable feature representation was defined, were not robust and do not meet the challenges of analysis and control of the parameter of the energy systems. Hence, recent years see the future in several deep learning based algorithmic frameworks such as Convolutional Neural Networks(CNNs) capable of learning good features automatically from the most complex representation of the hybrid energy system. This makes the detection task robust and faster in contrast to the methods which used a combination of mechanically engineered features and classifier within a sliding window approach. However, RCNNs were slow in the sense that they calculated convolutional features for each candidate region separately. Fast RCNN and Faster RCNN methods which use Object Proposal Methods (OPMs) to propose candidate regions to the classifier are among the top performers for object detection on common benchmark datasets-ANN and Faster R-ANN can also be explored for the same as they have much more advanced properties.

2.2 Genetic Algorithms

A Search algorithm is a algorithm which is designed to find out some parameters. But its result sometimes may not be to the point. Presently for many solutions heuristic algorithms are used, as they never give wrong results and are never slow. Always its results are near to optimum. There are various heuristic algorithms whose description is there in the literature. To put in nut shell, Genetic Algorithms are different from other normal optimization methods in four ways:-

* GAs handles a coding of the parameter set, and not individual the parameters themselves.
" GAs search from a huge population of points and not a single point.
* GAs use objective function information.
* GAs use probabilistic transition rules for execution and not deterministic rules.

The binary coded Genetic Algorithm works successfully in discrete search spaces and the performance largely it depends on the coding used to represent the problem variables. For solving of optimization problems having continuous search space, binary-coded Genetic Algorithms discretise the search space by coding the problem variables as binary strings. Anyhow the coding of real variables into finite length bit strings always leads to the inability to reach the solution with arbitrary precision. The problem of optimization in continuous search space is dealt with a number of Real coded Genetic algorithms whose crossover operators are found to have inadequate search power[5].

3. Energy Resources

Energy as it is known today is of broadly two types namely, Renewable energy resources and Non-Renewable Energy resources[6]. Non-Renewable Energy Resources are those when spent take a lot of time (Millions of years) to replenish and are made up of mostly decomposed biological wastes. Non-Renewable energy resources include coal, oil, natural gas, fossil fuels and nuclear energy. Renewable energy resources on the other hand are those which can be generated without depleting its source or causing any harm to the environment. These Resources include Sun, Water Bodies, Heat within the Earth taking form of different types of energies like Solar, Hydro and Geo-thermal[7][8][9].

3.1 Solar Energy

Solar Energy is a form of energy which comes under the category of renewable energy. Energy which comes to the Earth from the Sun in the form of Light energy when converted to Electrical Energy is known as Solar Power. This conversion is accomplished using two approaches, one is direct and the other is indirect. Direct approach is done by using Photovoltaic's (PV) while, the Indirect approach utilizes Concentrated Solar Power (CSP). Solar energy is very beneficial to for many good reasons. It is comparatively better to generate, it is pollution free and maintenance and wear costs are almost negligible. However, there are some short-comings to it too. Installation of the solar panels requires a lot of space. There's also an efficiency that is not 100% and one of the major disadvantages of the entire system is its dependence on weather. If it is very cloudy or it is night, there will be less to none production of solar energy.

3.2. Wind Energy

Wind when flows, contains a lot of kinetic energy in it. The energy generated by the conversion of this potential energy in wind into Electrical energy with the help of turbines is known as Wind Energy. The rotor blades of the turbines are made up of Glass fibre or Reinforced carbon-fibre plastics. Wind energy is very helpful in a lot of ways. Windmill uses the flow of wind to run its turbine and hence making it an evergreen and technically an unlimited resource. There are no harmful residues left by the generator and therefore causing no harm to the environment. Naturally, there are short-comings to this type of energy generation too. It is a very erratic form of energy because it solely depends on the regularity and direction of the wind. Also, there are cases in which many birds have flown into the turbines and died.

3.3. Hydro Energy

When water is stored in place of higher land it possesses potential energy, this potential energy and make the water fall on the blades of the turbine, the turbines start moving rapidly. These fast-moving turbines are used to generate electricity and the generated energy is called as Hydro energy. Hydro energy is a very clean and green fuel. It causes almost no pollution, the hydro plants run for a very long economic span and the maintenance costs are very cheap. The short-comings for Hydro energy are the risks that come up with putting up a plant. There are always chances for the wall-bursts, destabilization of natural flora and fauna also, sediment is retained on the other side of the dam (or the plant), causing soil erosion.

4. Hybrid Energy

The energy generated by the combination of two or more than two energy generation devices is known as Hybrid Energy. Combining of these devices is done to overcome barriers for the individual devices. Currently hybrid energy systems are gaining popularity on the global level. There are many benefits to the Hybrid systems which make it a more desirable system than the individual ones.

4.1. Hybrid Solar and Wind Energy

Hybrid solar and wind energy is the energy we generate by making a system of the combination solar panels and wind turbines. These kinds of systems tackle the problems faced by the individual systems. For example as demonstrated in Figure 1,2 and 3.

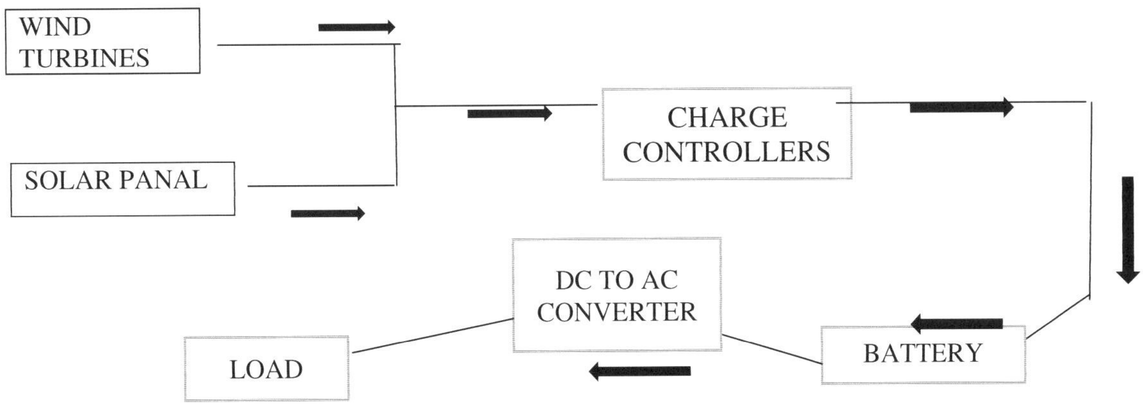

Figure 1: Hybrid wind and solar system

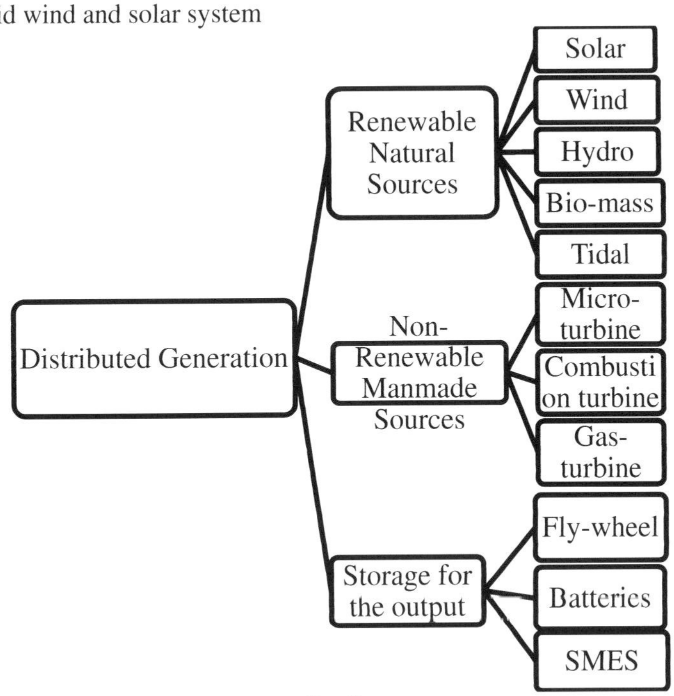

Figure 2 generation of energy and ways to distribute.

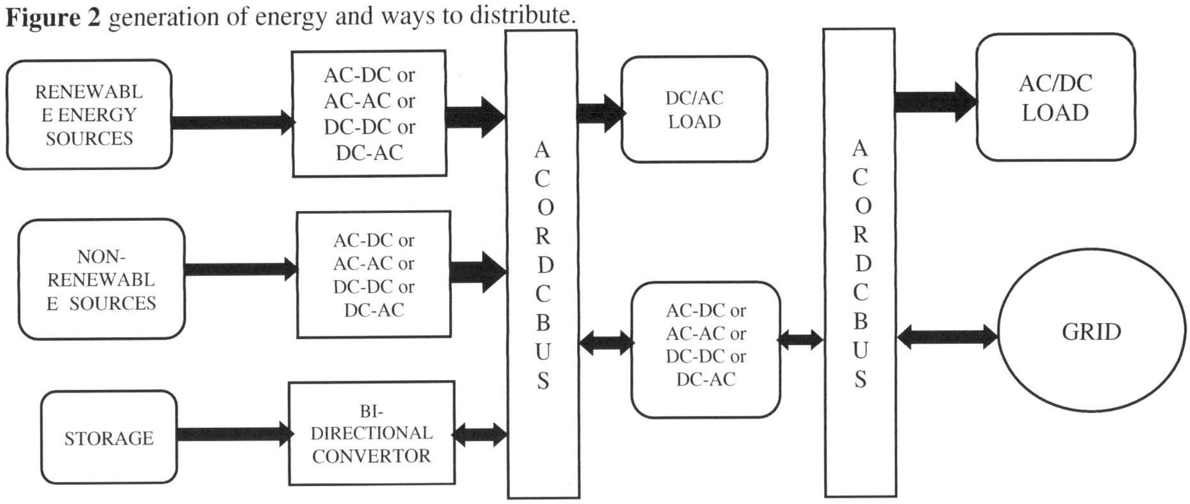

Figure 3 Basic components of hybrid system

The adjustment for the constant value can be achieved by following analysis
if $V_{dc} \geq V_{dc.ref}$ then $k = 1$;
if $V_{dc} \geq V_{dc.ref}$ then $k = -1$;
$i_{ac.ref}(n+1) = i_{ac.ref}(n) + (k)\Delta i$
where V_{dc} is the measured dc bus voltage;
$V_{dc.ref}$ is the reference value of the dc bis voltage;
$i_{ac.ref}(n)$is amplitude of current command for the present 60 Hz cycle;
k is the constant value which is proportional to value of injected ac current and
Δi is the current command adjustment.

5. Discussions and Conclusions

Findings of Literature Survey:
1. So broadly, as per the above study, Hybrid Energy architecture comprises of two parts: an ancient algorithm to generate candidate regions as input to the conventional systems, and the second part is using hybrid system accompanied by semantic algorithms.
2. Therefore, in a nutshell, from the thorough study of the literature, it can be concluded that to get in place high power and energy generation at the output, robust and AI based system, are needed so that improvements are made both at the coefficient generation method level and adaptive algorithmic level.
3. As per the study, hybrid energy based power generation proposal methods are way better than conventional approach, which unnecessarily increases the number of steps to be processed in the input space, thus increasing computational time.
4. Out of the proposal methods, Genetic algorithm based methods are the best available method based on deep learning based framework which gives superior performance with little adaptation.
5. On the other hand, Faster RCNN is the best detector on common benchmark datasets, but does not gives superior performance on hybrid systems computational complexity.
6. Although the resolution of handcrafted features based methods is higher, but deep learning based methods are computationally very fast.
7. So, it is desirable to achieve good detection performance, without compromising the speed.
8. After extensive literature survey, the remedy proposed for this, could be using an object detection system comprising of RPN and boosted classifiers as proposed in for pedestrian detection.
9. The challenge of missing small instances could be solved by pooling features of (Real Coded Genetic Algorithms) RCGA.

References

[1] T. Gonzalez, “Clustering to minimize the maximum intercluster distance,” Theoretical Computer Science, vol. 38, pp. 293-306, 1985.
[2] T.H. Cormen, C. E. Leiserson, R. L. Rivest and C. Stein, Introduction to Algorithms, MIT Press, 2001.
[3] S. Shmuel and O. Schwartz, “On the complexity of approximating TSP with neighborhoods and related problems,” Computational Complexity, 2006.
[4] A, Dumitrescu and J. Mitchell, “Approximation algorithms for TSP with neighborhoods in the plane,” ACM Symp. Discrete Algorithms, 2001.V. Smil, Energy in World History. Boulder (USA): Westview Press, 1994.
[5] Kalyanmoy Deb, "Optimization for Engineering Design Algorithms and Examples", Prentice Hall, 1995. M. Srinivas and L. Patnaik, "Genetic algorithms: A survey", IEEE Computers, Vol. 27, No. 6, pp. 17-26, July 1994.
[6] J. Perlin, Let it Shine: The 6,000-year Story of Solar Energy. San Francisco (USA): New World Library, 2013.
[7] G.N. Tiwari, S. Dubey, Fundamentals of Photovoltaic Modules and Their Applications. Cambridge (UK): Royal Society of Chemistry, 2010.

[8] Hybrid Renewable Energy System: A Review International Journal of Electronic and Electrical Engineering. ISSN 0974-2174,Volume 7, Number 5(2014), pp. 535-542
[9] A review on hybrid renewable energy systems K. Shivarama Krishna, K. Sathish Kumar School of Electrical Engineering, VIT University, Vellore-632014, Tamilnadu, Indiahttp://dx.doi.org/10.1016/j.rser.2015.07.187 1364-0321/ & 2015 Elsevier Ltd.

IOP Conf. Series: Earth and Environmental Science **168** (2018) 012005 doi:10.1088/1755-1315/168/1/012005

A hybrid energy system based on renewable energy for the electrification of low-income rural communities

Lucero Gaslac [1], Sebastian Willis[1], Grimaldo Quispe[1], Carlos Raymundo[2]

[1]Ingenieria Industrial, Universidad Peruana de Ciencias Aplicadas (UPC), Lima-Peru.
[2] Dirección de investigación, Universidad Peruana de Ciencias Aplicadas (UPC).

Email: u201114035@upc.edu.pe ; u201210010@upc.edu.pe, grimaldo.quispe@upc.edu.pe ; carlos.raymundo@upc.edu.pe

Abstract. Electrification of low-income rural areas that have a limited connection or no access to electrical grids is one of the most demanding challenges in developing countries such as Peru. The international commitment to stop global warming and the reduction in the cost of renewable sources of energy have reduced the prices of fossil fuels in some cases. This has opened the way to the current research which proposes a hybrid energy system (HES) based on the use of renewable sources of energy. Therefore, a renewable electricity system (HRES) was set up at the village of Monte-Catache in the Cajamarca region, which is one of the poorest areas of Peru. Surveys and field studies were used to evaluate the socioeconomic characteristics, availability of renewable energy resources, and energy demand of this region. Potential energy sources were evaluated, and isolated photovoltaic systems with a battery bank were found to be the most appropriate according to the results obtained in the simulation with HOMER. This proposal constitutes an interesting contribution for future energy solutions in isolated and low-income rural areas.

1. Introduction

Electric power is considered to be a variable for measuring the regional competitiveness of a country. At present, approximately 80% of the global demand for electric power is met by burning fossil fuels, which have negative impacts on the environment [1].

Conventionally, electric power in Peru is generated in thermoelectric and hydroelectric plants, following which the voltage is raised to a range of 110–480 kV to reduce losses in transmission systems during transportation, and eventually the power reaches the consumer via medium- and low-voltage distribution systems [2] [3].

Considering that 46% of the electric power in Perú is produced in thermoelectric plants and given the proven correlation of fuel combustion with greenhouse gases, renewable energies in electricity and hydrocarbons are deemed a state policy and increasing their share in the electricity generation as a whole and also mitigating the effects of climate change must be sought after [4].

To address the obstacles related to conventional electrification methods and to meet the current energy demand, the development of hybrid energy systems (HES) based on renewable energy sources is an economical and eco-friendly solution.

Renewable energy sources are virtually so abundant that they could supply more energy than the current global demand [5] [6]. Despite this incredible growth potential, renewable energy sources have not been fully exploited because of barriers such as technological, economic, and resource availability.

In Peru, the cost of supplying electricity to rural towns is excessively high because of the difficulty of access, geographical dispersion, and in some cases their proximity to protected environmental reserves. This is why it is difficult to connect the towns to the existing electricity distribution grids. Therefore, both the initial investment and the high costs of operation and maintenance of the grids make it virtually impossible for citizens to pay for electricity.

Due to these facts, generation technologies based on local renewable energy—known as isolated systems—have become a suitable option for these regions. In recent decades, both technological improvements and government policies to promote the use of renewable energy resources have resulted in a significant reduction in costs, which makes the proposed electrification of remote populations through renewable isolated systems an economically viable alternative [7] [8] [9].

In Peru, according to Legislative Decree No. 1002, such government policies imply that daily delivery of electric load by the Committee of Economic Operation of the System, a variable cost of production equal to zero, accelerated depreciation for the purposes of Income Tax and early recovery of the General Sales Tax [10] [11] [12] must be prioritized. Moreover, in 2011, the Ministry of Energy and Mines passed the Electricity Short-Term Market Regulation to promote effective competition in the electricity generation market of Peru through the inclusion of projects based on renewable energy resources (RER).

However, providing a constant supply of energy is a difficult task for most RERs because of the intermittent and unreliable atmospheric conditions, to which the production of energy is directly related. Therefore, in an effort to overcome the production variability of renewable energy-based systems and to provide a reliable power supply, such systems can be complemented with non-renewable generation systems, known as hybrid electric systems.

2. State of the art

Vast research has been conducted on isolated HSE and hybrid electrical systems, which combine two or more types of renewable technologies (e.g. photovoltaic energy, wind-turbines, mini-hydro generators, biomass, etc.) [12].

Due to the reduced manufacturing costs, photovoltaic systems are becoming an economical and technologically viable option. Studies have compared different hybrid systems that include photovoltaic technology as the main source of energy and other sources (battery banks, wind-turbines, and microhydraulic systems) as backup systems because of the intermittent nature of the sun and little irradiation at certain times of the year [13] [14]. This study indicated that with an increase in demand, the solar/batteries model becomes more inefficient and requires the inclusion of new sources of renewable generation. Other research established that hybrid energy schemes are more sustainable in terms of supplying electricity to a rural center compared with isolated photovoltaic systems because of the lack of solar irradiation in the months of winter. This is due to the fact that trusting a single technology generally results in the oversizing of the system, which increases the initial capital cost of the plant.

The multiple combinations of renewable energy technologies with non-renewable sources, as well as the reliance on several factors such as the demand by population, the availability of renewable energy sources (RES), and the component costs, compound to the complexity of a proper design of these combinations [15]. The HOMER software models each system configuration by performing hour-by-hour simulations of the different technology options, component costs, and resource availability. Then, they analyze the technical feasibility of the configuration and calculate the total cost of the implementation and operation of the system. A summary of studies that used HOMER for the sizing of isolated systems and with relevant characteristics for this study is presented in table 1.

However, the research conducted has not delved into the impact of the cost of electricity in rural villages that are difficult to access. Therefore, the purpose of the research is to design an HES that will meet the electricity demand of the population with the best available combination of RES and at the lowest cost possible to supply electricity in a reliable and sustainable manner [16] [17].

3. Proposed methodology

The proposed methodology is divided into five stages: selection of the village or community for the project, estimation of the energy demand of the populated center, determination of the optimal model for sizing the system, development of financial analysis, and implementation. These stages are represented in the flow chart in figure 1.

Table 1. Summary of important studies of isolated power generation systems.

Authors	Technology considered	Types of needs met	Sensitivity variables	Important topics
Beck and Schott [10].	•Small hydroelectric plants, wind-turbines, photovoltaic solar units, bio-diesel generators	•Domestic •Commercial and industrial •Medical Center •School	•Biodiesel price •Wind speed •Design flow	•Pre and post HOMER analysis
Hafez and Bhattacharya [3].	•Only diesel, fully-renewable, mix of diesel and renewable, connected to the external network	•Domestic thermal and electrical needs of a hypothetical rural community	•Diesel price, shortage of maximum allowable capacity	•The importance of an unmet load
Lee, Soto and Modi [15].	•Solar photovoltaic units	•Conservation of domestic and community food	•Not considered	•Algorithm to measure energy deficit •Probability
Ma, Yang and Lu [14].	•Solar/Wind Hybrid	• Hypothetical load of 250 kW h/day	•Solar Radiation •Wind speed •Photovoltaic, wind and battery capacity	•The importance of combining the demand for power and generation to minimize the dumping energy
Murphy, Twaha and Murphy [16].	•Diesel, solar energy, Solar-diesel hybrid	•Hypothetical loads	•Not considered	•Cost of reliable electricity and modeling a reliable grid
Alzola et al. [17].	•Photovoltaic solar energy, wind-turbines	•Domestic, industrial loads, water pumping loads	•Lack of capacity	•Framework study for calculating energy demand
Patil, Saini and Sharma [18].	•Micro-hydraulic energy power, photovoltaic solar energy, biomass gasification, and hybrid systems	•Domestic agriculture •Community •Rural industries	•The cost of biomass fuel	•Reliability parameters

3.1. Selection of the adequate community/village for the project

The first step in estimating energy demand is the formulation of criteria for the selection of the village or community. The framework for energy demand assessment developed by Camblong was used, and the selection criteria are as follows:

- No connectivity present to the grid and relative long time for future grid connectivity.
- Degree of grouping of dwellings.
- Presence of telecommunications towers.
- Presence of premises and small business that require energy.

3.2. Calculation of electricity demand

Calculation of the demand for electricity in small rural villages is based on their maximum power demand. For this, the energy demand method was developed, and it requires the following information about the activities of the populated center [18] [19].

IOP Conf. Series: Earth and Environmental Science **168** (2018) 012005 doi:10.1088/1755-1315/168/1/012005

- Activities in the home; times of going to bed and wake up, food preparation, etc.
- Commercial and industrial activities: agriculture, commerce, tourism, etc.
- Public services: schools, telecommunications, health, weather and sanitation.

The information obtained is recorded in the format described in table 2, and the power demand is a result of the power of the electrical equipment (P), the simultaneity coefficient (SC), and the number of users (No.) as is shown in equation 1. Thus, the total of the potential demand in a period will be the sum of all the potential demands by the village.

$$Dp = P \times CS \times N^{c} \quad (1)$$

3.3. Determination of the optimal model for sizing the system

For this purpose, several software techniques, and optimization methods exist to maximize/minimize the objective functions according to the constraints of the HES; however, all of them require the following data as input [20]:

3.3.1. Weather data. Information on wind speed, solar radiation, temperature, and speed of the village flow was collected to calculate the power output of the system.

3.3.2. Energy load required. The energy load or demand by the population is the most important factor to make simulations and optimize HES because these elements restrict the design of HES. This can be determined with existing data, forecasts, or field studies.

3.3.3. Characteristics of the equipment. The efficiency of the operation was determined by the characteristics of the equipment used in the HES. For this purpose, a table indicating the main characteristics such as power and voltage was used.

3.3.4. Search space. The HES involves diverse equipment and technology; therefore, certain degrees of slack must be considered in the simulation.

3.3.5. Economic data. For the case studied, various systems were used such as wind-turbines, photovoltaic panels, microhydraulic turbines, diesel generators, and battery banks, which are presented in a table that identifies their technical characteristics and initial capital costs, maintenance, and operation of the useful life of the HES.

3.3.6. Technical data. Technical data are required such as distribution strategies and operational reserves.

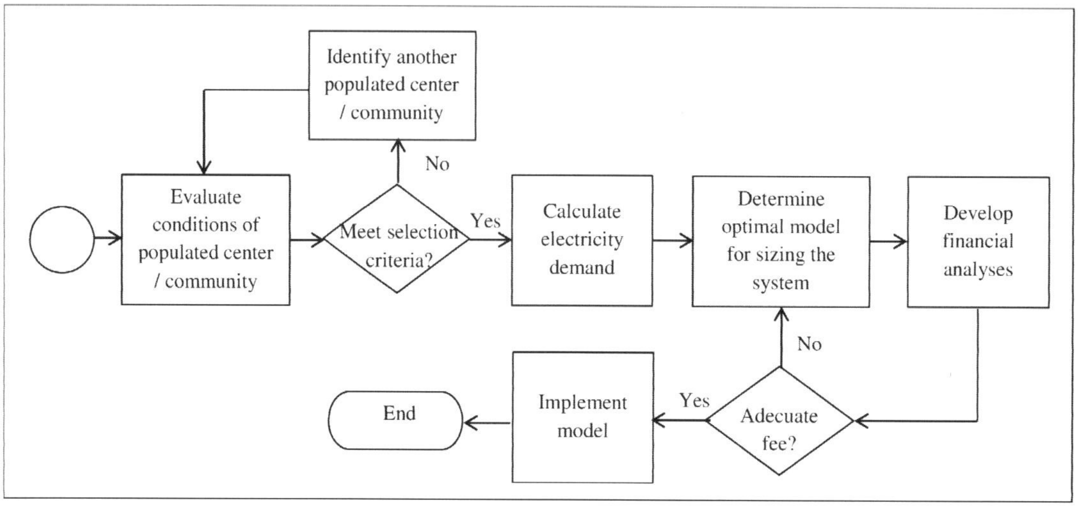

Figure 1. Proposed methodology for the implementation of a renewable HES

3.4. Financial analyses

The financial analysis comprises determining the period for recovery of the investment, which is the period of time required for the net cash flow of the investment to be recovered till their initial value. For this, the net cash flow of the project was set, and the initial investment of the project was estimated in period zero. The cash flow corresponding to the following periods is described in table 3.

Finally, the period in which the total investment was recovered was counted to determine the investment recovery term.

Table 2. Form for the evaluation of the current electricity demand of the village

Utility	Power (W)	SC (%)	No.	Hours of use	kWh Day	kWh Year
Residential						
Lighting	40,0	50%	100,0	4,0	8,0	2.920,0
Food preservation	400,0	98%	1,0	5,0	2,0	715,4
Recreation and amenities	300,0	50%	25,0	3,0	11,3	4.106,3
Other uses	1.300,0	50%	25,0	1,0	16,3	5.931,3
Total residential	2.040,0				37,5	13.672,9
Industrial and commercial						
Motors	5.000,0	2,0	8,0	40,0	14.600,0	2,0
Refrigerators	1.000,0	2,0	8,0	8,0	2.920,0	2,0
Ovens	0,0	0,0	0,0	0,0	0,0	0,0
Lighting	40,0	30,0	8,0	4,8	1.752,0	30,0
Others	3.000,0	30,0	8,0	360,0	131.400,0	30,0
Total industrial and commercial	9.040,0			412,8	150672	
Public services						
Health	2.500,0	100%	1,0	16,0	40,0	14.600,0
Education	3.000,0	100%	1,0	8,0	24,0	8.760,0
Communications	3.000,0	100%	1,0	8,0	24,0	8.760,0
Street lighting	2.750,0	100%	1,0	8,0	22,0	8.030,0
Other public services	2.000,0	100%	1,0	16,0	32,0	11.680,0
Total public services	13.250,0				142	51830
Total	24.330,0				592,3	216.174,9

4. Description of the selected rural area, components and scenarios

4.1. Selection of the populated center
In Peru, of the 3 million Peruvians without access to electricity, 78% come from rural areas. According to the General Directorate of Rural Electrification, Cajamarca is the region with the lowest coefficient of electrification because only 55% of the population has access to electricity [21]. Factors such as the climatic, geographical, economic, and cultural features, as well as the political context of villages, are the causes of the low degree of electrification observed. One of the districts of this affected region is the district of Catache, in the province of Santa Cruz, which auctioned the electrification project for its rural areas in 2008 and included 21 villages. The the village of El Monte (latitude: −6.6758°, longitude: −79.032016°) was chosen because of the presence of RES and its difficult access to existing distribution grids.

4.2. Calculation of energy demand
The results of the surveys state that 30% of houses in the village hold trade activities; this is why the use of household appliances and lamps is high, while the remaining 70% houses are used as dwellings. This reduces their electrical consumption to only two (02) lamps and audiovisual equipment. [22] [23] [24]. The load profile is obtained via the form shown in table 2.

4.3. Weather data
The historical database of the National Meteorology and Hydrology Service of Peru (SENAHMI) and the coordinates of the village were used to determine the average values of solar radiation and wind speed, as shown in figures 2 and 3, respectively. The flow velocity of the Carhuaquero River was calculated using the equations developed by Beck and Schott and presented in figure 4.

Table 3. Net flows of the project

		Year	**0**	**6**	**7**
Cost	Soles/kWh		0.62	1.85	2.22
Annual cost village	kWh			8,491	8,576
Total investment	Soles/house		-347,432	-	-
Income				66,058	72,431
Sale of energy	Soles			13,100	15,877
Fixed monthly charge per house	Soles	7.07		205.09	205.09
Cost of installation for supply	Soles	500.00	-	-	-
CO2 credits	Soles/ tCO2	28.95	-	1,613.61	1,629.75
Social benefit to the population	Soles	-	-	51,139	54,719
Costs (Operation Maintenance cost)	Soles			−131.00	−158.78
Expenses				−1,084	−1,139
Profit before income tax		-	-	63,626	69,835
Profit after income tax		-	-	63,626	69,835
Economic flow			347,432	63,626	69,835

Net financial flow	347,432	63,626	69,835
Cumulative flow	347,432	−42,758	27,077

4.4. Characteristics and economic data of the equipment used
Considering the products available in the Peruvian market, a list of the components was selected and their prices were found through different national distributors and manufacturers. The selected components and their corresponding costs are shown in table 4.

4.5. Simulation and optimization of the system´s model
The tool used to optimize the sizing of the HES is the Hybrid Optimization Model for Electric Renewables (HOMER). HOMER software was used to achieve economic optimization of the different configurations for the El Monte community. To do this, six different scenarios were analyzed for the best possible configuration for El Monte: diesel generator (case 1), FV + batteries (case 2), wind turbine (case 3), solar/wind hybrid (case 4), solar + microhydraulics hybrid (case 5), FV + diesel hybrid (case 6) and solar + wind + diesel hybrid (case 7). Figure 5 shows the configurations of the scenarios in HOMER.

4.6. Financial analysis
The scenario chosen by HOMER was an isolated solar system with a battery bank (case 2) whose component cost is \$347,432.00 Peruvian Soles. It was established that the payment by inhabitants for domestic consumption would be 0.62 Soles per kilowatt/hour, and this value that corresponds to the average cost in the region. In the first month a payment of 500 Soles will be made for activation of the service. Currently, on an average, each family spends 104.00 Soles per month on inefficient energy sources such as kerosene and candles. With these costs, the recovery period of the isolated solar electric system and batteries is 6.61 years.

5. Description of the selected rural area, components and scenarios
Figure 6 presents the costs of each configuration modeled using HOMER. The case 2 has photovoltaic panels supported by battery banks and meets the energy demand of the farm with the lowest cost per kW/h: 0.62 Soles kW/h. This increases the probability that 100% of homes can access the electricity supply service. Additionally, this configuration presents one of the lowest operating costs (26,151 Soles/year) because of the technology used.

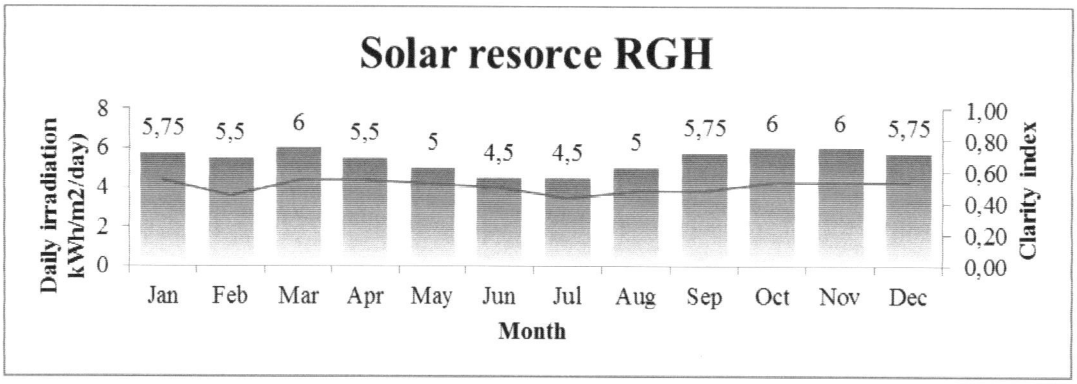

Figure 2. Average solar irradiation in El Monte Village, 5.38 kWh/m2

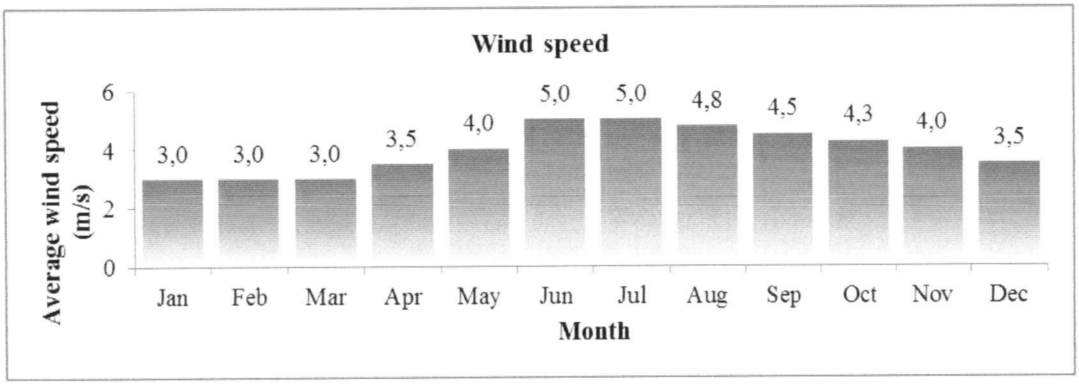

Figure 3. Average wind speed in El Monte Village, 3 m/s

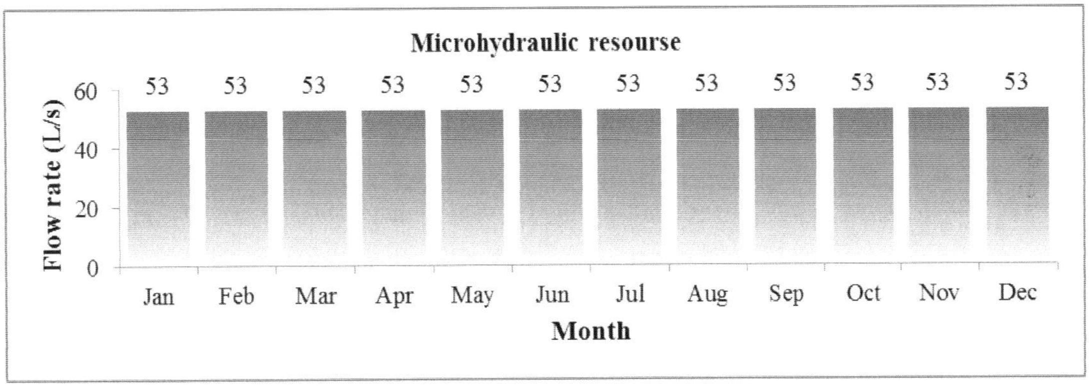

Figure 4. Carhuaquero river flow rate in El Monte Village, 53 L/s

On the contrary, case 3 presents the highest cost of kW/h in the simulation (1.65 Soles kW/h), and this confirms that relying only on a renewable energy technology results in a high cost of electric power because of the oversizing of the system and the high initial investment costs of the wind-turbines.

Case 4 was simulated numerous times because in seasons where there was less solar irradiation, wind speed was usually higher, which makes it an efficient system. However, for the case study, the wind speed was only 3 m/s, so there was an oversizing in panels to meet the demand, which translates into a higher initial cost (633,017 Soles).

Table 4. Characteristics of components and costs

Component	HOMER suffix	Model	Characteristics		Capital cost (Soles)	Operation and maintenance cost (Soles)
Wind turbine	AH10	AEOLOS - H 10kW 3 blades	Average power: Rotor diameter: Wind speed: Noise: Height:	10 kW 8 meters 10 m/s 45 db 12–18 m	93.000,00	24.64
Micro hydro turbine	TURBINE	MONO FLOAT 3 blades	Power: Rotational speed: Weight: Rotor diameter	250–5000 W 90–230 rpm 380 kg 1000 mm	47,820.00	1,460.00

Photovoltaic panel	FV	HONEY 60-cell module	Power: Maximum efficiency: Maximum tension:	265–280 W 17.1% 30.8 V	2,000.00	2,025.00
Inverter	INVERTER	Isolated system inverter	Voltage: Power: Output voltage:	10.5–15 V 300 W 210–240 V	15,000.00	14,400.00
Battery	H3000	Stationary battery	Capacity: Temperature:	3000 Ah −20°C–45°C	7,380.00	7,117.00
Diesel Generator	Generator	Gasoline generator set	Average output: Voltage: Motor power:	10 kW 110–220 V 20 HP	20,400.00	17,000.00

It is thus proven that it is possible to design an HES that can guarantee electricity access to the rural populations without access to the energy distribution system, and this HES is economically sustainable.

6. Impacts

6.1. Economic impact

As defined in the financial analysis, the cost will be 0.62 Soles /kWh as seen in table 3. The expense is 104.75 Soles/month. When compared against the renewable system, the former system would cost approximately 5.58 soles/month, after the first month in which a payment of Soles .500 is made to activate the service.

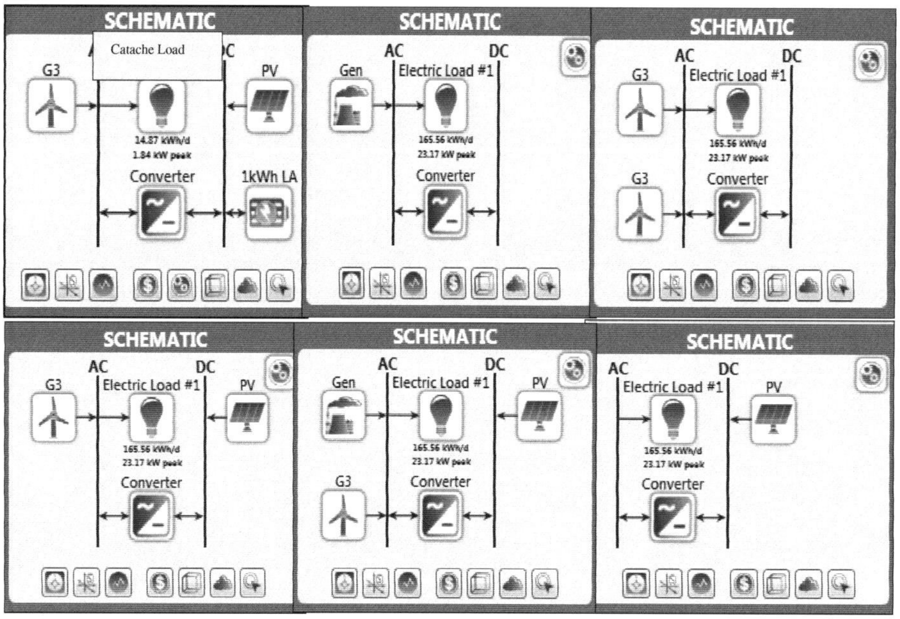

Figure 5. Sets of isolated configurations analysed using HOMER

6.2. Social impact

For the development of projects with a considerable social impact such as rural electrification, community participation is an active process where users not only receive the benefits of the project but also influence the direction and execution of the project.

6.2.1. Social and cultural sustainability in renewable project. Socio-cultural and political issues are part of the earliest stage of project development; for example, the careful selection of users and the appropriate mode of communication with different segments of a community will determine the level of success that a program can hope to achieve. It should not be assumed that a community requires a particular technology and that they will accept it as soon as they have access to it. The introduction of new technology can generate different reactions in different communities.

6.2.2. Focus on the energetic needs of users. The demand for energy technologies in rural areas goes through a long range of problems in which supplying lighting needs is only a part of the scenario. A successful rural electrification program should allow the diversified application of this energy, including money income, entertainment, and education.

6.2.3. Understanding the community. The first step to involve the community is to understand it because the level of participation in a project depends on the context in which it happens. Context may include, for example, population, economic conditions, religious traditions, health, and nutritional benefits, and these factors vary depending on the country. The appropriate development strategies in one country are not necessarily appropriate in another because the context differs.

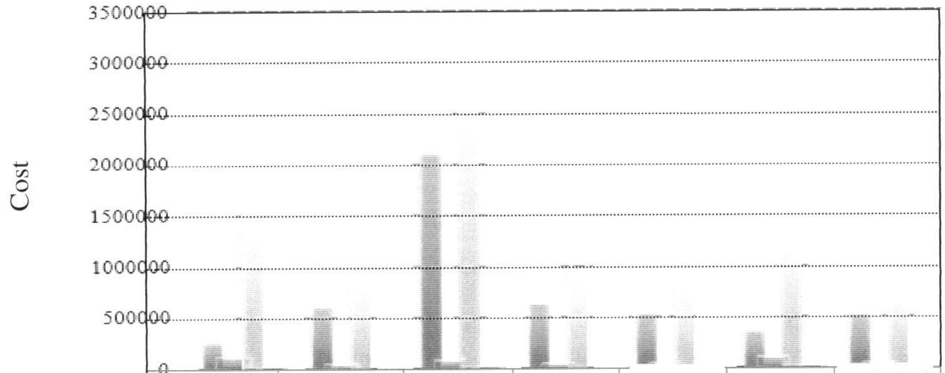

	Case 1 (Diesel Generator)	Case 2 (FV + Batteries)	Case 3 (Wind Turbine)	Case 4 (Wind and Solar Hybrid)	Case 5 (Solar and Microhydraulics hybrid)	Case 6 (FV + and Diesel Hybrid)	Case 7 (Solar, Wind and Diesel Hybrid)
Cap. Initial (S/.)	257820	347432	2109700	633017	529950	552332	521078
Oper. Cost (S/.)	99840	26151	64458	27552	24969	83702	24652
Total Netto (S/.)	1534112	931744	2933694	985220	839135	1422327	836210
COE (S/./kWh)	0.868	0.620	1.659	0.757	0.890	0.805	1.473

Figure 6. Systems of associated costs for each configuration

7. Conclusion

The design of an HES has been presented to provide electricity to a population that has difficult access to electricity and is not connected to a power grid. This strategy is based on global engineering systems stipulated by the IEEE and best practices.

The cost per kW/h achieved by the HES was 0.620 Soles, which equals the cost of electricity supply offered by the conventional grid. This means that such cost is competitive in the market and has the potential to reduce its costs due to the technological improvements and to the population growth expected, because more than 80% of the costs of implementation of the HES are fixed costs.

The solar-hybrid and battery combination is the most profitable for remote rural areas and allows a period for recovery of the investment. However, the cost analysis shows that an increase in energy demand would make the solar microhydraulics and battery bank scenario the most optimal.

It is estimated that the amount of emissions will be reduced by 87,852 $kgCO_2$-eq/KWh for the generation of 228,186 KWh in a period of 25 years. Therefore, it is recommended that future research is conducted on the relationship between the implementation of the HES and the environmental conditions to be faced by populations from these rural areas in the coming year.

References

[1] RK Akikur, R. Saidur, *Comparative study of stand-alone and hybrid solar energy systems suitable for off-grid rural electrification: a review, Renew Sustain*. Energy Rev. 27 (2013) 738-752.

[2] E.D. Giannoulis, D.A. Haralambopoulos, *Distributed generation in an isolated grid: methodology of case study for Lesvos e Greece*, Appl. Energy 88 (2011) 2530e2540

[3] O. Hafez, K. Bhattacharya, *Optimal planning and design of a renewable energy based supply system for microgrids*, Renew. Energy 45 (2012) 7e15.

[4] World Bank. "*Household electrification rate*" 2015:http://worldbank.270a.info/classification/indicator/EG.ELC.HOUS.ZS.html, Date of consultation: June 10, 2015.

[5] Pujarini, S. "*Dharnai: the story of one solar village*" 2015:http://www.greenpeace.org/international/e n/news/Blogs/makingwaves/Dharnai- Solar-Village/blog/53566/,Date of consultation: July 20, 2015.

[6] Ministry of Energy and Mines (MEM). "*PNER: National plan for rural electrification period 2015 - 2024*" 2014: dger.minem.gob.pe/ArchivosDger/PNER_ 2015-2024/F4-PNER-Dptos-2015-24.pdf, Date of consultation: June 15, 2016.

[7] Alliance for Rural Electrification. "*Hybrid mini-grids for rural electrification: Lesson learned*" 2015: http://www.ruralec.org\fileadmin\DAT\.

[8] International Monetary Fund. "*World Economic Outlook: Adjusting to lower commodity prices*", 2015. Documents/06_Publications/Position_pape rs/ARE_Mini-grids_-_Full_version.pdf, Date of consultation: June 10, 2015.

[9] Cota, R. "*Study and evaluation of Micro Networks for isolated communities*" 2015: https://prezi.com/eg8aitw9havv/estudio-y- evaluación-de-micro-redes-para- comunidades-aisladas/, Date of consultation: June 2, 2015.

[10] PERU. Congress of the Republic (2008) Decree Law 1002: Legislative decree to promote investment for the generation of electricity with the use of renewable energies.

[11] ENERGYGOV "*How Microgrids Work*": http://energy.gov/articles/how-microgrids- work, Date of consultation: June 17, 2015.

[12] General Electric. "*About mini grids*" 2015: http://www.generalelectric.com/about- mini-grids, Date of consultation: May 2014.

[13] GVEP International. (2011). *Mini Grid Development in Africa*. Retrieved on July 31, 2015, from Reeep's website: http://www.reeep.org/mini-grid- development-africa.

[14] Salvador, J. "*Access to energy in Peru: balance and policy options*" 2013 http://www.osinergmin.gob.pe/newweb/pa ges/Publico/CongresoInternacional/archivos/JUEVES_30/CTI/1.%20Acceso%20a% 20la%20Energia%20en%20el%20Peru-Julio%20Salvador.pdf, Date of consultation: May 30, 2013.

[15] Shervinzakeri, M. "*Technical Journal of Engineering and Applied Sciences Review on Renewable Energy, Sustainable Energy and Clean Energies*" (2014) ResearchGate, 121-122.

[16] Abdilahi Abdirahman Mohamed, Yatim Abdul Halim Mohd, Mustafa Mohd Wazir, Khalaf Omar Tahseen, Shumran Alshammari Fahad, NorFaizah Mohamed. *Feasibility study of renewable Energy based microgrid systemin Somaliland urban centers*. Renew Sustain EnergyRev2014; 40: 1048-59.

[17] Ngan MeiShan, TanCheeWei. *Assessment of economic viability for PV/wind/ diesel hybrid Energy system in southern peninsular Malasia*. RenewSustain Energy Rev2012;16 (1):634–47.

[18] General Directorate of Rural Electrification (2011) Rural Electrification in Peru. Power Point presentation. Cajamarca.

[19] Shaahid SM, Al-Hadhrami LM, Rahman MK. *Review of assessment of hybrid photovoltaic-diesel-battery power systems for residential loads for different provinces of Saudi Arabia.* Renew Sustain EnergyRev2014; 31: 174-81.

[20] Dursun Bahtiyar. *Determination of theoptimum hybrid renewable power generating systems for Kavakli campus of Kirk lareli University, Turkey.* Renew Sustain Energy Rev 2012;16(8):6183–90.

[21] Dávila, C.; Vilar, D., Villanueva, G., Quiroz, L. Brochure for the evaluation of demand, hydraulic resources, design and installation of micro hydroelectric power plants. Lima: Practical Solutions, 2010.

[22] Ghasemi Abolfazl, Asrari Arash, ZarifMahdi, AbdelwahedSherif. *Techno- economic analysis of stand-alone hybrid photovoltaic-diesel-battery systems for rural electrification in Eastern part of Iran.* RenewSustainEnergyRev 2013;28:456–62.

[23] Fadaeenejad M, RadziMAM, AbKadirMZA, HizamH. *Assessment of hybrid renewable power sources for rural electrification in Malaysia.* RenewSustain Energy Rev2014;30:299–305. Koussa Djohra Saheb, Koussa Mustapha. *A feasibility and cost benefit pro- spection of grid connected hybrid power system.* Renew Sustain Energy Rev 2015;50:628–42.

IOP Conf. Series: Earth and Environmental Science **168** (2018) 012006 doi:10.1088/1755-1315/168/1/012006

Dynamic Economic Scheduling Strategy for a Stand-alone Microgrid System Containing Wind, PV Solar, Diesel Generator, Fuel Cell and Energy Storage:- A Case Study

Lili Wang[1], Linjuan Zhang[1], Changqing Xu[1], Abinet Tesfaye Eseye[2,3], Jianhua Zhang[2] and Dehua Zheng[3]

[1] State Grid Henan Electric Power Company Economic and Technological Research Institute, Songshan South Road, Zhengzhou, Henan, China; 894759965@qq.com (L.Wang); zlinj@163.com (L.Zhang); zzplus@126.com (C.Xu)
[2] North China Electric Power University, Changping District, Beijing, 102206, China; abinet.tesfaye@gmail.com (A.T.Eseye); jhzhang001@ncepu.edu.cn (J.Zhang)
[3] Goldwind Science and Etechwin Electric Co., Ltd; BDA, Beijing, 100176, China; zhengdehua@goldwind.com.cn (D.Zheng)

Abstract. Efficient dynamic economic scheduling for a microgrid is essential to ensure optimal energy utilization and sustainability. In this paper, a day-ahead optimal dynamic scheduling for a stand-alone microgrid containing wind, PV solar, fuel cell, diesel generator and energy storage system is implemented. The primary objective of the dynamic economic scheduling is to minimize the energy production cost, maximize the energy storage system economic benefit and enhance the utilization of the renewables in the microgrid. The Genetic Algorithm (GA) optimization approach is proposed to solve the economic scheduling problem. Fluctuations of the load demands and renewables in the microgrid are considered and relevant predictions have been made to surmount these fluctuations. The proposed economic scheduling strategy has been tested on a case study microgrid in stand-alone mode (Goldwind Microgrid System, Beijing, China). Simulation results have demonstrated that the proposed approach can solve the day-ahead scheduling problem in a reasonably fast computation time. To validate and compare the performances of the proposed strategy, simulation results were also obtained using Pattern Search (PS) optimization technique. Comparisons of simulation results demonstrate the effectiveness of the proposed GA-based dynamic economic scheduling in attaining a minimum total cost of energy production within a short computation time.

1. Introduction

Microgrid is an aggregate of interconnected loads and distributed energy resources (including microturbines, fuel cells, diesel generators, energy storage, renewable energy resources, and all other kinds of distributed energy resources) at distribution level with distinct electrical boundaries that has black start capacity and can operate either in isolated or non-isolated mode. The transmission and distribution over capacity and expensiveness to secure or to expand, the electric and fuel gas prices fluctuations, the power quality, efficiency and reliability issues, cyber and physical security attacks, technological innovations of renewable energy resources are some of the influencing factors for the current shift of the world's energy industry from large interconnected and centralized power systems to the decentralized small power systems called microgrids. Advanced control technologies that can

IOP Conf. Series: Earth and Environmental Science **168** (2018) 012006 doi:10.1088/1755-1315/168/1/012006

combine several generation systems and energy storage units together in self-contained manners as microgrid architecture are emerging to offer electricity customers the opportunity to access reliable and secured electricity locally.

Wind and photovoltaic solar (PV) as the main renewable energy resources, and conventional generation sources such as microturbine (MT), fuel cell (FC) and diesel engines (DE) are the primary distribution generation (DG) resources used in microgrids. These energy sources together with energy storage system (ESS) can support each other to compensate for the intermittent nature of the renewables, and thereby enhance the system reliability and energy sustainability as much as possible. Microgrids can be operated in self-contained manner (isolated microgrids) for electrifying off-grid remote areas, islands, military bases, institutions and industries that are far from the main grid, [1], [2]-[5], or to supply a group of loads in cities, in grid-connected mode of operation [6] and [7].

Robust economic dispatching strategy for microgrids is essential to ensure economic energy utilization and energy sustainability in the smart grid context. As a result of the alternating nature of the renewable energy resources used in microgrids and the electricity price fluctuations in the main grid, economic dispatching is a complex software structure that needs accurate predictions of the load demands and renewable energy resources within the microgrid. The conventional and classical optimization methods are not efficient and fast enough to be used for real-time optimization of this complex economic dispatching problem. Consequently, at the present time, the implementation of intelligent and modern optimization methods for economic dispatching of microgrids has become the key research aim in the field [1], [6]-[12].

This paper intends to apply a heuristic modern optimization method, Genetic Algorithm (GA) [13], for day-ahead economic scheduling of an isolated microgrid, which runs once in a day followed by heuristic hourly decision adjustments based on real-time renewable generations and load demands in every hours of the day. A real industrial microgrid (Goldwind Smart Microgrid System), in Beijing, China, in stand-alone mode, is considered to deliver the power demand requirements of the various loads within the company (Goldwind Science and Etechwin Electric. Co., Ltd.), shown in Fig. 1.

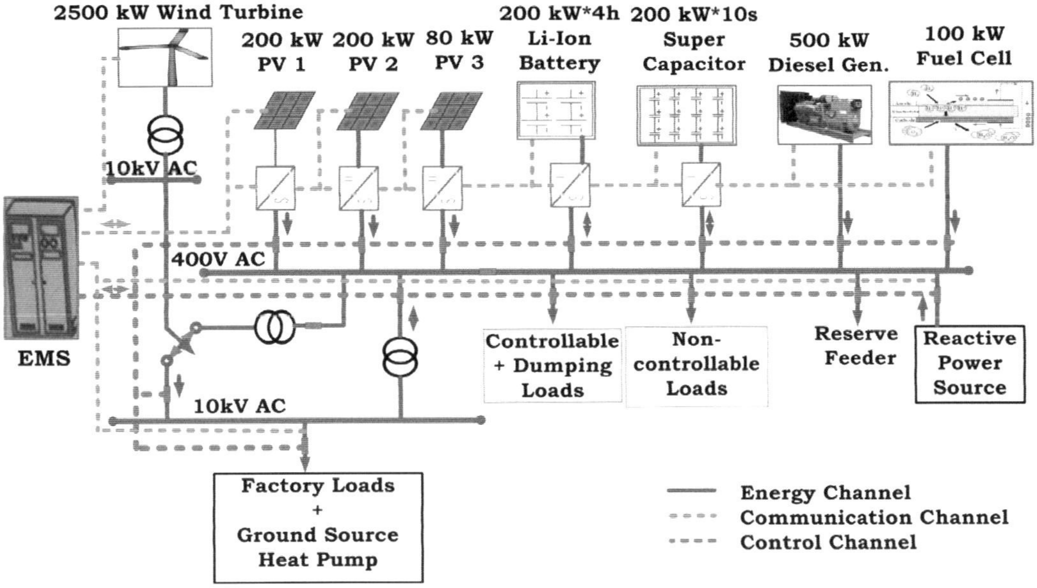

Figure 1. Configuration of the stand-alone microgrid under study.

The main components of the system are wind energy conversion system (WECS) that utilizes a permanent magnet synchronous generator (PMSG), three solar PV systems, fuel cell, diesel generator, a lithium-ion battery (Li-Ion) as ESS. The actual ratings of the components as shown in Fig. 1 are used in this study.

Besides, another direct search optimization method, Pattern Search (PS) was also implemented to validate the dispatch results obtained using GA. Neural networks based day-ahead forecasting of the wind power, PV solar power and the load demands have been carried out using data obtained from the microgrid supervisory control and data acquisition (SCADA) system and numerical weather prediction (NWP) weather prediction results around the microgrid installation site. These forecasting values were used as input variables in the economic scheduling simulation studies, using MATLAB/Simulink as the simulation environment.

The objective function for both optimization methods (GA and PS) is modified in every hour, during which the load demands and generations are supposed to be invariant. The Li-Ion storage unit was used to ensure long-term energy sustainability and to supply the power demanded during each hourly time interval to support the energy imbalance caused by sudden changes in load, wind speed, solar irradiation or other weather parameters. In this paper, a 24-hour simulation studies were carried out to show the work done on dynamic economic dispatching for isolated microgrid. The rapid convergence and global optimality of the GA [13], and the simulation results obtained indicate the robustness of the proposed economic dispatching strategy for microgrids.

This paper is organized as follows. Section 2 presents the objective and constraint function formulations. In Section 3, GA is briefly discussed and the proposed economic scheduling strategy is presented. The case study simulation results are discussed and comparison is provided in Section 4, and finally the paper is concluded in Section 5.

2. Optimization Model for the Dynamic Economic Scheduling Problem

The objective problem of the optimization model for dynamic dispatching in isolated microgrid with the prescribed setting, shown in Fig. 1, is formulated in this section. The microgrid, considered in this research paper, operates in isolated mode which can neither send power to the main grid nor receive power from the main grid, and thus the load demands in the microgrid are satisfied by its local generations. During all the operational periods, the microgrid is expected to minimize the energy production cost within it, maximize the economic benefits of the Li-Ion energy storage unit and ensure the renewable energy utilization to the maximum possible extent.

This objective function is subjected to four decision variables: the power operating points of the diesel generator and the fuel cell, the charging or discharging power of the energy storage unit and its state of charge. The associated constraint functions considering the DGs' capacity and operational limits, the energy storage unit charging and discharging power limits and its state of charge (SOC) conditions, and all other technical requirements within the microgrid are also built in this section.

2.1. Objective Function Formulation

Some prerequisite information for a day-ahead economic scheduling in microgrids should be known in advance [14] and [15]. This information is as follows:

- Day-ahead hourly load forecast
- Day-ahead hourly wind power forecast
- Day-ahead hourly PV power forecast
- Conventional DGs' cost functions and power limits
- ESS charging/discharging power limits and SOC conditions

In this scheduling optimization model, the simulation time interval is taken as one hour; thus the 24-hours ahead renewable generation and load demand forecasts are also given in one hour time interval.

Objective function:

$$Min \sum_{t=1}^{n} \sum_{i=1}^{m} \left(F_i\left(P_i(t)\right)\tau_i(t) + SC_i(t) \right) \tag{1}$$

IOP Conf. Series: Earth and Environmental Science **168** (2018) 012006 doi:10.1088/1755-1315/168/1/012006

Where, n is the number of time intervals for a scheduling day; m indicates the number of all types of conventional and dispatchable units within the microgrid.

$\tau_i(t) = 1$, if the *i*th dispatchable unit is in operation;

$\tau_i(t) = 0$, if the *i*th dispatchable unit is OFF at time t;

The conventional DGs cost functions are given as follows:

$F_i(P_i(t)) = b_iP_i(t) + c_i$, Fuel cell

$F_i(P_i(t)) = a_iP_i(t)^2 + b_iP_i(t) + c_i$, Diesel engine

Here, a_i, b_i and c_i are the corresponding cost function parameters of each conventional DG in the microgrid. $SC_i(t)$ is the start-up cost function of each conventional DG and is given by:

$SC_i(t) = sc_i$, if $\tau_i(t) - \tau_i(t-1)= 1$

SCi(t) = 0, otherwise

Where, sci is the start-up cost of the *i*th conventional DG unit.

2.2. Constraint Functions Formulation

The objective function formulated above is subjected to the following constraints including DGs and ESS capacity and operational limits, and all other technical requirements within the microgrid;

- Power output of the ith unit at time t:

$$P_i^{\min} \le P_i(t) \le P_i^{\max} \tag{2}$$

- Load demand and generation balance:

$$\sum_{i=1}^{m} P_i(t) = P_{load}(t) - P_{wind}(t) - P_{pv}(t) - P_{ess}(t) \tag{3}$$

- ESS charging/discharging power output:

$$P_{ess}^{\min} \le P_{ess}(t) \le P_{ess}^{\max} \tag{4}$$

$P_{ess}(t) > 0$, ESS is discharging power;

$P_{ess}(t) < 0$, ESS is charging power;

$P_{ess}(t) = 0$, ESS is inactive/not in operation

- ESS dynamic operation performance:

$$SOC(t+1) = SOC(t) - \frac{\eta_t \cdot P_{ess}(t)}{C_{ess}} \tag{5}$$

$$SOC^{\min} \le SOC(t+1) \le SOC^{\max} \tag{6}$$

Where, η_t is the charging or discharging efficiency of the ESS; C_{ess} denotes the rated capacity of the ESS; $P_{load}(t)$, $P_{wind}(t)$ and $P_{pv}(t)$ respectively represent the predicted load demands, wind generation and PV solar generation at time t. Thus, the decision variables that need to be determined are the power outputs from the dispatchable DGs $P_i(t)$ (for i =1, 2, …, m), the ESS charging or discharging power output $P_{ess}(t)$ and its state of charge SOC(t) at the same time t.

3. Proposed Optimal Scheduling Strategy

The proposed economic scheduling strategy for the isolated microgrid, with the prescribed setting above, has the following main goals:

1. Minimize the total cost of energy production;
2. Maximize the renewable energy resources utilization;
3. Maximize the economic benefit of the Li-Ion storage;
4. Maximize the Li-Ion life by monitoring its SOC and charging/discharging power;

5. Electrify dumping and testing loads when the load demands are fully satisfied and the Li-Ion is fully charged in case of excess renewable generations;
6. Reduce the use of the fuel fired conventional DGs by enhancing the renewable DGs utilization for less carbon emission.

Fig. 2 illustrates the information flows of the proposed scheduling system, which a scheduling center converts to output commands for dispatchable DGs output power, ESS optimal charging/discharging power and load monitoring action (shed, shift or dump). Decisions are based primarily on renewable generation and load demand forecasts. These decisions are generated in this paper for a day-ahead time horizon on one hour interval basis.

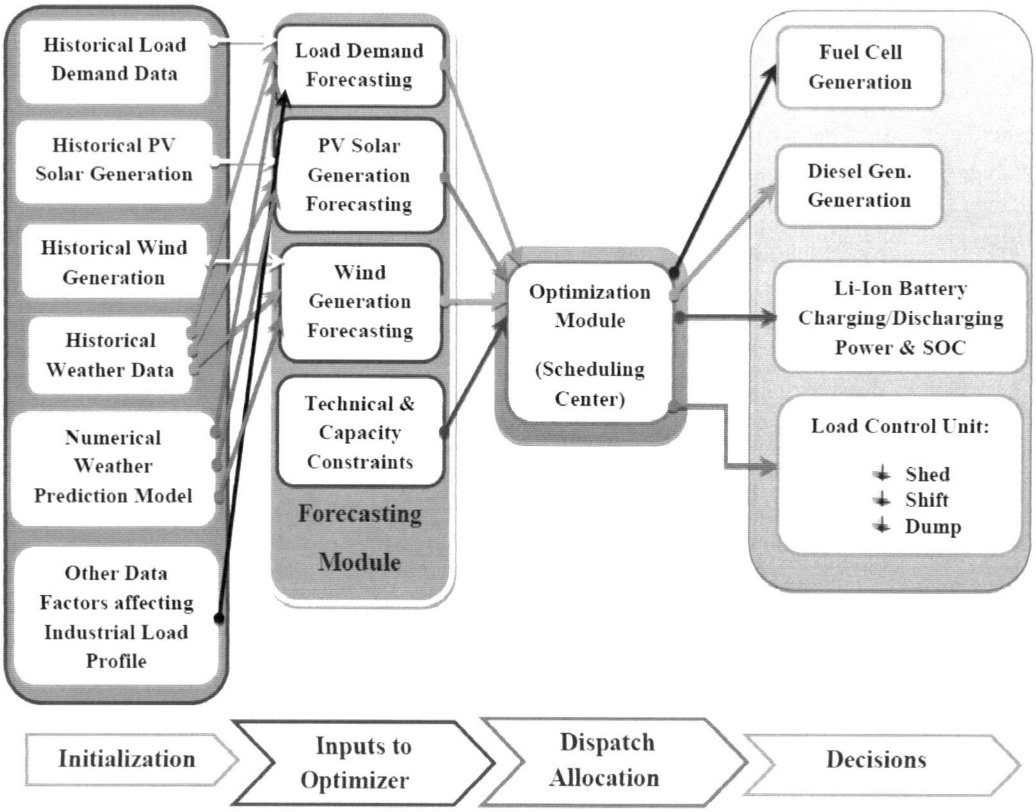

Figure 2. Information flow of the proposed dynamic scheduling strategy.

3.1. Genetic Algorithm (GA)

Most practical optimal design engineering problems are described by mixed continuous-discrete variables, and discontinuous and con-convex design domains. If traditional and standard nonlinear optimization techniques are implemented for this sort of engineering problem they will be inefficient, computationally expensive, and, in most cases, find a relative optimal solution that is closest to the starting point [16]. Genetic algorithms (GAs) are more suitable for solving such engineering problems, and in most cases they are able to find the global optimal solution with a high degree of probability. GAs were first presented systematically by Holland [17]. The GA basic ideas of analysis and design based on the concepts of biological evolution can be found in the work of Rechenberg [18].

GAs were inspired by Darwin's theory of survival of the fittest. They are based on the principles of natural selection and genetics. The natural genetics fundamental elements—reproduction, crossover, and mutation—are used in the genetic search process. GAs differ from the classical techniques of optimization in the following aspects [16]:

1. A population of initial points is used for kicking off the procedure instead of a single design point, thus GAs are less likely to get trapped at a local optimum.
2. GAs do not use the derivatives of the objective function, only the objective function values are used in the search procedure.
3. The design variables in GAs are coded as strings of binary variables that correspond to the chromosomes in natural genetics. Thus the search technique is naturally suited for solving discrete and integer programming engineering problems. The string length can also be varied to attain any desired resolution in the case of continuous design variables.
4. In every new generation, a new set of strings (offspring) is produced by using probabilistic transition rules, not deterministic rules.

Fig. 3 shows a flowchart diagram of GA.

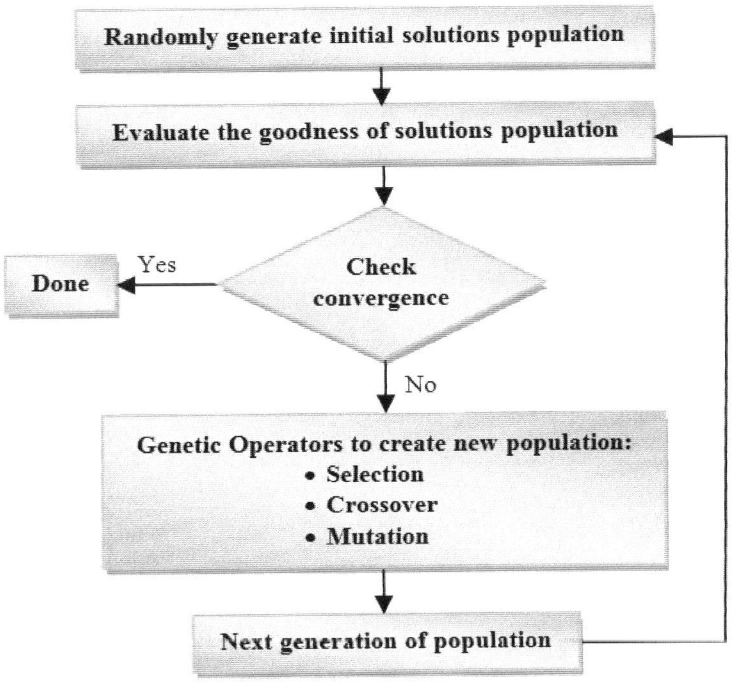

Figure 3. Flowchart of GA.

4. Case Study and Simulation Results

The stand-alone Wind-PV-DE-FC-Li-Ion microgrid in this research study is designed to deliver power to an industrial company, Goldwind Science and Etechwin Electric Co., Ltd., in Beijing, China. The generation capacity limits of the DGs in this microgrid are provided in Table 1, and the dispatchable DGs cost function parameters are also given in Table 2. The 24-hours ahead hourly load demand forecast for a typical day in the microgrid is shown in Fig. 4.

The predicted wind and PV generations, for the same typical day, are also shown in Figs. 5 and 6, respectively.

Fig. 7 shows the GA-based optimal energy scheduling for the same typical day, and the corresponding state of charge of the Li-Ion battery is shown in Fig. 8.

As depicted in Figs. 4 to 8, there is a high generation of wind energy and zero PV solar generation during the first six hours (12am - 6am) of the simulation period. In this period, the generation from the wind energy completely supplies the load demands and charges the Li-Ion battery which was at its minimum SOC (20%) before the simulation started, and all the conventional DGs are off (zero power) to minimize the fuel cost. The Li-Ion battery continuously charges (shown in Fig. 7 as negative power) and reaches its maximum storage capacity (800 kWh or 100% maximum SOC), shown in Fig. 8, at 6am, and its charging power comes to zero (not in operation or inactive).

Table 1. DGs installed capacity.

DG	No. of Units/Systems	Unit/System Power Capacity (kW)		Total Power Capacity (kW)	
		Min Power	Max Power	Min Power	Max Power
FC	1	10	100	10	100
DE	1	10	200	20	500
	1	10	300		
PV	4	0	100	0	480
	1	0	50		
	1	0	30		
Wind	1	0	2500	0	2500
LI-Ion battery*	1	-200	200	-200	200

* Li-Ion: SOC_{min} = 20%, SOC_{max} = 100%, C_{ess} (Capacity) = 800 kWh, η = 1

Table 2. Dispatchable DGs cost function parameters.

DG	Cost Function Parameters			Start-up Cost ($)
	a ($/kWh)2	b ($/kWh)	c ($/h)	
FC	0	0.0164	3.68	18
DE	0.00025	0.0156	0.3312	23

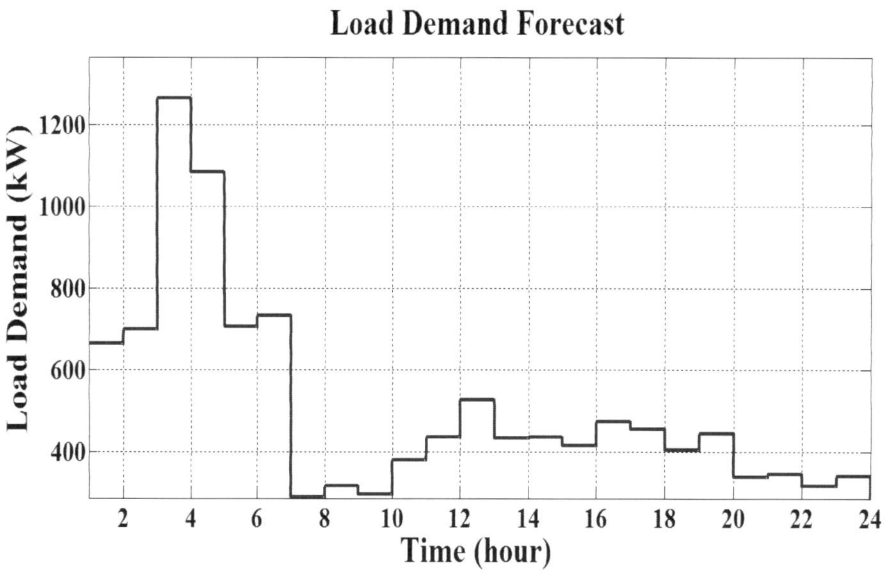

Figure 4. Forecasted load demand.

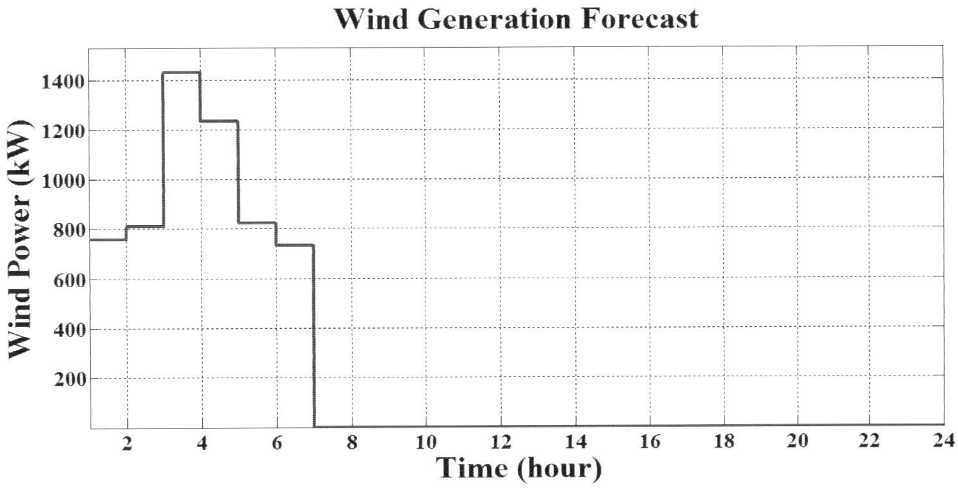

Figure 5. Forecasted wind power generation.

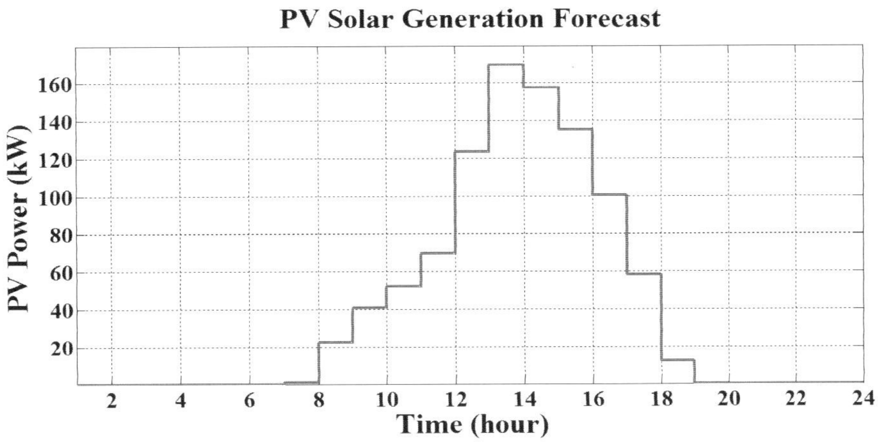

Figure 6. Forecasted PV solar power generation.

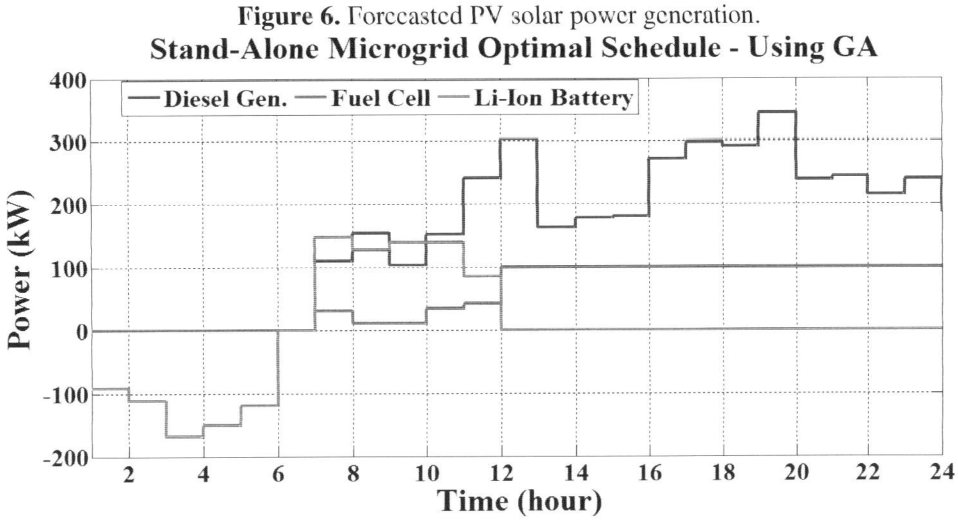

Figure 7. Dynamic economic scheduling using GA.

IOP Conf. Series: Earth and Environmental Science **168** (2018) 012006 doi:10.1088/1755-1315/168/1/012006

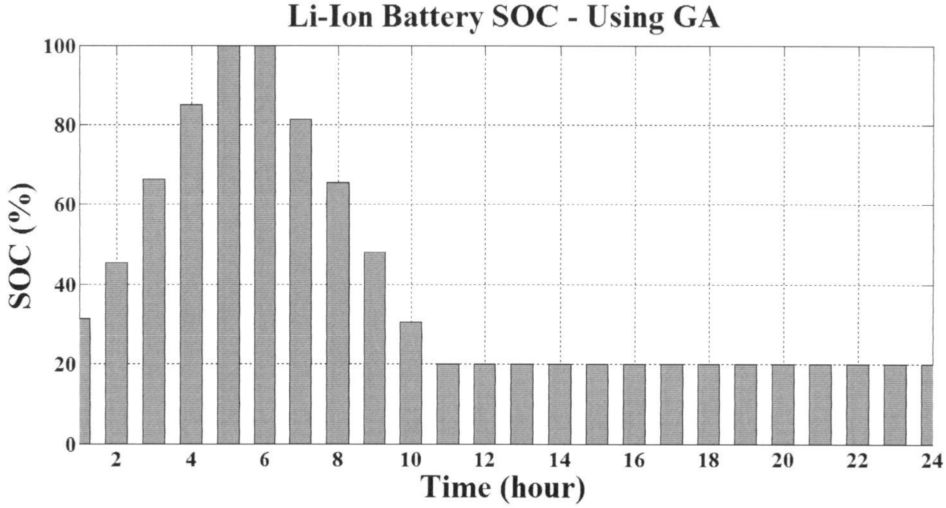

Figure 8. SOC of Li-Ion battery obtained using GA.

Since 7am, the wind generation gets down to zero while the PV source starts generating power. As the power generation from the renewables is insufficient to supply the load demands during this time, the DE and FC start operation to support the PV source. The Li-Ion battery sends or discharges power to the microgrid during this period to reduce the fuel cost. The Li-Ion battery continuously discharges (shown in Fig. 7 as positive power) and reaches its minimum storage capacity (160 kWh or 20% minimum SOC), shown in Fig. 8, at 12pm and its discharging power becomes zero.

To minimize the cost of energy production, the Li-Ion battery is not in operation or inactive (shown in Fig. 7 as zero power) since 12pm; until it will be charged again by an available excess renewable generation in the microgrid and its SOC is kept at its minimum value of 20%. The PV source together with the conventional DGs supplies the load demands from 12pm to 7pm and becomes zero at 7pm. The load demands are solely supplied by the DE and FC from 7pm to 12am as shown in Fig. 7.

In order to validate and compare the performances of the proposed GA-based dynamic economic dispatching strategy, a PS-based dynamic economic dispatching for the same microgrid and system parameters was obtained. Fig. 9 shows the PS-based optimal dynamic energy scheduling simulation result for the same typical day, and the Li-Ion battery state of charge using this approach is also shown in Fig. 10.

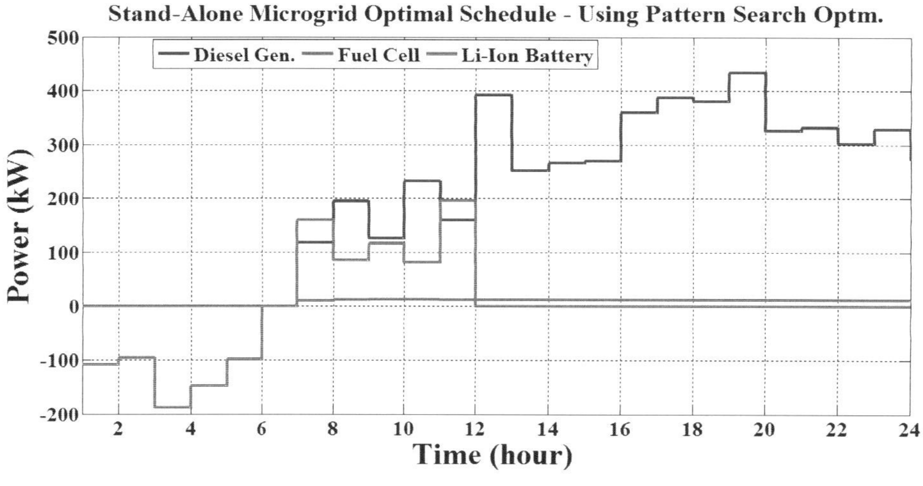

Figure 9. Dynamic economic scheduling using PS.

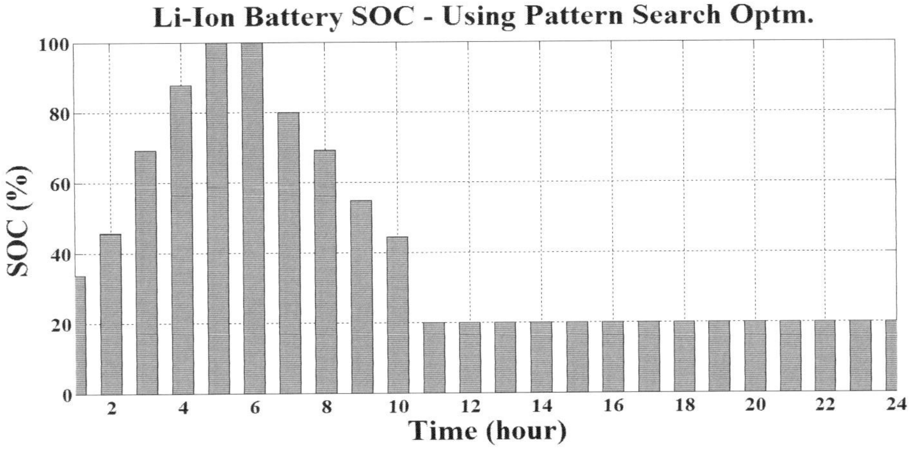

Figure 10. SOC of Li-Ion battery obtained using PS.

The hourly energy production costs obtained using both optimization approaches are shown in Fig. 11. As seen in this figure, the energy production cost is zero until 6am as there is enough generation from the renewables and the fuel fired conventional DGs are off; moreover, it's clearly seen that the GA obtains the minimum energy production cost almost for all of the operation hours throughout the day.

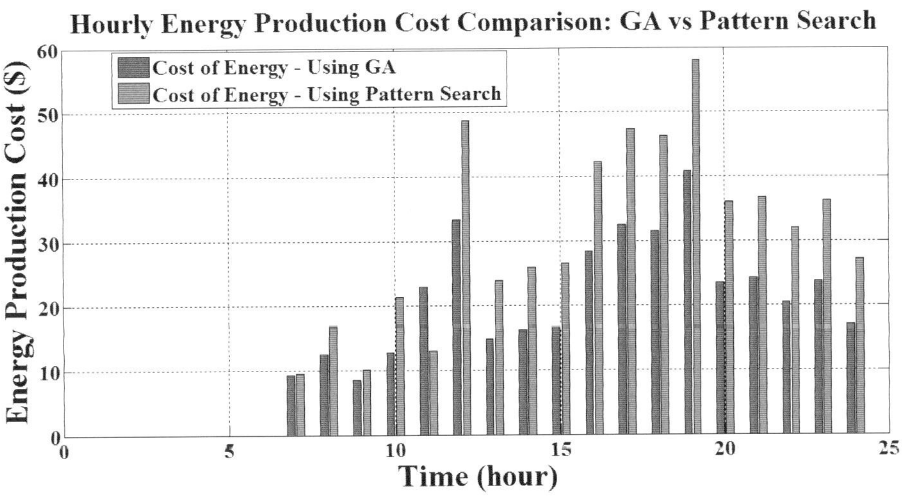

Figure 11. Comparison of hourly energy production cost.

The total energy production cost for the whole day and the hourly average cost by both methods are also presented in Table 3. As shown in Table 3, the GA-based economic scheduling strategy has given a lower cost of energy production than the PS-based dynamic scheduling approach.

Table 3. Cost of energy comparison.

Algorithm	Energy Production Cost ($)	
	Hourly Average Cost	Daily Total Cost
GA	16.2225	389.3402
PS	23.0382	552.9169

Table 4 gives the total computation time taken by both dynamic scheduling optimization approaches in a Matlab/Simulink simulation environment (using Intel (R) core (TM i5-5200 CPU, 2.20GHz processor, 4GB RAM, and 64-bit operating system computer).

Table 4. Computation time.

Algorithm	Total (for 24-hour ahead scheduling simulation) Computation Time (second)
GA	1.6824
PS	3.9452

The GA-based strategy has allocated the dynamic economic schedule within a short period of time compared to the PS-based scheduling approach as given in Table 4 above.

5. Conclusions

GA-based dynamic economic scheduling strategy for a stand-alone microgrid containing wind energy conversion system, PV solar, diesel generator, fuel cell and Lithium-ion battery was presented in this paper. The proposed economic scheduling strategy takes into account the variation of load demands and intermittency of renewable energy resources in the microgrid, and thus appropriate day-ahead forecasts have been made to withstand these fluctuations. Simulation results confirm the suitability and possible advantages of the developed dynamic economic scheduling model in minimizing the total daily energy production cost, maximizing the energy storage economic benefits, and enhancing the utilization of renewable energy resources in the microgrid. To test and validate the proposed economic scheduling strategy performances, simulation results were also carried out for the same microgrid system using pattern search (PS) optimization technique. Comparison of simulation results demonstrate the robustness of the proposed GA-based dynamic economic scheduling in obtaining minimum cost of energy production. Furthermore, the proposed approach results global optimum solutions and converges rapidly that shows its capability for real time dynamic dispatching of microgrids.

References

[1] A.T.Eseye, Jianhua Zhang, Dehua Zheng, and Dan Wei, "Optimal Energy Management Strategy for an Isolated Industrial Microgrid Using a Modified Particle Swarm Optimization," *IEEE 2016 International Conference on Power and Renewable Energy*, pp. 494 - 498, Shanghai, China, October 21-23, 2016.

[2] C. Wang, M. H. Nehrir, C. M. Colson, and J. Li, "Power management of a stand-alone hybrid wind-microturbine distributed generation system," in *Proc. Power Electronics and Machines in Wind Applications (PEMWA 2009)*, Jun. 24–26, 2009, pp. 1–7.

[3] R.-J. Wai, C.-Y. Lin, R.-Y. Duan, and Y.-R. Chang, "High-efficiency power conversion system for Kilowatt-level stand-alone generation unit with low input voltage," *IEEE Trans. Ind. Electron.*, vol. 55, no. 10, pp. 3702–3714, Oct. 2008.

[4] F. Valenciaga and P. F. Puleston, "Supervisor control for a stand-alone hybrid generation system using wind and photovoltaic energy," *IEEE Trans. Energy Convers.*, vol. 20, no. 2, pp. 398–405, Jun. 2005.

[5] C.Wang and M. H. Nehrir, "Power management of a stand-alone wind/photovoltaic/fuel cell energy system," *IEEE Trans. Energy Convers.*, vol. 23, no. 3, pp. 957–967, Sep. 2008.

[6] Abinet Tesfaye Eseye, Dehua Zheng, Han Li, and Jianhua Zhang, "Grid-price Dependent Optimal Energy Storage Management Strategy for Grid-connected Industrial Microgrids", *9th Annual IEEE Green Technologies Conference (GreenTech)*, Denver, Colorado, USA, March 29-31, 2017.

[7] Han Li, Abinet Tesfaye Eseye, Jianhua Zhang, and Dehua Zheng, "Optimal Energy

Management for Industrial Microgrids with High-Penetration Renewables", *Journal of Protection and Control of Modern Power Systems (PCMP)-Springer*, vol. 2, no. 1, April, 2017.
[8] V. Miranda and H. P. Sio, "Economic dispatch model with fuzzy wind constraints and attitudes of dispatchers," *IEEE Trans. Power Syst.*, vol. 20, no. 4, pp. 2143–2145, Nov. 2005.
[9] C. L. Chen, "Simulated annealing-based optimal wind-thermal coordination scheduling, generation," *Transmission & Distribution, IET*, vol. 1, no. 3, pp. 447–455, May 2007.
[10] Abinet Tesfaye Eseye, Jianhua Zhang, and Dehua Zheng, "Optimal Operational Planning for Microgrids under Grid-connected and Islanded Operation Modes", *CSEE Journal of Power and Energy Systems (JPES)*. (Accepted for publication)
[11] C. Chun-Lung, "Optimal wind-thermal generating unit commitment," *IEEE Trans. Energy Convers.*, vol. 23, no. 1, pp. 273–280, Mar. 2008.
[12] J. Hetzer, D. C. Yu, and K. Bhattarai, "An economic dispatch model incorporating wind power," *IEEE Trans. Energy Convers.*, vol. 23, no. 2, pp. 603–611, Jun. 2008.
[13] Yiwei Ma, Ping Yang, Hongxia Guo and Yuewu Wang, "Dynamic Economic Dispatch and Control of a Stand-alone Microgrid in DongAo Island," *J Electr Eng Technol (JEET)*, vol. 10, pp. 30-40, Oct, 2015.
[14] Zheng Zhao, "Optimal Energy Management for Microgrids", PhD Dissertation, Electrical and Computer Engineering, Clemson University, Clemson, south Calorina, USA, January 2012.
[15] Alberto Borghetti, Mauro Bosetti and Samuele Grillo, "Short-term scheduling and control of active distribution systems with high penetration of renewable resources," *IEEE Systems Journal*, vol. 4, No. 3, September, 2010.
[16] Singiresu S. Rao, *Engineering Optimization: Theory and Practice, Fourth Edition*, John Wiley & Sons, Inc., 2009.
[17] J. H. Holland, *Adaptation in Natural and Artificial Systems*, University of Michigan Press, Ann Arbor, MI, 1975.
[18] I. Rechenberg, *Cybernetic Solution Path of an Experimental Problem*, Library Translation 1122, Royal Aircraft Establishment, Farnborough, Hampshire, UK, 1965.

Acknowledgments
This work is supported financially and technically by the State Grid Henan Electric Power Company Economic and Technological Research Institute, North China Electric Power University, and Goldwind Science and Etechwin Electric Co., Ltd.

IOP Conf. Series: Earth and Environmental Science **168** (2018) 012007 doi:10.1088/1755-1315/168/1/012007

The optimal operation strategy and simulation for CCHP system

Zhenglong Zhao[1], Bin Zou[1,2]

1. School of Mechatronic Engineering and Automation, Shanghai University, Shanghai, China
2. Shanghai Key Laboratory of Power Station Automation Technology, Shanghai China

Abstract. This paper summarizes the typical operation strategy for CCHP, and the operation problem is modeled as nonlinear programming in which the relationship between generating and heating and the generating efficiency change with the power generating are considered as the constraints. The KKT condition of the optimal operation model are investigated for the typical operation strategy, and the condition that the typical operation strategy is optimal operation strategy are obtained. A CCHP simulation algorithm based on optimal operation strategy is presented in this paper. The numerical results demonstrated that the operation cost based the presented operation strategy reduces significantly comparison with the traditional operation strategy.

1. Introduction

The distributed energy system with 80% primary energy utilization and environmental benefits have be become one of the important means to solve the city's energy supply [1]. The most significant feature of the distributed energy system is the recovery of waste heat from power generation to supply heat energy for users around. There is a tight relationship between generating and heating in combined heating power system which is the key equipment in the distributed energy system, and the generating and heating efficiency change with the power generating [2].

The operation strategy of combined heating power system to meet the load profile is an important issue. The load profiles are not match the relationship between generating and heating of CHP system, so the operation strategy to dispatch the PGU and other equipment to meet the load demand will affect directly the primary energy ratio, economic and environmental benefits [3]. The two famous operation strategy are following thermal load (FTL) and following electric load (FEL). The previous studies show that it is a good strategy that the CHP system operates under FEL when the thermal load is comparatively larger than the electric load, and under FTL when the electric load is larger [4][5]. The operation strategy is called Hybrid electric–thermal load (HET).

The operation strategy of combined cooling, heating, power system (CCHP) is more complex than CHP's. The absorption chiller and electric chiller are used cooling, and the equivalent thermal electric profile change with the various chiller's operation profile. Using the absorption chiller to supply the cooling, [6] show that the performance of CCHP in primary energy consumption, operation cost and environment is better than that of independent suppling system, [7] and [8] presented a new operation strategy dispatch the chillers, the results are better than HET, but do not know whether it is optimal.

The operation strategy of CCHP system is also included in many researches on optimal operation of distributed generations in micro-grids and the energy hub. The linear programming, nonlinear programming, dynamic programming, mixed integer linear programming and mixed nonlinear

IOP Conf. Series: Earth and Environmental Science **168** (2018) 012007 doi:10.1088/1755-1315/168/1/012007

programming are used to model the operation problem of CCHP system [9]-[12].

This paper summarizes the typical operation strategy for CCHP, and the operation problem is modeled as nonlinear programming in which the relationship between generating and heating, the generating efficiency change with the power generating are considered. The KKT condition of the optimal operation model are investigated for the typical operation strategy, and the condition that the typical operation strategy is optimal operation strategy are obtained. A CCHP simulation algorithm based on optimal operation strategy is presented in this paper. The numerical results demonstrated that the operation cost based the presented operation strategy reduces significantly comparison with the traditional operation strategy.

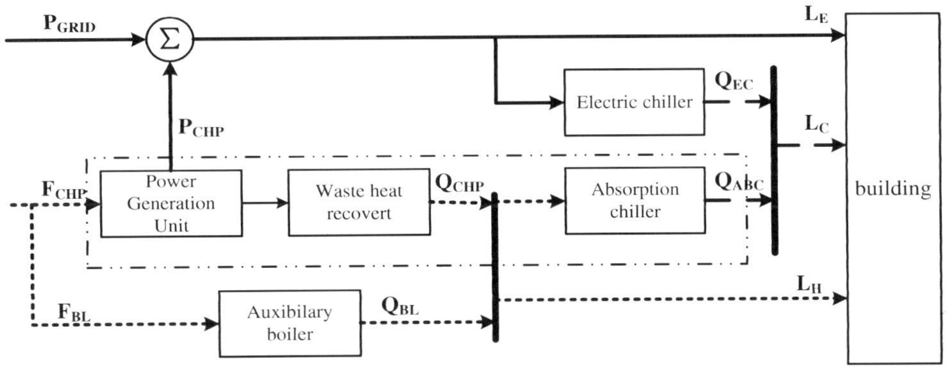

Fig.1 Schematic of the CCHP system

2. Operation Strategy of the Combined Heating Power System

A schematic of the CCHP system is shown in Fig.1. There are three energy bus in the system, names thermal bus, cooling bus and power bus. The energy demands of consumers are served in three ways: 1) power is provided by power generation unit (PGU) or/and power grids; 2) heat is provided by waste heat recover from PGU or/and auxiliary boiler; 3) cold is provided by electric chiller or/and absorption chiller.

It is important features that the thermal recover is dependent on the power generation by the PGU. The relationship between the production electricity energy by PGU and the thermal recover are represent by equation (1)-(4).

$$P_{CHP}(t) = F_{CHP}(t) \cdot \eta_{CHP.E} \tag{1}$$

$$Q_{CHP}(t) = F_{CHP}(t)(1-\eta_E) \cdot \eta_{CHP.H} \tag{2}$$

$$f = P_{CHP} / P_{CHP.MAX} \tag{3}$$

$$\eta_{CHP.E} = a + b \cdot f + c \cdot f^2 \tag{4}$$

By the equation (1)-(4), the production electricity energy by PGU and the thermal energy from waste recover is a function that is shown in Fig.2 as Q-E curve. If the production electricity energy known, the thermal energy from the waste heat recover can be determined by the function, in turn which can also be found.

In the system, the load balance can be expressed as (5)-(7).

$$L_E(t) - P_{CHP}(t) - P_{GRID}(t) + Q_{EC}(t) / COP_{EC} = 0 \tag{5}$$

$$L_H(t) + Q_{ABC}(t) / COP_{ABC} - Q_{CHP}(t) - Q_{BL}(t) = 0 \tag{6}$$

$$L_C(t) - Q_{ABC}(t) - Q_{EC}(t) = 0 \tag{7}$$

2.1 *The FTL and FEL operation strategy*

The Q-E curve of the generation and waste heat recover separates the space into two parts in Fig.2.

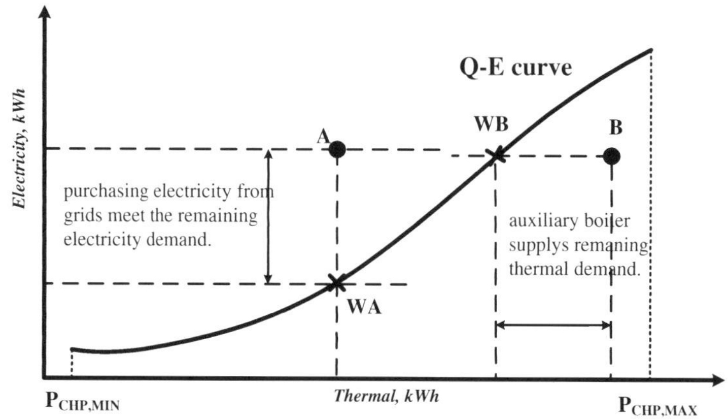

Fig.2 the Q-E curve of generating energy and recovered waste heat

The heat and electric load profiles in the left of the Q-E curve, known as the larger electric load, like as point A in Fig.2, the case can expressed as

$$\frac{L_E}{L_H} > \frac{L_E}{Q_{CHP}(L_E)} \tag{8}$$

The $Q_{CHP}(L_E)$ means that recovered heat when production electricity energy is L_E, which follows the function determined by Q-E curve.

The case that the heat and electric load is in the right of the Q-E curve known as the larger thermal load, like as point B in Fig.2, that is,

$$\frac{L_E}{L_H} < \frac{L_E}{Q_{CHP}(L_E)} \tag{9}$$

In the larger electric load, the following thermal load strategy (FTL) known as good strategy. The load profile point A in Fig.2, the work point of CHP is WA in Fig.2 when the FTL strategy is used, the thermal energy recovered from waste heat by PGU generation meets the thermal demand, but the generation by PGU is less than the electric load. The insufficient demands are provided by purchasing from generation grids.

In the larger thermal load, the following electric load (FEL) strategy is considered to be a good choose. The load profile B point in Fig.2, the work point is WB according to the FEL strategy, the electricity generated by PGU is equal to the electric load, and the recovered waste heat from the PGU is not enough to meet the thermal load, additional heat must be provided by the auxiliary boiler.

There are some research investigated the advantage through simulation, As far as I know, the optimality of this operation strategy has not been confirmed. In fact, there are other feasible operation strategy, for example, selecting the work point between the WA and WB when the load profile is the point A.

2.2 *The optimal model of combined thermal power system*

For certain thermal and electric load profile, an optimal model can be used to determine the output of the equipment. In the section, the cold load is zero, the absorption chiller and the electric chiller are not considered. The optimal operation model can be expressed as

$$\min \rho_{CHP} \cdot F_{CHP}(t) + \rho_{BL} \cdot F_{BL}(t) + \rho_{GRID} \cdot P_{GRID}(t) \tag{10}$$

s.t.

$$P_{CHP}(t) = F_{CHP}(t) \cdot \eta_{CHP.E} \tag{11}$$

$$Q_{CHP}(t) = F_{CHP}(t)(1-\eta_E) \cdot \eta_{CHP.H} \tag{12}$$

$$f = P_{CHP} / P_{CHP.MAX} \tag{13}$$

$$\eta_{CHP,E} = a + b \cdot f + c \cdot f^2 \tag{14}$$

$$L_E(t) - P_{CHP}(t) - P_{GRID}(t) = 0 \tag{15}$$

$$L_H(t) - Q_{CHP}(t) - Q_{BL}(t) = 0 \tag{16}$$

$$Q_{BL}(t) = F_{BL}(t) \cdot \eta_{BL} \tag{17}$$

$$P_{CHP,MIN} \leq P_{CHP}(t) \leq P_{CHP,MAX} \tag{18}$$

$$Q_{BL,MIN} \leq Q_{BL}(t) \leq Q_{BL,MAX} \tag{19}$$

$$P_{GRID,MIN} \leq P_{GRID}(t) \geq P_{GRID,MAX} \tag{20}$$

The objective function of the model is operation cost (10), the decision variable are the generation energy by PGU, the buying electricity energy from power grids, and the thermal energy produced by auxiliary boiler. The constraints include combined heating power production (11)-(14), load balancing (15) and (16), auxiliary boiler production (17) and the maximum and minimum limit on output of each device's (18)-(20). Without loss of generality, assume all minimum limit is zero.

The Lagrange function of the model is

$$\begin{aligned}\mathscr{L} = {} & \rho_{CHP} \cdot F_{CHP}(t) + \rho_{BL} \cdot F_{BL}(t) + \rho_{GRID} \cdot P_{GRID}(t) \\ & + \lambda_E\left(L_E(t) - P_{CHP}(t) - P_{GRID}(t)\right) \\ & + \lambda_H\left(L_H(t) - Q_{CHP}(t) - Q_{BL}(t)\right) \\ & + \mu_{CHP}^{+}\left(P_{CHP}(t) - P_{CHP,MAX}\right) + \mu_{CHP}^{-}\left(P_{CHP,MIN} - P_{CHP}(t)\right) \\ & + \mu_{BL}^{+}\left(Q_{BL}(t) - Q_{BL,MAX}\right) + \mu_{BL}^{-}\left(Q_{BL,MIN} - Q_{BL}(t)\right) \\ & + \mu_{GRID}^{+}\left(P_{GRID}(t) - P_{GRID,MAX}\right) + \mu_{GRID}^{-}\left(P_{GRID,MIN} - P_{GRID}(t)\right)\end{aligned} \tag{21}$$

The optimal solution of the model must meet the KKT condition. The KKT condition of the model include three part. The one is the derivative of the Lagrange function equal to zero, that is,

$$\frac{\partial \mathscr{L}}{\partial P_{CHP}} = \rho_{CHP} \cdot \frac{\partial F_{CHP}(t)}{\partial P_{CHP}} - \lambda_E - \lambda_H\left(\frac{\partial Q_{CHP}(t)}{\partial P_{CHP}}\right) + \mu_{CHP}^{+} - \mu_{CHP}^{-} = 0 \tag{22}$$

$$\frac{\partial \mathscr{L}}{\partial Q_{BL}} = \rho_{BL} \cdot \frac{1}{\eta_{BL}} - \lambda_H + \mu_{BL}^{+} - \mu_{BL}^{-} = 0 \tag{23}$$

$$\frac{\partial \mathscr{L}}{\partial P_{GRID}(t)} = \rho_{GRID} - \lambda_E + \mu_{GRID}^{+} - \mu_{GRID}^{-} = 0 \tag{24}$$

The detailed expression of the $\partial F_{CHP}/\partial P_{CHP}$ and $\partial Q_{CHP}/\partial P_{CHP}$ see appendix.

The second part of the KKT condition are the constraints of model. The third part is KKT multipliers of the inequality constraint greater than or equal to zero and the complementary slackness, that is,

$$\mu_*^{(*)} \geq 0 \tag{25}$$

$$\mu_*^{(+)}(x - x_{\max}) = 0 \tag{26}$$

$$\mu_*^{(-)}(x_{\min} - x) = 0 \tag{27}$$

If the operation strategy is optimal, the KT condition mast be meet by the results of the operation strategy. According to the load profile and operation strategy, the constraint and complementary slackness, the KKT multipliers can be determined, then the solvability condition of equation (22)-(24) are investigated.

2.3 *The optimal operation strategy*

Assume the heat and electric load profile meet the inequality (8), and the FTL strategy is used, then the recovered waste heat meet the thermal demand, the electric load is meet by PGU generation and purchasing power grid. It means that the output of PGU, purchasing electricity is between their maximum and minimum limit, the output of the auxiliary boiler is equal to the minimum output, is zero. Based on formula (26) and (27), the KKT multipliers of the constraint are $\mu_{CHP}^{(+)} = \mu_{CHP}^{(-)} = \mu_{GRID}^{(+)} = \mu_{GRID}^{(-)} = \mu_{BL}^{(+)} = 0$ and $\mu_{BL}^{(-)} > 0$. Substituting these KKT multiplier into equation (22), (23) and (24), then $\lambda_E = \rho_{grid}$ and

$$\lambda_H = \frac{\rho_{CHP} \cdot \frac{\partial F_{CHP}(t)}{\partial P_{CHP}} - \rho_{GRID}}{\frac{\partial Q_{CHP}(t)}{\partial P_{CHP}}} < \frac{\rho_{BL}}{\eta_{BL}} \tag{28}$$

If the condition (28) is meet, then the FTL is optimal strategy.

Other strategy in electric load lager always is not the optimal strategy. When the work point of combined heat power system is over WA, the load balance is not meet, so the KT condition is not meet. And when the work point is at the bottom of the point WA, the generation electricity by PGU and the recovered heat all are not meet the load, purchasing electricity and auxiliary boil should be run at same time. It means the KT multiplier are$\mu_{CHP}^{(+)} = \mu_{CHP}^{(-)} = \mu_{GRID}^{(+)} = \mu_{GRID}^{(-)} = \mu_{BL}^{(+)} = \mu_{BL}^{(-)} = 0$, substituting these KKT multiplier into equation (22), (23), and (24), the (22) is not meet. In the larger electric load, the FTL is the optimal operation strategy.

Assume the heat and electric load profile meet the inequality (9), and the FEL strategy is used, the electric load is meet by the generating electricity by PGU, the thermal load is meet by recovered waste heat and product thermal by auxiliary boil. The output of the combined heat power system and the auxiliary boiler is between the maximum and minimum, the purchasing electricity is equal to zero. The KKT multipliers of the constraint are$\mu_{CHP}^{(+)} = \mu_{CHP}^{(-)} = \mu_{BL}^{(+)} = \mu_{BL}^{(-)} = \mu_{GRID}^{(+)} = 0$ and $\mu_{GRID}^{(-)} > 0$. By (23) can know$\lambda_H = \rho_{BL}/\eta_{BL}$, substituting these value into (22), and considering the (23), there are

$$\lambda_E = \rho_{CHP} \cdot \frac{\partial F_{CHP}(t)}{\partial P_{CHP}} - \left(\frac{\rho_{BL}}{\eta_{BL}}\right)\left(\frac{\partial Q_{CHP}(t)}{\partial P_{CHP}}\right) < \rho_{GRID} \tag{29}$$

When the inequality (29) is true, the results of the FTL operation strategy meet the KkT condition. In normal price and the efficiency of the combined heat power system, the inequality (29) is true.

Follow the same logic, it can be proved that the other operation strategy in lager thermal load are not an optimal.

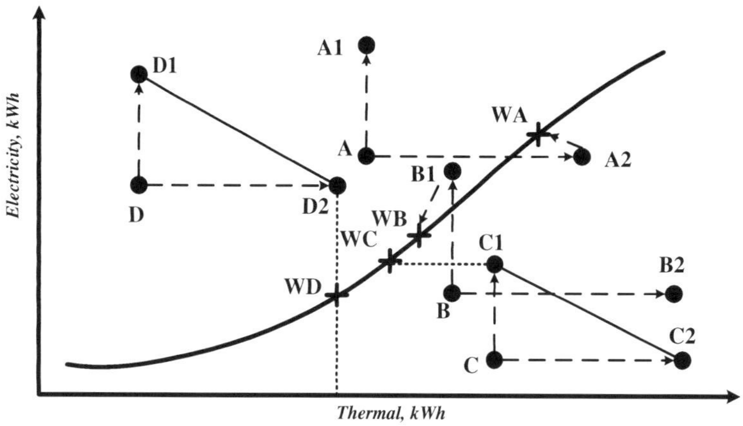

3. The operation strategy of the combined cooling heating power system

3.1 *The typical operation strategy*

When the cold demand is considered, the operation strategy is more complicated than only thermal and electric demand. In the section, nine typical strategy are summed up as alternative strategies for optimal operation strategy. The cold, thermal and electric demand profile are divided into four region. For the convenience of description, let us define the equivalence thermal load and the equivalence electric load.

$$L_{E,E}(t) = L_E(t) + Q_{EC}(t)/COP_{EC} \tag{30}$$

$$L_{H,E}(t) = L_H(t) + Q_{ABC}(t) / COP_{ABC} \tag{31}$$

According to the load profile and the equivalence load profile, four states are defined that covers all load space.

State 1. When the thermal and electric demand meet the inequality (8) and the equivalence thermal load and electric load meet the inequality (9) while the cold demand is all supplied by absorption chiller.

State 1 can be explained as shown in Fig.3, the point A in which the thermal and electric load profile meet the inequality (8), and the point A2 is the equivalence load profile that the cold demand is met by absorption chiller, and the equivalence thermal load and electric load meet the inequality (9). A1 is the equivalence load profile that the cold demand is met by electric chiller, and the thermal load and the equivalence electric load meet the inequality (9). The points on the line connecting point A1 and point A2 represent the equivalent load while the absorption chiller and electric chiller work at same time to supply the cooling.

In state 1, there are three typical operation strategies. The first is called EC-FTL (Electric Cooling Following the thermal load, EC-FTL), while the cold demand is all met by electric chiller. The equivalence load profile located in point A1, then the thermal load is supplied by recovered waste heat, and the equivalence load is supplied by PGU and purchasing electricity from grids. The second typical operation strategies is called HC-FEL (Thermal Cooling Following the electric load, HC-FEL), while the cooling demand is all met by absorption chiller. The equivalence load profile located in point A2, the electric load is met by PGU, and the equivalence thermal load is supplied by the recovered waste heat and the auxiliary boiler.

The third typical operation strategy is called TEWM (Thermal Electric While Meeting, TEWM). The absorption chiller and electric chiller are run at same time to supply the cold demand in the operation strategy, and the equivalence load located in point WA that is the intersection of the Q-E curve and segment A1-A2. Although any point on segment A1-A2 corresponds to an operation strategy, the point WA is the most likely optimal strategy, because of the cooling heating and electricity load are all supplied by CCHP system.

State 2. When the thermal and electric demand meet the condition (9) and the equivalence electric load and thermal load meet the inequality (8) while the cold demand is all met by electric chiller.

For example, the thermal and electric demand profile located at point B in Fig.3, the equivalent load profile located at point B1 while electric chiller supplies the cold demand, and the point B1 and point B are located at above and below the Q-E curve.

In state 2, the typical operation in state 2 is same in state 1.

State 3. When the thermal and electric demand meet the inequality (8) and the equivalence thermal load and electric load also meet the inequality (8) while the cold demand is all met by absorption chiller.

All the equivalence load used various operation strategy in State 3 satisfy the inequality (8), it means that the typical operation strategy for the equivalence load is always FTL. There are three typical way to produce the equivalence load. The first typical way is that electric chiller supplies the cold demand, the second typical way is that absorption chiller met the cold demand, and the third is that two type chiller work at same time to supply the cold demand. So the typical operation strategies in state 3 are the Electric Cooling Following the thermal load (EC-FTL), Thermal Cooling Following the thermal load, HC-FTL and Joint Cooling Following the thermal load (JC-FTL).

The point D in Fig.3 represent the load profile, and point D1 represents the equivalence load profile while electric chiller meet the cold demand, point D2 presents the equivalence load profile while absorption chiller meet the cold demand. The points on the segment D1-D2 represent that the absorption chiller and the electric chiller supply the cooling at same time.

State 4. When the thermal and electric demand meet the inequality (9) and the equivalence thermal and electric load meet the inequality (9) while the cold demand is all met by electric chiller.

A load profile which is stat 4 is shown in Fig.3 as point D. when the cold demand is all met by electric chiller, the point D1 represents the equivalence load profile, while point D2 represents the equivalence load profile that the absorption chiller supply the cold demand. The segment D1-D2 represent the equivalence load profile that the two type chiller work at same time to supply the cooling.

The equivalence load profiles meet the inequality (9), so the operation strategy for the equivalence load is FEL.

There are three typical operation strategies in state 4, that is, the Electric Cooling Following the electric load (EC-FEL), Thermal Cooling Following the electric load, HC-FEL and Joint Cooling Following the electric load (JC-FEL).

3.2 *The optimal operation model of CCHP*

Considering the cooling demand, the optimal operation model (10)-(20) is modified, the load balance equation (15) and (16) are replaced by (5), (6) and (7). And two new constraints are added as

$$Q_{EC,MIN} \le Q_{EC}(t) \le Q_{EC,MAX} \tag{32}$$

$$Q_{ABC,MIN} \le Q_{ABC}(t) \le Q_{ABC,MAX} \tag{33}$$

The adding new decision variable are the producing cooling $Q_{ABC}(t)$ by absorption chiller and producing cooling $Q_{EC}(t)$by electric chiller.

The KKT condition add two new equation in addition to equation (22)-(24), that is,

$$\frac{\partial \mathscr{L}}{\partial Q_{ABC}(t)} = -\lambda_C + \lambda_H \frac{1}{COP_{ABC}} + \mu_{ABC}^{+} - \mu_{ABC}^{-} = 0 \tag{34}$$

$$\frac{\partial \mathscr{L}}{\partial Q_{EC}(t)} = \lambda_E \frac{1}{COP_{EC}} - \lambda_C + \mu_{EC}^{+} - \mu_{EC}^{-} = 0 \tag{35}$$

3.3 *The optimal condition of the operation strategy*

A. the optimal strategy in state 1 and state 2

Using TEWM strategy in state 1, the PGU, absorption chiller and electric chiller work between their maximum and minimum limit, and the auxiliary boil's output and the purchasing electricity are aero that is their minimum limit. So the KKT multipliers are $\mu_{CHP}^{(+)} = \mu_{CHP}^{(-)} = \mu_{EC}^{(+)} = \mu_{EC}^{(-)} = \mu_{ABC}^{(+)} = \mu_{ABC}^{(-)} = \mu_{BL}^{(+)} = \mu_{GRID}^{(+)} = 0$; $\mu_{BL}^{(-)} > 0$; $\mu_{GRID}^{(-)} > 0$. Substituting these KKT multiplier into equation (22)-(24) and (34)-(35), when inequality (37) is true, the operation strategy is optimal strategy because the KKT condition is true.

$$\lambda_H = \frac{\rho_{CHP} \cdot \dfrac{\partial F_{CHP}(t)}{\partial P_{CHP}}}{\dfrac{\partial Q_{CHP}(t)}{\partial P_{CHP}} + \dfrac{COP_{EC}}{COP_{ABC}}} \tag{36}$$

$$\lambda_H < \frac{\rho_{BL}}{\eta_{BL}}, \lambda_H < \rho_{GRID} \frac{COP_{ABC}}{COP_{EC}} \tag{37}$$

The result of the EC-FTL strategy is that the PGU, electric chiller run between their maximum and minimum limit, the purchasing electricity from grids is also between the maximum and minimum, the absorption chiller and auxiliary boil are not work, their output is their minimum value. The KKT multipliers are $\mu_{CHP}^{(+)} = \mu_{CHP}^{(-)} = \mu_{EC}^{(+)} = \mu_{EC}^{(-)} = \mu_{GRID}^{(+)} = \mu_{GRIS}^{(-)} = \mu_{BL}^{(+)} = \mu_{ABC}^{(+)} = 0$; $\mu_{BL}^{(-)} > 0$; $\mu_{ABC}^{(-)} > 0$. Substituting these KKT multiplier into equation(22)-(24) and (34)-(35), when following inequality (39) is true, the EC-FTL is optimal strategy.

$$\lambda_H = \frac{\rho_{CHP} \cdot \left(\frac{\partial F_{CHP}(t)}{\partial P_{CHP}(t)}\right) - \rho_{GRID}}{\left(\frac{\partial Q_{CHP}(t)}{\partial P_{CHP}(t)}\right)} \tag{38}$$

$$\rho_{GRID} \frac{COP_{ABC}}{COP_{EC}} \leq \lambda_H \leq \frac{\rho_{BL}}{\eta_{BL}} \tag{39}$$

Comparing (37) and (39), when the EC-FTL is optimal strategy, the TEWM is not, Vice versa vice versa.

The result of EC-FEL can be investigated using KKT condition, the condition that EC-FEL is optimal strategy is

$$\lambda_E = \rho_{CHP} \cdot \frac{\partial F_{CHP}(t)}{\partial P_{CHP}} - \frac{\rho_{BL}}{\eta_{BL}} \cdot \left(\frac{\partial Q_{CHP}(t)}{\partial P_{CHP}}\right) \tag{40}$$

$$\frac{\rho_{BL}}{\eta_{BL}} \cdot \frac{COP_{EC}}{COP_{ABC}} < \lambda_E < \rho_{GRID} \tag{41}$$

The COP_{EC} is larger than COP_{ABC}, the inequality (41) is not true in general. In state 1, the optimal strategy may be WETM or EC-FTL according to the condition are (37) and (39).

The strategy in state 2 is same as that in state 1, the conclusion are also same as that in state 1.

B. the optimal strategy in state 3

When the HC-FTL strategy is used in state 3, the PGU, the absorption chiller, and the purchasing electricity will between their maximum and minimum, and the producing energy of the electric chiller and the auxiliary boil are zero. The KKT multipliers are $\mu_{CHP}^{(+)} = \mu_{CHP}^{(-)} = \mu_{ABC}^{(+)} = \mu_{ABC}^{(-)} = \mu_{GRID}^{(+)} = \mu_{GRID}^{(-)} = \mu_{BL}^{(+)} = \mu_{EC}^{(+)} = 0$; $\mu_{BC}^{(-)} > 0$; $\mu_{BL}^{(-)} > 0$. Substituting these KKT multiplier into equation (22)-(24) and (34)-(35), when λ_H is calculated according to (38), and following inequality (42) is true, the operation strategy is optimal.

$$\lambda_H \leq \frac{\rho_{BL}}{\eta_{BL}}, \lambda_H \leq \rho_{GRID} \frac{COP_{ABC}}{COP_{EC}} \tag{42}$$

The EC-FTL is an optimal strategy when the inequality (39) is true. Compare (42) with (39), if the EC FTL is the optimal strategy, then the HC-FTL is not, and vice versa vice versa

JC-FTL is not able to be an optimal strategy. The result of JC-FTL are that the both chiller is work between their minimum and maximum output, and so are the PGU and the purchasing electricity. The auxiliary boil's output is zero in its minimum value. The multipliers are $\mu_{CHP}^{(+)} = \mu_{CHP}^{(-)} = \mu_{ABC}^{(+)} = \mu_{ABC}^{(-)} = \mu_{GRID}^{(+)} = \mu_{GRID}^{(-)} = \mu_{EC}^{(+)} = \mu_{EC}^{(-)} = \mu_{BL}^{(+)} = 0$, $\mu_{BL}^{(-)} > 0$. Substituting these KKT multiplier into equation (22)-(24) and (34)-(35), the KKT condition (22)-(24) and (34)-(35) are not true, the JC-FTL is not optimal strategy. In state 3, the optimal strategy is the one of the EC-FTL and FC-FTL according to the condition (39) and (42).

C. the optimal strategy in state 4

In the HC-FEL, the purchasing electricity and absorption chiller's output are zero in their minimum value, and the output of PGU, electric chiller and the auxiliary boil are between the maximum and minimum. According to the complementary slackness, The KKT multipliers are $\mu_{CHP}^{(+)} = \mu_{CHP}^{(-)} = \mu_{EC}^{(+)} = \mu_{EC}^{(-)} = \mu_{BL}^{(+)} = \mu_{BL}^{(-)} = \mu_{ABC}^{(+)} = \mu_{GRID}^{(+)} = 0$; $\mu_{ABC}^{(-)} > 0$; $\mu_{GRID}^{(-)} > 0$. Substituting these KKT multiplier into equation (22)-(24) and (34)-(35), the condition that HC-FEL is optimal strategy is that the λ_E is calculated according to (40), and following inequality (43) is true.

$$\lambda_H \leq \frac{\rho_{BL}}{\eta_{BL}}, \lambda_H \leq \rho_{GRID} \frac{COP_{ABC}}{COP_{EC}} \tag{43}$$

The optimal condition of the EC-FEL strategy's in state 4 is same as in state 1, that is, (41) need be true.

JC-FEL is not optimal operation strategy for the same reason as JC-FTL in state 3.

4. The Simulation Algorithm Based On Optimal Operation Strategy

Based on the optimal strategy condition discussed in section 3, a chronological simulation algorithm is presented in the section.

Table 1 The parameters of the equipment

equipment	Maximum kW	Minimum kW	Parameter
Power generation unit	900	0	The parameters of generating efficiency $a=-0.2, b=0.4, c=0.1$ recovered waste heat efficiency $\eta_{CHP.H}=0.8$
Absorption chiller	3000	0	$COP_{ABC}=0.75$
Electric chiller	3000	0	$COP_{EC}=3$
Auxiliary boil	1000	0	$\eta_{BL}=0.8$

Table 2 the price data

Item	value
low heat values of natural gas, LHV, kWh/m^3	9.86
normal price of natural gas, ρ_{gas}, ￥/m^3	3.99
favorable price of natural gas, ρ_{gas}, ￥/m^3	2.43
electricity price on peak, ρ_{GRID}, ￥/kWh	1.032
electricity price on peak off ρ_{GRID}, ￥/kWh	0.474

The chronological simulation includes three main procedures:

1) According to hourly cold, heating and electricity load, determine that the load profile belongs to which state in state 1- state4;

2) Calculate the output of various equipment based on the load balance (10) and the operation strategy. According to the optimal strategy condition, choice the operation strategy.

3) Based on gas consumption of PGU and auxiliary boiler, purchasing electricity energy, and the price, calculate the operation.

5. Case study

This section presents the calculation results based the actual data. The parameters of PGU, the efficiency of the waste heat recover, and the coefficients of performance of electric chiller and absorption chiller are shown in table 1, which is come from [12]. The electricity price and gas price are shown in table 2 that come from the Shanghai development and Reform Commission Web site [17].

The hourly cooling, thermal and electricity load is used by a hospital in Shanghai[12], Four load profiles in these hours are selected as the typical load profiles which fall within the state 1 to 4 respectively, while the electricity price is electricity price on peak and natural gas is the normal price. The typical four cooling, thermal and electricity load are shown in table 3.

The results based on various operation are shown in table 4. The first column in table 4 is the state that means the load profiles in table 4, and the 2nd column gives the operation strategies, column 3rd to column 8th gives the output of the devices, the last column is the cost corresponding operation strategies. According to the optimal condition, the optimal operation strategy is TEWM in state 1 and

2 while the electricity price is electricity price on peak and the normal price of natural gas, the cost in state 1 and 2 are the lowest. Similarly, the optimal strategy in state 3 and state 4 is HC-FTL and EC-FEL respectively, their cost are the lowest in optional operation strategies.

The annual production energy to meet the annual load based on the presented simulation algorithm are shown in table 6.

A simulation algorithm based on traditional strategy is used as the comparison of the presented method. In the comparison operation strategy, all clod demand is met by absorption chiller, and then the FTL strategy is used if the equivalent load meet inequality (8) or the FEL strategy is used if the equality (9) is met by the equivalent load. The operation strategy is HET.

The output of the combined heat and power system are higher in the presented algorithm than that in the HET, and the output of the auxiliary boiler and purchasing electricity from grid in the presented algorithm are less than that in the HET. Because that the efficiency of the CHP is high than the auxiliary boiler, and the electricity cost is less than the electricity price, the cost based optimal strategy is less than the cost based on the HET. In fact, the TEWM is employed 2298 hours that is one-fourth total hour number, the cost based optimal strategy reduce 20% compared to the HET.

Table 3 the typical load profiles of the four state

STATE	Load		
	cold kWh	thermal kWh	electricity kWh
1	270	1559	852
2	292	1923	955
3	180	363	691
4	337	2017	940

Table 4 the results of the various operation strategies

state	strategy	P_{CHP} kWh	Q_{CHP} kWh	Q_{EC} kWh	Q_{ABC} kWh	P_{GRID} kWh	Q_{AB} kWh	cost ￥
1	TEWM	1006	1925	142	195	0	0	407
	EC-FTL	801	1665	337	0	253	0	586
	HC-FEL	970	1876	0	337	0	239	489
2	TEWM	1023	1950	272	20	0	0	414
	EC-FTL	1004	1923	292	0	24	0	431
	HC-FEL	955	1856	0	292	0	456	571
3	HC-FTL	121	603	0	180	570	0	637
	EC-FTL	62	363	180	0	674	0	720
	JC-FTL	89	483	90	90	624	0	680
4	EC-FEL	1024	1951	337	0	0	65	441
	HC-FEL	940	1836	0	337	0	630	635
	JC-FEL	982	1892	169	169	0	349	539

Table 5 the results of the Optimal strategy simulation and the HET operation strategy

The item	HET	Optimal Strategy
P_CHP, kWh	3624705.345	3848172.466
Q_CHP, kWh	9448685.233	9661266.466
Q_EC, kWh	354883.5476	687787.5376
Q_ABC, kWh	6014920.133	5741943.57
P_GRID, kWh	2663269.39	2117012.585
Q_BL, kWh	1049456.831	21807.60121
COST, ￥	4338968.516	3440631.979

6. Conclusion

The solution of the optimal operation model for CCHP system are different from the traditional operation strategy, the operation simulation algorithm need to be improved to reflect the characteristics of the optimal operation for estimating the performance of CCHP system. This paper investigated the typical operation strategy and their optimal condition based on KKT condition, and presents a simulation algorithm according to the optimal operation strategy. The research obtained following conclusion:

1) The optimal operation strategy change with the extrinsic price, like as electricity price and gas price, and intrinsical equipment efficiency, like as COP of electric chiller and absorption chiller.

2) The optimal condition of operation strategy presented in the paper can be used as selecting operation strategy in simulation. The simulation algorithm given a realistic estimation of CCHP system that operation based optimal operation model.

Nomenclature

CCHP	combined cooling, heating and power	P	The electricity energy, kWh
COP	coefficient of performance	Q	The thermal energy or cooling energy, kWh
EC-FEL	Electric Cooling Following the electric load	x	The one of the P and Q
EC-FTL	Electric Cooling Following the thermal load	ρ	The unit operation cost, ￥/kWh
FTL	following the electric load	η	efficiency
FEL	following the thermal load	$\mu^{(+)}$	The KT compliers corresponding to the $\leq$ constraint
HC-FEL	Thermal Cooling Following the electric load	$\mu^{(-)}$	The KT compliers corresponding to the $\geq$ constrain
HC-FTL	Thermal Cooling Following the thermal load	*Subscripts*	
HC-HET	Thermal Cooling Hybrid electric–thermal load	ABC	The absorption chiller
HET	hybrid electric–thermal load	BL	The auxiliary boiler
JC-FEL	Joint Cooling Following the electric load	C	The cooling demand
JC-FTL	Joint Cooling Following the thermal load	CHP	The combined heating, power system
TEWM	Thermal Electric While Meeting	E	The electricity demand
OS	The simulation based optimal strategy	EC	The electric chiller
Symbols		H	
a, b, c	The coefficient of the generation efficiency	MAX	The maximum output limit
F	The gas consumption, kWh	MIN	The minimum output limit
L	The load demand, kWh	*	The one of the CHP, ABC, EC, BL

Reference

[1] Wang J-J, Jing Y-Y, Zhang C-F, Zhai ZJ. Performance comparison of combined cooling heating and power system in diffident operation models, Appled Energy, 2011, 88(12): 4021-4631.

[2] Pnghui Gao, Yanjun Dai, YenWah Tong, Pengwei Dong. Energy matching and optimization analysis of waste to energy CCHP system with exergy and energy level. Energy 2015, 79: 522-535.

[3] Mago P, Chamra L. Analysis and optimization of CCHP systems based on energy, economical, and envioronmental considerations. Energy and Building, 2009, 41(10):1099-1106.

[4] Alan Moran, Redro J.Mago and Louay M. Comra. Thermo economic modeling of micro-CHPfor small commercial applications. Int.J.Energy Res., 2008, 32: 808-823

[5] R/J.Mago,N.Fumo, L.M.Chamra. Perforance analysis of CCHP systems operationting following the thermal and electric load. Int. J. Energy Res., 2009, 33: 852-864.

[6] Mingxi Liu, Yang Shi, Fang Fang. A new operation strategy for CCHP systems with hybrid chillers, Applied Energy, 2012, 95: 164-173.

[7] Wang J J, Sui J, Jin H G. An improved operation strategy of combined cooling heating and power system following electrical load. Energy, 2015, 85: 654. 666.

[8] Fang F, Wang Q H, Shi Y. A novel optimal operational strategy for the CCHP system based on two operating modes. IEEE Transactions on Power Systems, 2012, 27(2): 1032. 1041.

[9] X.Q.Kong, R.Z.Wang, X.H.Huang. Energy optimization model for a CCHP system with available gas turbines. Applied Thermal Engineering, 2005, 25: 377-391
[10] Zhanbo Xu, Xiaohong Guan, Qing-shan Jia, Jiang Wu, Dai Wang, Siyuaaaaaaaan Chen. Performance analysis and comparison on energy storage devices for smart building energy management. IEEE Tans. on Smart Grid, 2012, 3(4): 2136-2147.
[11] Evins R, Orehounig K, Dorer V, Carmeliet J. New formulations of the energy hub model to address opera-tional constraints. Energy 2014;73:387-398.
[12] Sashirekha A, Pasupuleti J, Moin NH, Tan C. Combined heat and power (CHP) economic dispatch solved using Lagrangian relaxation with surrogate subgradient multiplier updates. Electr Power Energy Syst 2013, 44:421-30.
[13] Iman Gerami Moghaddam, Mohsen Saniei, Elaheh Mashhour. A comprehensive model for self-scheduling an energy hub to supply cooling, heating and electrical demands of a building. Energy 94 (2016) 157-170
[14] Hossein Karami, Mohammad Javad Sanjari, Seyed Hossein Hosseinian, G. B. Gharehpetian. An Optimal Dispatch Algorithm for Managing Residential Distributed Energy Resources. IEEE Trans. on Smart Grids, 2014, 5(5): 2360-2367
[15] Bischi A, Taccari L, Martelli E, Amaldi E, Manzolini G, Silva P, Campanari S, Macchi E. A detailed MILP optimization model for combined cooling, heat and power system operation planning. Energy, 2014, 74: 12. 26
[16] C.Y.Zheng, J.Y.Wu, X.Q.Zhai, R.Z.Wang. A novel thermal storage strategy for CCHP system based on energy demands and state of storage tank. Electrical power and energy systems, 2017, 85:117-129.
[17] http://www.shdrc.gov.cn/main?main_colid=319&top_id=312&main_artid=19986.

Appendix

According to (1)

$$\frac{\partial F_{CHP}(t)}{\partial P_{CHP}(t)}=\frac{1}{\eta_{CHP.E}}+P_{CHP}(t)\cdot\left(\frac{1}{\eta_{CHP.E}}\right)^2\cdot\frac{\partial\eta_{CHP.E}}{\partial P_{CHP}} \tag{A.1}$$

According to (3),(3)

$$\frac{\partial\eta_{CHP.E}}{\partial P_{CHP}}=b\cdot\frac{1}{P_{CHP.MAX}}+2\cdot c\cdot f\cdot\frac{1}{P_{CHP.MAX}} \tag{A.2}$$

Substituting (A.2) into (A.1)

$$\frac{\partial F_{CHP}(t)}{\partial P_{CHP}(t)}=\frac{1}{\eta_{CHP.E}}-\left(\frac{1}{\eta_{CHP.E}}\right)^2\cdot\left[b\cdot f+2\cdot c\cdot f^2\right] \tag{A.3}$$

According to (2)

$$\frac{\partial Q_{CHP}(t)}{\partial P_{CHP}}=\eta_{CHP.H}\cdot\left(\left(\frac{\partial F_{CHP}(t)}{\partial P_{CHP}}\right)(1-\eta_{CHP.E})-F_{CHP}(t)\cdot\left(\frac{\partial\eta_{CHP.H}}{\partial P_{CHP}}\right)\right) \tag{A.4}$$

Substituting into (A.2), (A.3)

$$\frac{\partial Q_{CHP}(t)}{\partial P_{CHP}}=\eta_{CHP.H}\cdot\left(\frac{1-\eta_{CHP.E}}{\eta_{CHP.E}}-\frac{1}{(\eta_{CHP.E})^2}\left(b\cdot f+2\cdot c\cdot f^2\right)\right) \tag{A.5}$$

Research on Multi-objective Optimization Linear Programming Problem of Power Generation under Carbon Emission Trading

Guoying Wu[1], Erfeng Xu[3], Dunnan Liu[3], Yingjing He[2]

1 State Grid Zhejiang Electricity Power Company, Hangzhou, Zhejiang 310007, China
2 State Grid Zhejiang Economic Research Institute, Hangzhou, Zhejiang 310018, China
3 North China Electric Power University, Changping District, Beijing 102206, China

E-mail: 18810905481@163.com

Abstract. Carbon emissions trading is an important means to achieve the target of low carbon and energy saving, the construction and promotion of which has been widely concerned in recent years. As carbon emitters, electricity industry is shouldering more pressure to reduce emissions and will definitely participate in carbon emissions trading. Energy saving dispatching is one of the most important ways to realize energy saving and emission reduction for electricity power companies. One kind of multi-objective optimization linear programming model for generating units under carbon emissions trading is constructed in this paper. The model is verified by a practical case analysis, which ensures the optimal allocation of power resources with multi-objective of Energy saving, emission reduction and economy, and sustainable development of electricity power companies.

1. Introduction

As a means of market regulation, the introduction of carbon emissions trading will effectively promote the development of energy-saving and emission reduction technologies in power industry. However, with the introduction of carbon emissions trading system in power industry, carbon emissions trading has brought the great impact on the traditional trading and dispatching mode of power industry, which makes the mode of traditional power resource allocation and the mechanism of operation scheduling dispatching need to be further improved. In addition to considering the emission of sulfuric dioxide, nitrogen oxides and solid particles, it is critical to for power industry to optimize the allocation of power resources with carbon constraints.

So far, there have been a lot of research results on the power generation scheduling under carbon emissions trading. For example, literature [1] has established three optimization models of inter-regional generation scheduling including clean energy generation, taking minimum regional generation cost and minimum comprehensive cost as objective functions respectively and the models considered the issues of economic efficiency, carbon emission and transmission constraints. Literature [2] has constructed a day-ahead dispatching scheduling model which takes minimum synthetically generation cost as optimization objective. Literature [3-4] has established the thermal power unit scheduling optimization models and multi-objective model according to the cost, energy consumption and emission quantity. And these models are compared with the traditional electricity distribution model. Literature [5-6] has set up low carbon power generation scheduling model for generation planning the day before. Then optimization model of carbon generation scheduling is transformed into a single-objective optimization model for power system energy consumption and carbon emissions as the goal of minimizing the total cost. Literature [7-8] has established optimization models based on equal utility efficiency, minimum coal consumption, minimum carbon emission, least generation cost and least total cost to analyse and evaluate the result of generation permit substitution with different allowance allocation.

The optimal allocation of generation resources is a problem of generator unit commitment planning substantially. This paper studies the problem of priority scheduling strategy for generator units and established the multi-objective optimization linear programming model of power generation under carbon emissions trading, which is applied to a real case. Finally, electricity purchasing optimization strategy for electricity power companies is obtained in the mode of carbon emissions the transaction.

2. Priority strategy of generator units dispatching under carbon emission trading

The equal incremental principle refers to every generator unit in power system operating according to the same incremental rate of consumption. Therefore, the total energy loss in power system will at the lowest level. Similarly, if each generator unit in power system operates with the same consumption level, which includes total input among various type of resources such as primary energy consumption, the use of emission rights and capital investment, the total resources loss in power system would be at the lowest level while the system operation mode will be in most economical level accordingly.

The process of constituting the priority strategy of generator units dispatching under carbon emission trading is mainly divided into four steps.

First of all, it is necessary to collect the data of power supply, coal consumption, emissions of all pollutants and electricity purchase capital consumption in recent years. Through regression analysis based on the data obtained, the function mapping relation between unit power supply and coal consumption, emissions of all pollutants and electricity purchase capital consumption is put forward correspondingly.

Secondly, according to the function mapping relation obtained, solve the derivative function of each function relation to power supply and get the relationship between the micro variations of power supply and the micro variations of coal consumption, emissions of all pollutants and electricity purchase capital consumption among each generator unit.

Then, the weight of each unit coal consumption, emissions of all pollutants and electricity purchase capital consumption is calculated and multiplied by the micro-increment relationship obtained in the previous step. So that an integrated consumption of the micro-increment relationship can be received by the sum of the above results.

Finally, making integrated consumption micro-increment rate of generator unit equal to each other, incremental value interval of each generator unit can be calculated combining with a series of constraint conditions under the equal incremental principle, and then the optimal generator unit dispatching program can be put forward according to the range of micro-increment rate.

3. Multi-objective optimization linear programming model of power generation under carbon emissions trading

3.1. Construction of linear programming model

Power trading center promotes energy saving and emission reduction in generator units through bilateral transactions. Therefore, the primary goal of the linear programming model is energy conservation, followed by emission reduction, the final consideration of electricity prices. Assuming that there are N generator units in the region, objective function can be obtained combining with energy-saving emission reduction targets.

- Target of minimum coal consumption.

$$MinM = \sum_{i=1}^{n} u_i \times q_i \quad (1)$$

- Target of minimum emissions of all pollutants.

IOP Conf. Series: Earth and Environmental Science **168** (2018) 012008 doi:10.1088/1755-1315/168/1/012008

$$MinC=\sum_{i=1}^{n}Ec_i\times q_i,MinS=\sum_{i=1}^{n}Es_i\times q_i,MinN=\sum_{i=1}^{n}En_i\times q_i,MinG=\sum_{i=1}^{n}Eg_i\times q_i \quad (2)$$

- Target of minimum power purchasing cost.

$$MinU=\sum_{i=1}^{n}P_i\times q_i \quad (3)$$

Combining with the constraint conditions, a set of constrained inequalities for multi-objective linear programming problems can be proposed.

- Restriction of generator unit output power.

$$0\le q_i\le Qr_i \quad (4)$$

- Restriction of power supply and demand.

$$\sum_{i=1}^{n}q_i\le Q_d \quad (5)$$

- Restriction of emissions of all pollutants.

$$\sum_{i=1}^{n}Ec_i\times q_i\le W_c,\sum_{i=1}^{n}Es_i\times q_i\le W_s,\sum_{i=1}^{n}En_i\times q_i\le W_n,\sum_{i=1}^{n}Eg_i\times q_i\le W_g \quad (6)$$

If electricity power companies promote energy saving and emission reduction in power generation companies through energy saving dispatching, the smaller the objective function values of formula (1) – (3) are, the better emission reduction effect is. Therefore, the model is to obtain the generating capacity of each generator unit when the objective function value is smallest based on the inequality constraint conditions (4) - (6), which is regarded as optimal program under linear programming. However, it is quite difficult to solve the problem of multi-objective linear programming directly. In this way, it is much easier to transform multi-objective linear programming problem into single-objective linear programming problem.

3.2. Solution to linear programming model

First of all, under the constraints conditions of the inequalities, the optimal value of each objective function is obtained by solving the single-objective linear programming problem, which is written as M_0, C_0, S_0, N_0, G_0 and U_0.

At the same time, because the objective function is the minimum, and the direction of optimization is the same, we can introduce a super-objective function.

$$Z=\alpha M+\beta\left(\beta_c C+\beta_s S+\beta_n N+\beta_g G\right)+\gamma U \quad (7)$$

$$\alpha+\beta+\gamma=1 \quad (8)$$

$$\beta c+\beta s+\beta n+\beta g=1 \quad (9)$$

Among them, α, β, γ, βc, βs, βn and βg are all weight coefficient equal to or more than 0.

In addition, compromise constraint condition is constructed.

$$\beta_c\left(C-C_0\right)=\beta_s\left(S-S_0\right) \quad (10)$$

$$\beta_s\left(S-S_0\right)=\beta_n\left(N-N_0\right) \quad (11)$$

$$\beta_n\left(N-N_0\right)=\beta_g\left(G-G_0\right) \quad (12)$$

$$\alpha(M-M_0)=\beta\left[\beta_c(C-C_0)+\beta_s(S-S_0)+\beta_n(N-N_0)+\beta_g(G-G_0)\right] \quad (13)$$

$$\alpha(M-M_0)=\gamma(U-U_0) \quad (14)$$

After constructing the super-objective function and the compromise constraints, a new linear programming problem can be presented.

$$MinZ=\alpha M+\beta(\beta_c C+\beta_s S+\beta_n N+\beta_g G)+\gamma U \quad (15)$$

Among them, the constraint conditions are inequalities (4) to (6) and compromise constraints are inequalities (10) to (14).
Finally, compromise solution set can be obtained by solving the linear programming problem above with the method of solving single-objective linear programming problem, which the non-inferior solution of the original multi-objective linear programming problem. At this point, the values of which is obtained are the optimal bilateral trading generator units and generating capacity under the multi-objective linear programming.

4. Model application

As a means of market regulation, carbon emissions trading will effectively promote the development of energy saving and emission reduction technologies in power industry and contribute to the realization of carbon emission reduction targets in China. However, with the introduction of carbon emissions trading system in power industry, carbon emissions trading has brought great impact on traditional trading and dispatching mode of power industry. So, the mode of power resource allocation and the mechanism of scheduling dispatching should be further improved. It is critical for power industry to optimize the allocation of power resources with carbon constraints.

4.1. Restriction of generator unit output power

Restrictions of 24 generator units' output are shown in Table 1.

Table 1. Restriction of generator unit output power.

Generator unit	**Installed capacity (MW)**	**Maximum utilization hours (h)**	**Upper limit of output power (MWh)**	**Subhead integrated plant power consumption rate (%)**	**Limit of generator unit output power (MWh)**
1	1000	7889	7888720	4.50%	7533728
2	1000	7780	7779750	4.66%	7417214
3	600	8129	4877298	4.06%	4679280
…	…	…	…	…	…
22	300	7230	2169135	6.05%	2037902
23	315	7510	2365518	5.65%	2231866
24	315	8176	2575431	5.65%	2429919

4.2. Restriction of power supply and demand and emissions of all pollutants

Based on installed capacity and maximum utilization hours of each generator unit, upper limit of output power can be calculated. The remaining part of the output power is the maximum annual power supply, which is deducted from plant power consumption and power loss during power transmission.
Restrictions of each generator unit' s power supply and pollutants emissions are shown as Table 2.
Based on multi-objective linear programming model above, power supply, pollutants emissions and purchasing cost of generator unit can be calculated. The results obtained by MATLAB program are shown in Table 3.

Compared to generator units dispatching with single objective including minimum coal consumption or minimum emissions of various pollutants, the difference among the operation modes of generator units with different targets is shown in Figure 1.

Table 2. Restriction of power supply and pollutants emissions.

Generator unit	**Power supply (MWh)**	**CO_2 emission factor (t/MWh)**	**SO_2 emission factor (t/MWh)**	**NO_x emission factor (t/MWh)**	**Solid particulate emission factor (t/MWh)**
1	q_1	0.542463	0.000169	0.000462	0.000091
2	q_2	0.543552	0.000152	0.000531	0.000073
3	q_3	0.521601	0.000262	0.00203	0.000087
…	…	…	…	…	…
22	q_{22}	0.644812	0.00018	0.003175	0.000058
23	q_{23}	0.683169	0.000264	0.00071	0.000116
24	q_{24}	0.68486	0.000256	0.000963	0.000099
Constraint	74916344	42548000	20420	96300	7700

Table 3. Result of solution to multi-objective optimization linear programming.

Generator unit	**Power supply (MWh)**	**Coal consumption (t)**	**CO_2 emission (t)**	**SO_2 emission (t)**	**NO_x emission (t)**	**Solid particulate emission (t)**	**Power purchasing cost (RMB)**
1	7533728	2392915	4086769	1273	3481	686	263731
2	7259256	2268050	3945783	1103	3855	530	272450
3	4679280	1463408	2440717	1226	9499	407	171661
…	…	…	…	…	…		
22	2037902	648732	1314064	367	6470	118	77389
23	2231866	708060	1524742	589	1585	258	86257
24	2429919	768483	1664154	622	2340	241	86662

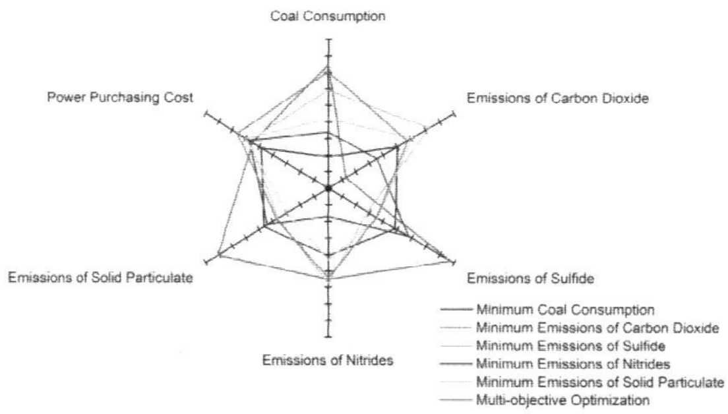

Figure 1. The difference among operation modes of generator units with different targets.

As the figure illustrated, power purchasing cost has not been greatly improved between dispatching model with multi-objective and single-objective. However, all kinds of pollutants emissions and coal consumption have decreased significantly. Therefore, multi-objective optimization linear programming model proposed in this paper can reduce coal consumption and pollutant emissions as much as possible on the premise of economic consideration, which will conducive to energy saving and emission reduction in power industry.

5. Conclusion

In this paper, and one kind of multi-objective optimization linear programming model for generating units under carbon emissions trading is constructed. Meanwhile, the model is solved by a practical case. The result obtained is considered as a compromise solution for all targets, which provides a solution for electricity power companies to the optimal allocation of power resources with multi-objective of energy saving, emission reduction and economy, and ensure scheduling operation of power system with multi objectives and sustainable development of electricity power companies.
In order to solve the problems of energy shortage and environmental pollution, there will be a large-scale renewable energy power. At this time, the constraints of power system dispatching are no longer limited to coal consumption and pollutant emissions, as well as more technical and environmental constraints. How to realize economic operation management of power system with a variety of constraints, will be the research direction of the new energy power system in the future.

References

[1] Zhang Lihui, Xiong Jun, Ju Liwei, Wu Hongliang and Tan Zhongfu, “An optimization model for inter-regional energy allocation under consideration of clean energy,” Journal of Hunan University (Natural Sciences), vol. 42, Apr. 2015, pp. 71-74.
[2] Zhang Chengfei, Yuan Yue, Zhang Xinsong, Cao Yang and Zhao Min, “Day-ahead dispatching scheduling for power grid integrated with wind farm considering influence of carbon emission quota,” Power System Technology, vol. 38, Aug. 2014, pp. 2114-2120.
[3] Shen Yangshuo, Research on approach analysis and optimization model of co2 emission reduction of power generation side in China. Beijing, North China Electric Power University, 2013.
[4] Wang Rui, Study on mode and price mechanism of low-carbon generation dispatch. Beijing, North China Electric Power University, 2013.
[5] He Nan, Models for the impact of carbon trading on generation permits trade in China. Beijing, North China Electric Power University, 2013.
[6] Mei Tianhua, Bian Qiaoyan, Xie jun,Gan Deqiang, “Low-carbon power dispatching and benefit allocation considering carbon emission allowance,” Automation of Electric Power Systems, vol. 40, Dec. 2016, pp. 49-55.
[7] Zhan Yi, Zuo Zhongqiu, “Energy-saving generation dispatching model about power dispatching under carbon emission trading mechanism,” Engineering Journal of Wuhan University, vol. 45, Oct. 2012, pp. 662-666.
[8] Zhang Zhanan, Cai Xingguo, “ Short-term energy-saving schedule with carbon emission trading considered,” Power System and Clean Energy, vol. 28, Jan. 2012, pp. 60-66.

Acknowledgments

This work is supported by National Natural Science Foundation of China *Research on the Coordinating Mechanism and Benefit Assessment Models between Power Trading and Carbon Trading* (71401055) and National Power Grid Corp science and technology project *Research and Application of Key Technologies for Power Grid Planning, Investment and Purchasing Power Optimization* (Contract No. SGZJJY00SJJS1600031).

IOP Conf. Series: Earth and Environmental Science **168** (2018) 012009 doi:10.1088/1755-1315/168/1/012009

Modelling and Application of Output Current of the Solar Array for Satellite in Sun Synchronous Orbit

Y S Bian[1,2], **C Xu**[2], **J F Li**[2] and **X J Wang**[2]

[1]State Key Laboratory of Astronautic Dynamics, Xi'an 710043, China

[2]Key Laboratory for Spacecraft In-Orbit Fault Diagnosis and Maintenance, Xi'an 710043, China

E-mail:bys02201@163.com

Abstract. The output current of the solar array for satellite in sun synchronous orbit is a critical parameter for performance in-orbit analyzing. In order to analysis the rules of the solar array output current accurately, some influential factors, such as the incidence angle of the sun, the temperature of the solar array, the earth-sun distance factor and the fixed offset angle to the sun of solar wing, are comprehensively researched. Then a model of output current is established. Results in-orbit show that the model has high accuracy, which can provide a reference for optimization design of satellite solar array.

1. Introduction

The sun synchronous orbit satellite needs sunlight to provide energy, and it can use the solar array for transforming light energy into electrical energy. So the output power of solar array is a very important basis for energy budget and balance 0, which decides the work condition of the spacecraft. The output power of the solar array is reflected by the output current, which is mainly affected by the incidence angle of the sun, the earth-sun distance factor, the solar array temperature and the attenuation factor of the solar array.

Johnson et al 0 obtained the decay rules of the solar cell with the time of 15 months' in-orbit operation by the direct measurement and calculation of the in orbit equipment. However, the study focused on the attenuation characteristics. Zhang et al 0 considered the influence of the solar incident light intensity and the working temperature of the solar array on the output power of the solar array on solar aircraft, whose research had certain reference value to the satellite. Fabio and Fabrizion 0 analyzed the UNISAT-3 solar array in-orbit performance. Their study focused on the decaying rate of different solar array materials. Jing et al 0 used the normalization method to obtain the influence rules of earth albedo, stars occlusion, attenuation factors on the output power of the solar array. But the lack of precise modelling of the output current makes it can only reflect the general trend. In addition, the existing studies did not consider the influence of the fixed offset angle to the sun of solar wing on the output current in the actual projects.

In this paper, the relationship between the solar array output current, the incidence angle of the sun, the temperature of the solar array, the earth-sun distance factor, the attenuation factor of the solar array and the fixed offset angle to the sun of solar wing is comprehensively analyzed. Then the Poisson series is used and improved to fit their relationship. Finally the model of the solar array output current is established, and it is verified in practical application.

2. Output power of the solar array

On the basis of in-orbit data statistics, the range of the bus voltage of the sun synchronous orbit satellite is very small. The output voltage of the solar array can be regarded as a constant. So, the variation rule of the output power of the solar array is reflected by the variation of the output current of the solar array 0. The output current of the power supply array of the solar array is directly affected by the illumination change of the sun, which is used for characteristic analysis in this paper0.

The output current of the solar array I is related to the incidence angle of the sun, the temperature of the solar array, the earth-sun distance factor and the attenuation factor, which is given by 0:

$$I=\left[I_{BOL}+\eta(T-25)\right]\cdot\cos\beta\cdot F_s\cdot F \tag{1}$$

where I_{BOL} (A) is the output current of the solar array with the incidence angle of 0° and the working temperature of 25°C at the beginning of life. η is the current-temperature coefficient. T is the temperature of the solar array. β is the incidence angle of the sun. F_s is the earth-sun distance factor. F is the attenuation factor of the solar array, which includes the attenuation factor of space particle radiation, the attenuation factor of ultraviolet radiation, the attenuation factor affected by the collision of meteoroid and the alternation of heat and cold, and et al 0. Besides, earth albedo and stellar occlusion can also influence the output power. I_{BOL} and η in equation (1) can be obtained by multi-group data at the beginning of satellite life using the least square method.

The angle of incidence of the sun's rays relative to the orbital plane, i.e., the angle between the sun's rays and the normal plane of the orbit, which is given by 00

$$u(t)=\arccos\left\{\cos i\sin\left[\arctan\frac{\sin(2\pi t/365)}{\cot\varepsilon}\right]+\sin i\cos\left[\arctan\frac{\sin(2\pi t/365)}{\cot\varepsilon}\right]\cdot\sin\left[\frac{t_0}{12}\pi-\Delta\alpha(t)\right]\right\} \tag{2}$$

where $t\in[0,365]$ and $t=0$ means vernal equinox. The sun moves eastward along the equator from the vernal equinox. t_0 is the local time of descending node of satellite in sun synchronous orbit, and $t_0\in[0,24]$. ε is the obliquity and $\varepsilon=23°26'$. $\Delta\alpha(t)$ is the accessional difference of the real sun and the mean sun. If the angle between the solar array and the orbit plane is κ, the angle between the solar ray and the normal vector of the solar array, i.e., $\beta=u(t)-\kappa$.

The earth-sun distance factor is given by 00:

$$F_s=1.00011+0.034221\cos\psi+0.00128\sin\psi+0.000719\cos 2\psi+0.000077\sin 2\psi \tag{3}$$

where $\psi=2\pi t/365$, and $t\in[0,365]$.

The attenuation factor of the solar array is fitted by 0:

$$F=ax^3+bx^2+cx+d \tag{4}$$

where a, b, c and d are polynomial variables. x is the working days of the in-orbit satellite.

When the fixed angle of the solar array to the sun is θ ($0\le\theta\le 90°$), the output current of the power supply array can be calculated by:

$$I(\theta)=\left[I_{BOL}+\eta(T-25)\right]\cdot\cos\beta\cdot F_s\cdot F\cdot\cos\theta \tag{5}$$

3. Poisson series fitting of output current

In this study, the temperature of the solar array T, the incidence angle of the sun β, the earth-sun distance F_s, the attenuation factor F and the fixed angle of the solar array to the sun θ are taken as variables, so equation (2) can be changed to:

$$\frac{I(\theta)}{I_{BOL}}=F\cdot F_s\cdot\cos\theta\cdot\cos\beta+\frac{\eta(T-25)}{I_{BOL}}\cdot F\cdot F_s\cdot\cos\theta\cdot\cos\beta \tag{6}$$

From equation (6) we can see that $I(\theta)$ is the combination function of polynomials (F_s, F and T) and trigonometric functions (θ and β). A special series is introduced to fit the output current $I(\theta)$, i.e., Poisson series. The combination of polynomials and trigonometric functions are used in the theory of perturbations of celestial mechanics, orbital mechanic, nonlinear mechanics and nonlinear differential equation, which is called Poisson series 00 and also called polynomial-trigonometric function. Its general form is given by:

$$P=\sum_{i_1,\ldots,i_n}\sum_{j_1,\ldots,j_m} C_{i_1,\ldots,i_n}^{j_1,\ldots,j_m} x_1^{i_1} x_2^{i_2}\cdots x_n^{i_n}\begin{pmatrix}\sin\\ \cos\end{pmatrix}\left(j_1\phi_1+\cdots+j_m\phi_m\right) \tag{7}$$

where $x_1,x_2,\cdots,x_n$ are polynomial variables. $\phi_1,\phi_2,\cdots,\phi_m$ are trigonometric variables. C_i^j are Poisson coefficients. $i_1,i_2,\ldots,i_n$ and $j_1,j_2,\ldots,j_m$ are all integers. There are three polynomial variables and two trigonometric variables in equation (6), so equation (7) can be simplified into a (3, 2) type of Poisson series:

$$P(x_1,x_2,x_3,\phi_1,\phi_2)=\sum_{i_1,i_2,i_3=0}^{n}\sum_{j_1,j_2=0}^{m} x_1^{i_1}x_2^{i_2}x_3^{i_3}\left[A_{i_1,i_2,i_3}^{j_1,j_2}\cos\left(j_1\phi_1+j_2\phi_2\right)+B_{i_1,i_2,i_3}^{j_1,j_2}\sin\left(j_1\phi_1+j_2\phi_2\right)\right] \tag{8}$$

where $x_1=F$, $x_2=F_s$, $x_3=\eta(T-25)/I_{BOL}$, $\phi_1=\theta$, and $\phi_2=\beta$. $A_{i_1,i_2,i_3}^{j_1,j_2}$ and $B_{i_1,i_2,i_3}^{j_1,j_2}$ are Poisson coefficients. n and m represent the order of polynomial and trigonometric variables respectively.

If the number of the output current $I(\theta)$ is N, it means that there are N sets of data $\{x_1(k), x_2(k), x_3(k), \phi_1(k), \phi_2(k), I(k)/I_{BOL}\}_{k=1}^{N}$. In order to calculate Poisson coefficients $A_{i_1,i_2,i_3}^{j_1,j_2}$ and $B_{i_1,i_2,i_3}^{j_1,j_2}$, the least square method is used to minimize the following objective unction:

$$F=\sum_{k=1}^{N}\left[I(k)/I_{BOL}-P(k)\right]^2 \tag{9}$$

Each set of data can be written into a linear equation about the unknown coefficient:

$$\begin{aligned}I(k)/I_{BOL}&=A_{0,0,0}^{0,0}+A_{0,0,0}^{0,1}\cos\left[\phi_2(k)\right]+\cdots+A_{0,0,0}^{0,m}\cos\left[m\phi_2(k)\right]+B_{0,0,0}^{0,1}\sin\left[\phi_2(k)\right]+\cdots+\\&B_{0,0,0}^{0,m}\sin\left[m\phi_2(k)\right]+\cdots+A_{n,n,n}^{m,0}x_1^n(k)x_2^n(k)x_3^n(k)\cos\left[m\phi_1(k)\right]+\cdots+\\&A_{n,n,n}^{m,m}x_1^n(k)x_2^n(k)x_3^n(k)\cos\left[m\phi_1(k)+m\phi_2(k)\right]+\\&B_{n,n,n}^{m,0}x_1^n(k)x_2^n(k)x_3^n(k)\sin\left[m\phi_1(k)\right]+\cdots+B_{n,n,n}^{m,m}x_1^n(k)x_2^n(k)x_3^n(k)\sin\left[m\phi_1(k)+m\phi_2(k)\right]+\varepsilon_k\end{aligned} \tag{10}$$

Equation (10) can be expressed in matrix equation with n=1 and m=1:

$$\mathbf{S}_{N\times1}=\mathbf{H}_{N\times49}\boldsymbol{\theta}_{49\times1}+\boldsymbol{\varepsilon}_{N\times1} \tag{11}$$

where

$$\mathbf{S}=\left(I(1),I(2),\cdots,I(N)\right)^{\mathrm{T}} \tag{12}$$

$$\boldsymbol{\varepsilon}=\left(\varepsilon_1,\varepsilon_2,\cdots,\varepsilon_N\right)^{\mathrm{T}} \tag{13}$$

$$\boldsymbol{\theta}=\left(A_{0,0,0}^{0,0},A_{0,0,0}^{0,1},B_{0,0,0}^{0,1},\cdots,A_{1,1,1}^{1,0},B_{1,1,1}^{1,0},A_{1,1,1}^{1,1},B_{1,1,1}^{1,1}\right)^{\mathrm{T}} \tag{14}$$

$$\mathbf{H}=\begin{bmatrix}1 & \cos\left[\phi_2(1)\right] & \cdots & x_1(1)x_2(1)x_3(1)\cos\left[\phi_1(1)+\phi_2(1)\right]\\ 1 & \cos\left[\phi_2(2)\right] & \cdots & x_1(2)x_2(2)x_3(2)\cos\left[\phi_1(2)+\phi_2(2)\right]\\ \vdots & \vdots & \vdots & \vdots\\ 1 & \cos\left[\phi_2(N)\right] & \cdots & x_1(N)x_2(N)x_3(N)\cos\left[\phi_1(N)+\phi_2(N)\right]\end{bmatrix} \tag{15}$$

So the estimated value of $A_{i_1,i_2,i_3}^{j_1,j_2}$ and $B_{i_1,i_2,i_3}^{j_1,j_2}$ can be calculated by:

$$\hat{\boldsymbol{\theta}} = \left(\mathbf{H}^{\mathrm{T}}\mathbf{H}\right)^{-1}\mathbf{H}^{\mathrm{T}}\mathbf{S} \tag{16}$$

There are 49 Poisson coefficients in equation (9), which makes the matrix in equation (16) is close to singular. Each variable in equation (8) is multiplied by a scaled factor, i.e., $x_1 = k_1 F$, $x_2 = k_2 F_s$, $x_3 = k_3 \eta (T-25)/I_{BOL}$, $\phi_1 = k_4\theta$, and $\phi_2 = k_5\beta$.

4. Experimental results and analysis

Some in-orbit satellite, which has been working normally for eighteen months, is selected for our experiment. 256 sets of in-orbit data in two months about the solar array output current, the incidence angle of the sun, the temperature of the solar array, the earth-sun distance factor, the attenuation factor of the solar array and the fixed offset angle to the sun of solar wing, are used to verify our model. The earth-sun distance factor, the incidence angle of the sun, and the attenuation factor of the solar array are showed in figure 1, figure 2 and figure 3, respectively. The fixed angle of the solar array to the sun is set to be 45° , i.e., θ=45° .

Table 1. Data statistics of different parameters.

Parameter	Numerical value		Amplitude of variation
	Max	Min	
F_s	1.035050	1.035020	2.65%
F	0.986039	0.985532	0.05%
cosβ	-18.05	-19.39	5.67%
T(℃)	54.86	11.37	79.27%
I(A)	47.69	22.34	53.16%

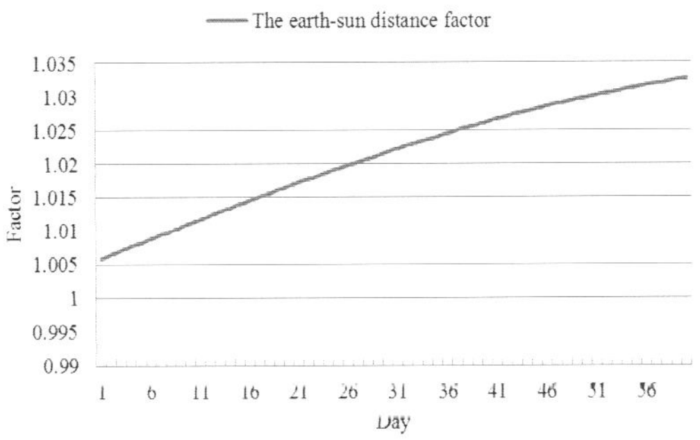

Figure 1. The earth-sun distance factor for two months.

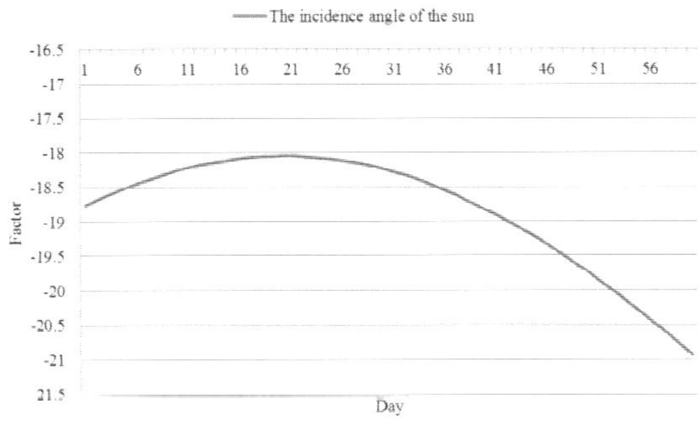

Figure 2. The incidence angle of the sun for two months.

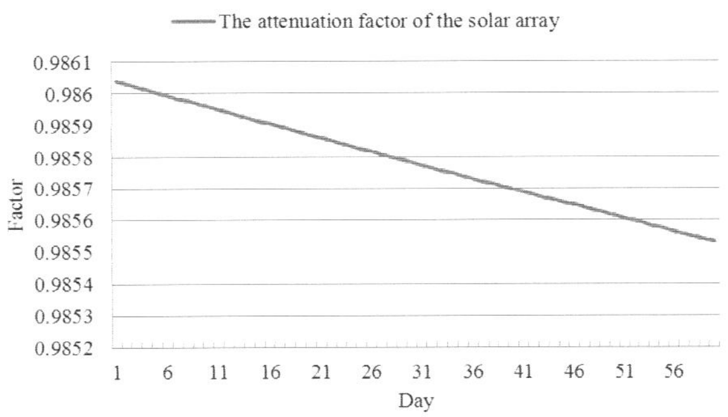

Figure 3. The attenuation factor of the solar array for two months.

The earth-sun distance factor increases progressively while the attenuation factor of the solar array decreasing progressively. However, their changes are very small, which are 2.65% and 0.05% respectively. As is showed in table 1, the temperature of the solar array has the largest influence on the output current compared with the other parameters. So, more attention should be paid on the change of the temperature of the solar array for in-orbit satellites.

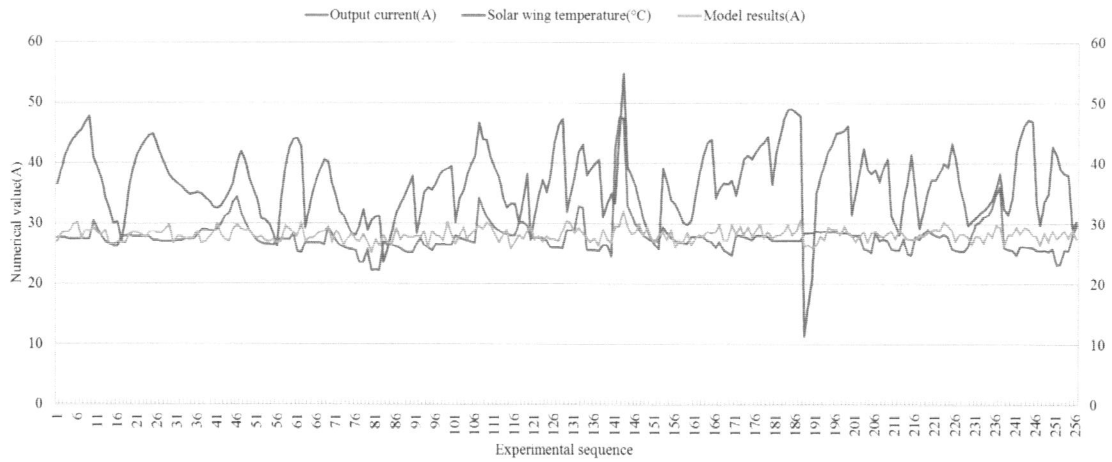

Figure 4. Model results compared to output current with different solar wing temperatures.

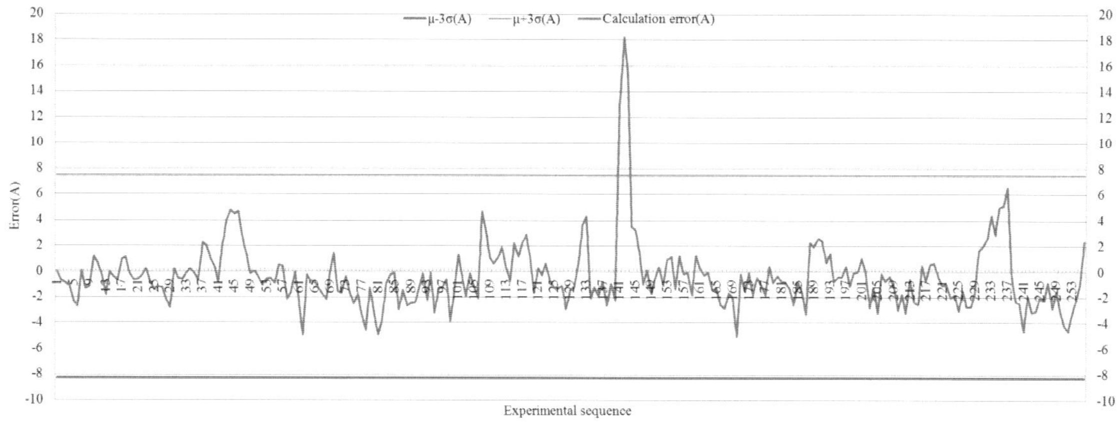

Figure 5. Error distribution of model calculation results.

The calculation results are showed in figure 4. Our model fits the trend of the output current well. The mean error is μ=-0.3711 A, and the standard deviation is σ=2.6208 A. The error distribution of model calculation results is showed in figure 5. The confidence interval of experimental results is 98.83% (3σ). The calculation results of our model show that the trend to be slightly fluctuated. However, there is a sharp increase point of the temperature (55 °C) in figure 4, which makes the calculation error of our model very big. The average value of the temperature is about 37 °C, which makes the sharp point to be a noise. Except that our model fits the output current well.

5. Conclusions

The relationship between the solar array output current, the incidence angle of the sun, the temperature of the solar array, the fixed offset angle to the sun of solar wing and so on is comprehensively analyzed. Then Poisson series is improved to fit their relational expression according to the in-orbit working condition. And then a model with multiple variables is established. Experimental results show that our model fits the output current well, which can be used in the related studies. Besides, the temperature of the solar wing has a great influence on the output current (i.e., the output power). Therefore, more attention should be paid on the change of the temperature in satellite management.

IOP Conf. Series: Earth and Environmental Science **168** (2018) 012009 doi:10.1088/1755-1315/168/1/012009

References

[1] Li Q, Sun X W, Lin L T, Gao C, Wang Q, and Hu X 2015 Earth Albedo Coefficient Estimation to LEO Satellite *J. Spa. Ene*. **3** 12-21
[2] Johnson M H and Ball J K 1992 Combined release and radiation effects satellite (CRRES) *J. Spa. Roc*. **29** 556-563
[3] Zhang J X, Liang X G, Wang G 2005 A New Method for Calculating Output of Space Solar Array *Ene. Con. Technol*. **23** 112-115
[4] Fabio S, Piergentili F 2008 Analysis of the UNISAT-3 solar array in-orbit performance *J. Spa. Roc*. **45** 142-148
[5] Jing Y L, Sun H T and Lei Y J 2013 Analysis on Performance of Solar Array for Satellite in Sun Synchronous Orbit *J. Spa. Ene* . **22** 61-66
[6] Tan W C and Hu J G 2009 *Spacecraft system engineering* (Beijing: China Science and Technology Press) p 105
[7] Peng M, Wang W W, Wu J and Li M 2011 Stydy on Attenuation Factor of Si Solar Array for Satellite in Sun Synchronous Orbit *J. Spa. Ene* . **20** 61-67
[8] Wang Y Q 1995 A method for calculating the incident angle of the sun relative to the orbital plane of the sun synchronous orbit *J. Spa. Ene* . **4** 65-73
[9] Ball J M, Stranks S D and Hörantner M T 2015 Optical properties and limiting photocurrent of thin-film perovskite solar cells *J. Ene. Env. Sci*. **8** 602-609
[10] Richard B E and Riordan C 1986 Simple solar spectral model for direct and diffuse irradiance on horizontal and tilted planes at the earth's surface for cloudless atmospheres *J. Cli. App. Met*. **25** 87-97
[11] Jimenez V A, Barrionuevo A and Will A 2016 Neural Network for Estimating Daily Global Solar Radiation Using Temperature, Humidity and Pressure as Unique Climatic Input Variables Smart *J. Gri. Ren. Ene*. **7** 94
[12] William J H 1970 A Fortran-based list processor for Poisson series *Cel. Mech. Dyn. Ast*. **2** 474-480
[13] Perminov A S and Kuznetsov E D 2015 Expansion of the Hamiltonian of the planetary problem into the Poisson series in elements of the second Poincare system Sol *J. Sys. Res*. **49** 430

IOP Conf. Series: Earth and Environmental Science **168** (2018) 012010 doi:10.1088/1755-1315/168/1/012010

Improving the performance of solar chimney by addressing the designing factors

Guomin Zhang*, Long Shi

Civil and Infrastructure Engineering Discipline, School of Engineering, RMIT University,Melbourne, VIC 3030, Australia

Email: kevin.zhang@rmit.edu.au

Abstract. Solar chimney is a reliable system totally based on solar energy to enhance natural ventilation in buildings, but challenge still exists to optimize its performance with the lowest cost. In this study, three designing factors, including configuration, installation conditions, and material usages, were reviewed to provide a technical guide for engineering applications. Regarding the configuration, the performance of solar chimney can be enhanced with a high cavity, an appropriate cavity gap (usually 0.2-0.3 m), equivalent inlet and outlet area, and height/gap ratio of 10-15. Regarding installation conditions, an optimum inclination angle of 45° was usually suggested, and large openings can enhance the performance while the increasing rate keeps decreasing. The principles of material usage are to maximum the heat absorption and reduce the heat losses, considering properties such as thermal conductivity, absorptivity, emissivity, transmissvity, and reflectivity.

1. Introduction

Solar chimney as a reliable renewable energy system has been largely utilized to conquer the global energy crisis. Buildings represent nearly 40% of total energy use in the U.S. and about 50% of this energy is used for heating, ventilating and cooling the space. Conventional heating and cooling systems are having a great impact on security of energy supply and greenhouse gas emissions [1]. Therefore, the application of renewable energy, such as solar, wind, tides, and geothermal energy, has become a critical topic [2].

To save the traditional energy, solar chimney has been largely used to enhance the natural ventilation in buildings. It is reliable system based on solar energy, which is basically a solar air heater, its position may be vertical or horizontal and according to the position it will be a part of a wall or roof and the classification of solar chimney can be varied according to different configuration or functions [3]. Experimental results showed that a solar chimney house could reduce the average daily electrical consumption of an air-conditioner by 10-20% [4].

Although solar passive features can add 0-15% to design and construction costs, paying this initial cost in return is long life energy saving. The challenge of designing solar chimney is to optimize the performance under the lowest construction cost. One can find different variations in solar chimney design, which is affected by a number of factors such as the location, climate, orientation, size of the space to be ventilated and the internal heat gains [5].

Therefore, three types of influencing factors were reviewed in this study to optimize the performance of soar chimney, including configuration, installation conditions, and material usage. The objectives of this study is to provide a comprehensive review about the designing factor, and then to provide a technical guide for solar chimney design in sustainable buildings.

2. Fundamentals of solar chimney

Solar chimney is a way of enhancing the natural ventilation in buildings based on passive solar energy. The basic driving mechanism of the air flow inside a solar chimney is thermal buoyancy, which is caused by the density variation of the air due to the temperature difference between the inside and outside of the building [6]. Solar chimney is basically a solar air heater, where its position may be vertical or horizontal as a part of wall or roof, shown in Fig. 1 [7].

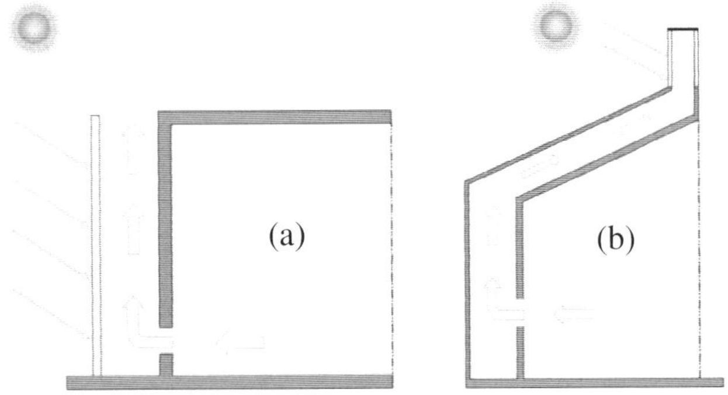

Fig. 1 Two typical solar chimney used in buildings [7]: (a) wall solar chimney; and (b) roof solar chimney

Fig. 1(a) shows a schematic of wall solar chimney (Trombe wall) for winter heating. The chimney is constructed by the external glazing and the inside wall. The external glazing allows the solar radiation penetrating into the chimney for heating up. The air in the chimney cavity then moves upward under the buoyancy after they are heated. Fig. 1(b) shows a typical roof solar chimney. The performance of solar chimney is much dependent on the temperature difference between inside and outside of building. A thermal storage layer below the chimney cavity is also used to extend the heating period for cloudy days or nights. Comparing to Trombe wall, the air flow in roof solar chimney will encounter resistance because of additional bends of duct, while its advantages and disadvantage can be seen in Table 1 [7].

Table 1. Advantages and disadvantages of two types of solar chimney [7]

Type	Advantages	Disadvantages
Vertical solar chimney	▪ The external glass gain sun radiation, solar collector is not needed; ▪ The air flow in chimney could go upward directly without bends; ▪ Easier to be control with inlet and outlet for different climatic condition; and ▪ Stack height is not restricted by roof height.	▪ Insulation is need to prevent direct heat transfer between chimney and interior room because of high temperature and high contact area; and ▪ Barriers are strictly prevented because the solar gained wall is lower than roof solar collector;
Roof solar chimney	• Very large collector areas easily achieved; • May be more aesthetically pleasing than a tower; • No additional towers needed; • Likely to be cheaper than a tower design; and • Easier to retrofit.	• Stack height is restricted by roof height; • Heat transfer between heated air and glass is higher than for a vertical surface; • Additional bends create greater pressure-losses; and • Incorporation of thermal mass may be more difficult.

3. Influencing factors of solar chimney

3.1. Configuration

It is quite sure that a higher cavity height can result in a better performance for wall solar chimney. Several reasons may be explained for this. The first is due to the pressure difference distribution enhanced by the high cavity, resulting in the rising of ventilation rate [8]. The increased heat gain is another reason where an increase in wall height by a quarter is equivalent to an increase in heat gains by three quarters [9]. The situation for roof solar chimney may be a little different. An experiment showed that there exists an appropriate cavity length, beyond which the obtained heat and natural ventilation rate cannot be increased drastically [10].

Cavity gap, also called chimney width or channel depth in the literatures, is the thickness of roof duct and the distance between the inside wall and external glazing for roof and wall solar chimney, respectively. The performance of both wall and roof solar chimney are highly influenced by the cavity gap. Studies have been carried out on finding out the optimum cavity gap. Many studied [11, 9, 12, 13] suggested a cavity gap of 0.2-0.3 m to achieve the maximum performance of solar chimney. This statement has been made based on a viewpoint of solar chimney design that the maximum effect with lower material costs is achieved under this cavity gap [13], or the stack effect that the stack effect is maximum when the cavity gap is within this range [14, 15].

Inlet and outlet areas are determining factors for the air entrance and escape. Within a certain range of cavity gap, both inlet and outlet areas show positive influences on the performance of solar chimney. An analytical analysis by Bassiouny and Koura [16] concluded that increasing the inlet size three times can improve the performance by almost 11%. Another experimental study also showed that the ventilation flow rate increased with an increase in ventilation inlet area [17]. Based on both experimental [18] and numerical [19] studies, it is noted that optimized ventilation flow rate can be obtained when the outlet area takes the same area as the inlet area. For unequal openings, the outlet area shows a higher importance than inlet area to enhance the air flow in the cavity, shown in **Fig. 2** [20, 18].

Chimney aspect ratio usually refers to the ratio between the cavity height and cavity gap thickness. Previous studies have been undertaken to find the optimum aspect ratio. A optimum ratio of 10 was obtained from both numerical modelling [19, 21] and experiments [22]. For roof solar chimney, Du et al. [23] obtained that the optimal length to width ratio of a chimney is 12 based on numerical modelling. For the solar collector on the top roof, the ratio of solar chimney's length to hydraulic diameter should be greater than 15 to ensure developed flow [24].

3.2. Installation conditions

Inclination angle for solar chimney generally refer to the angle between cavity and the horizontal. It is usually applicable to roof solar chimney. As the roof solar chimney should be assembled with roof, the inclination angle is much dependent on the inclination angle of the roof, shown in Fig. 1(b). The optimum inclination angle is one of the key influencing factors on the performance of solar chimney. Most of the previous studies obtained an optimum inclination angle of 45° [25, 26, 2, 27, 10]. Its airflow rate is about 45% higher than that for a vertical chimney at otherwise identical conditions [25, 26], which was explained by lowest pressure loss. Zhai et al. [10] explained this as a balance between the stack pressure and convective heat transfer coefficient.

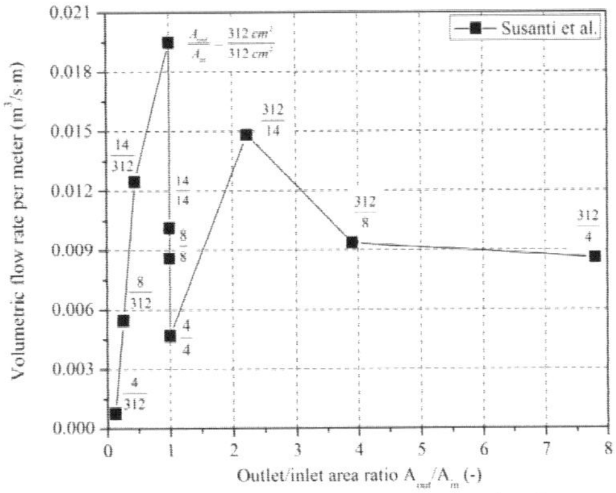

Fig. 2 The influence of outlet/inlet area ratio on volumetric flow rate per meter [20, 18]

Opening usually refers to the window, door, skylight, and other styles that can bring in fresh air in the buildings. The function of the opening of the room is to promote the circle of airflow, which benefits the natural ventilation and heat exchange in buildings. Priyadarsini et al. [28] found from experiments that even though the enhancement in the velocity is much higher when the doors are closed, they are more or less localized as the velocity increase is more evenly distributed when the doors are kept open.

In our previous study [29], a room coefficient (C_R) is proposed to describe the influence of both the room configurations on the performance of wall solar chimney. It can be seen from our numerical results that the inlet flow rate increases dramatically with the area of opening initially but the rate of increase then slows down at a later stage, shown in **Fig. 3**.

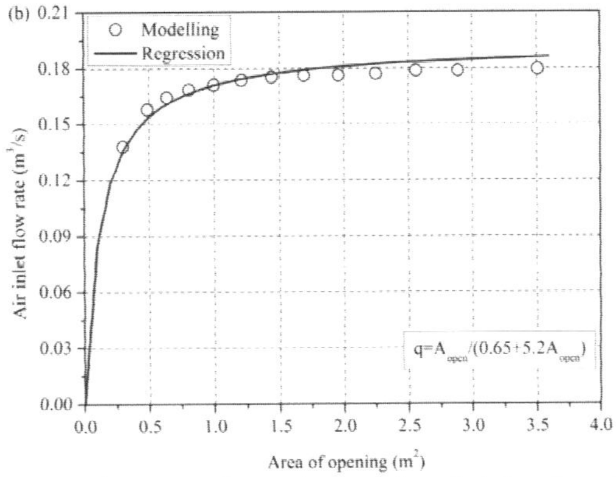

Fig. 3 The influence of opening on the air inlet flow [29]

The empirical model developed in our previous study can be expressed by [29]:

$$\dot{q}_{in} = C_R \cdot \left[h \cdot H_{sc} \cdot A_{hot} \cdot \left(T_{sc} - T_0\right)\right]^{1/3} \quad (1)$$

where C_R is the room configuration coefficient related to room openings, chimney inlet and outlet,

$$C_R = \frac{0.1 + 0.0016A_r - 0.003A_r^{-1}}{9.9 + 1.5A_{open}^{-1}}, \ where\ A_r = \frac{A_{in}}{A_{out}} \quad (2)$$

3.3. Material usages

Glazing is largely used in both wall and roof solar chimney, shown in Fig. 1(a) and (b). For the glazing, several properties may be important to the performance of solar chimney, such as transmissvity, reflectivity and absorptivity. An experimental study by Lee et al. [2] showed that a high transmissivity of glass can increase the outlet temperature and improve the performance of collectors, which showed a relatively higher importance than the reflectivity and absorptivity.

Double glazing shows its advantage in enhancing the performance of solar chimney. Hatami and Bahadorinejad [30] assessed six cases of air flow including: two cases for air heaters equipped with one glass cover and four cases for air heater with two glass covers. Maximum efficiency was obtained when air heaters had two glass covers and air could flow through all channels. Although double glazing shows advantage, it seems to be a good choice for winter heating but not for summer cooling [9, 5].

Materials have been also utilized in solar chimney for several special functions, such as thermal insulation, thermal storage and thermal absorption. It is known that a thermal storage layer with phase change materials under the cavity is to absorb as much heat as possible from the hot air in the cavity. It can release the heat at a later stage when the phase changes back. The bottom layer of insulation is to minimize the heat loss for the storage layer.

For solar collector, the solar absorptivity and emissivity play important role for the performance of roof solar chimney. Lee and Strand [31] reported the effect of absorptivity on ventilation enhancement and showed that improvement in air flow rate can be up to 57% by increasing the solar absorptance of the absorber wall of a solar chimney from 0.25 to 1.0. This was also confirmed by Liu et al. [32]. The numerical results [6] showed that the mass flow rate is an increasing function of surface emissivity and input heat flux and the mass flow rate is up to 59% higher with a surface emissivity of 0.9 than that with a zero surface emissivity.

Thermal insulation has been investigated regarding the performance of solar chimney. An experimental study showed that decreasing the heat transfer resistance in the airflow channel had the most significant effect on thermal efficiency enhancement, comparing to other four parameters, including height of a stagnant air layer, optical properties of a transparent cover, emissivity of an absorber plate, and conductive thermal resistance of a back plate on the thermal performance of a solar air collector [33].

The optimum thickness of wall was also investigated. A numerical study taken by Afonso and Oliveira [34] suggested that a thickness of 5 cm insulation wall is sufficient and no significant improvements can be achieved with thickness above 10 cm. It was mentioned that the optimum storage thickness for solar chimney depends on building use pattern. A small thickness was suggested for daytime ventilation, whereas, greater thickness was suggested for night time ventilation. An insulation thickness of 5 cm was considered optimum.

4. Conclusions

Solar chimney is a reliable system to enhance the natural ventilation in buildings totally based on solar energy, which plays an important role in release the global energy crisis. Although a large number of studies have been undertaken, the optimization design is still a challenge regarding solar chimney. Therefore, three important designing factors, including configuration, installation conditions, and material usages, were reviewed to provide a technical guide for engineering applications. In the aspect of the configuration, improved performance of solar chimney can be achieved for a high cavity, an appropriate cavity gap (usually 0.2-0.3 m), equivalent inlet and outlet area, a range of height/gap ratio (usually 10 for wall solar chimney and 12-15 for roof solar chimney). Regarding installation conditions, an optimum inclination angle of 45° was usually suggested, and opening of the room can enhance the performance but the increasing rate will slow down when it approaches to value. Materials used for solar chimney play an important role for the performance, such as thermal conductivity, absorptivity, emissivity, transmissvity, and reflectivity. Double glazing shows advantage, but it is a good choice for winter heating but not for summer cooling.

References

[1] Li HR, Yu YB, Niu FX, Shafik M and Chen B 2014 *Renewable Energy* **62** 468
[2] Lee DS, Hung TC, Lin JR and Zhao J 2015 *Appl Energy* **154** 651-662

[3] Lal S, Kaushik SC and Bhargav PK 2013 *International Journal of Development and Sustainability* **2** 1
[4] Khedari J, Rachapradit N and Hirunlabh J 2003 *Energy* **28** 1099
[5] Harris DJ and Helwig N 2007 *Appl Energy* **84** 135
[6] Khanal R and Lei CW 2011 *ANZIAM Journal* **52** 899
[7] Shi L and Chew MYL 2012 *Renewable Sustainable Energy Rev.* **16** 192
[8] Ding WT, Hasemi Y and Yamada T 2005 *Energy Build.* **37** 411
[9] Gan GH 1998 *Energy Build.* **27** 37
[10] Zhai XQ, Dai YJ and Wang RZ 2005 *Energy Build.* **37** 373
[11] Arce J, Jimenez MJ, Guaman JD, Heras MR, Alvarez G and Xaman J 2009 *Renewable Energy* **34** 2928
[12] Mathur J, Bansal NK, Mathur S, Jain M and Anupma 2006 *Sol. Energy* **80** 927
[13] Miyazaki T, Akisawa A and Kashiwagi T 2006 *Renewable Energy* **31** 987
[14] Balocco C 2002 *Energy Build.* **34** 469
[15] Barozzi GS, Imababi MSE, Nobile E and Sousa ACM 1992 *Build. Environ.* **27** 433
[16] Bassiouny R and Koura NSA 2008 *Energy Build.* **40** 865
[17] Spencer S 2001 *An experimental investigation of a solar chimney natural ventilation system* (Master's Thesis, Concordia University, Canada)
[18] Susanti L, Homma H, Matsumoto H, Suzuki Y and Shimizu M 2008 *Energy Build.* **40** 2196
[19] Li AG, Jones P, Zhao PG and Wang LP 2004 *JOurnal of Asian Architecture and Building Engineering* **3** 233
[20] Shi L, Zhang GM, Cheng XD, Guo Y, Wang JH and Chew MYL 2016 *Build. Environ.* **110** 115
[21] Wang LP and Li AG 2006 The 23rd Conference on Pasive and Low Energy Architecture (Geneva, Switzerland)
[22] Bouchair A 1994 *Building Services Engineering Research and Technology* **15** 81
[23] Du W, Yang QR and Zhang JC 2011 *Renewable Energy* **36** 265
[24] Tan AYK and Wong NH 2013 *Energies* **6** 145
[25] Chen ZD, Bandopadhayay P, Halldorsson J, Byrjalsen C, Heiselberg P and Li Y 2003 *Build. Environ.* **38** 893
[26] Halldorsson J, Byrjalsen C, Chen ZD, Bandopadhayay P and Heiselberg P 2002 The 8th International Conference on Air Distribution in Rooms (Copenhagen, Denmark: p 597
[27] Xu JL and Liu WH 2013 *Energy Build.* **66** 467
[28] Priyadarsini R, Cheong KW and Wong NH 2004 *Energy Build.* **36** 61
[29] Shi L and Zhang GM 2016 *Build. Environ.* **103** 250
[30] Hatami N and Bahadorinejad M 2008 *Sol. Energy* **82** 903
[31] Lee KH and Strand RK 2009 *Energy Build.* **41** 615
[32] Liu T, Lin WX, Gao W, Luo CX, Li M, Zheng QH, et al. 2007 *International Journal of Green Energy* **4** 601
[33] Yang M, Wang PS and Yang XD 2012 *Build. Environ.* **56** 361
[34] Afonso C and Oliveira A 2000 *Energy Build.* **32** 71

IOP Conf. Series: Earth and Environmental Science **168** (2018) 012011 doi:10.1088/1755-1315/168/1/012011

Experimental Study on a Low-Temperature Heat Pump Sludge Drying System

T Zhang[1,2] and Q Z Zhu[1]

[1]College of Energy and Mechanical Engineering, Shanghai University of Electric Power, 2103 Pingliang Road, Yangpu District, Shanghai, P.R.China

E-mail: zhtyn86@163.com

Abstract. The low-temperature sludge drying has become a kind of emerging hot drying technology. A small low-temperature heat pump sludge drying system is constructed in this paper. System operation characteristics are experimental studied. The results demonstrate that the set of an auxiliary fan is helpful to the system stable operation, but also leads temperature fluctuation of the main and auxiliary condenser. The inlet temperature and humidity of moist air in the drying closet and evaporator presents the opposite vibrate trend. The moisture contents in the above two devices both show a trend of increasing first and then decreasing, but the fluctuation range in drying closet is bigger. The wind resistances of the drying box and evaporator, which account for more than 90% of the whole system, both shows a certain extent fluctuation. The average specific power consumption of the system is 0.97 kWh/kg.

1. Introduction

Untreated sludge, when directly discharged to the environment, can cause secondary pollution in water, land and atmosphere. With the improvement of living standard, sludge drying treatment plays an important role in economic development and environmental protection. At present, there are several popular kinds of sludge dewatering treatment methods[1-3]: 1) Mechanical dewatering, that is, using mechanical force to pull water out of the sludge, but the treated sludge still has a high moisture content, which is around 80% to 85%. 2) Pressure filtration dewatering, that is, putting an instantaneous pressure on the sludge to assist dewatering. The sludge moisture content can be reduced to 70% to 75%. 3) Chemical reagent adding and pressure filtration; that is, adding chemical reagents for sludge modification, which can accelerate the water precipitating out. The sludge moisture content can reduce to 60-70% when combining with pressure filtration technology. 4) Sludge incineration, that is, incinerating the sludge or sludge and other combustible material mixture under a certain temperature and aerobic condition, the process can convert the organic matter in sludge into the corresponding burning gaseous substances; 5) Sludge heat drying, that is, employing the smoke, water vapor or other industrial waste heat as the heat source to vaporize the moisture in the sludge directly or indirectly. The sludge, which is processed by the first three methods, cannot directly discharge unless coupled with other disposition methods. Sludge incineration can realize the maximum extent of sludge decrease, but the system scale is normally tremendous [4]. The hot drying of sludge is the most significant sludge drying technology in recent decades. One can choose different drying methods according different sludge characteristics and heat sources [5].

Traditional sludge heat drying technologies has many disadvantages, such as large equipment investment, high energy consumption, high operation cost, explosion risk and easy causing secondary pollution. The low-temperature sludge drying, which is energy-saving, environmental protection, operational safety, has become an emerging drying technology [6]. In recent years, employing low-temperature heat pump for sludge drying has been one of the hot spots.

A small low-temperature heat pump sludge drying system was constructed in this paper, and performances of key components such as evaporator, condenser and drying closet during the test were experimental studied.

2. Low temperature heat pump sludge drying system setup

2.1. System structure and operation principle

The structure schematic diagram of the system is illustrated in Fig. 1. It is mainly composed of a heat pump system, drying closet, fans and wind pipes. Main and auxiliary condensers, which are linked in series, are set. There are also two fans set. The auxiliary fan only works when the drying hot air temperature is higher than the setting value. The air is under a closed cycle, so there is no exhaust gas emission, and the condensed sewage can be directly discharged after treatment.

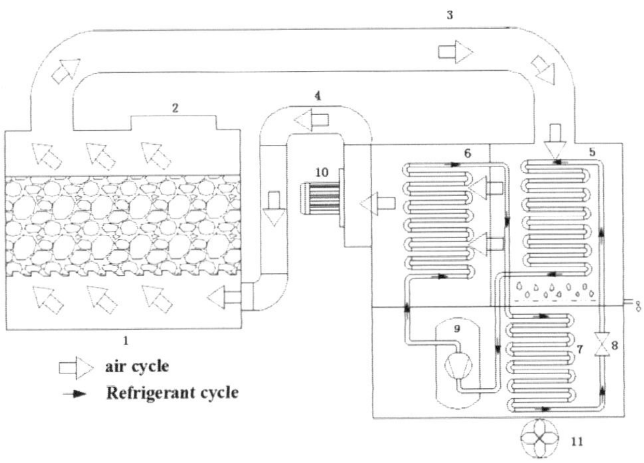

Fig1. schematic diagram of the system structure (1 drying closet;2 feeding port;3 air return pipe;
4 air inlet pipe;5 evaporator;6 main condenser;7 auxiliary condenser;8 throttle valve;9 compressor;
10 main fan;11 auxiliary fan)

To ensure system stable operation, the inlet air temperature of the drying closet is limited to no higher than 55℃. That is, the corresponding maximum condensing temperature is about 60℃. The limited temperature, which can avoid deflagration during the sludge drying process and guarantee compressor long-term effectively operation at the same time, helps with the high COP of system performance. Dewater cycle is as follows: with the help of main fan (10), high temperature and low humidity air is blown into drying closet (1), in where the hot air vaporizes the moisture in the sludge; there is humidity increasing and temperature decreasing of the hot air in the drying closet. Then the air will be blown into the evaporator (5) through the air return pipe (3), and be cooled and dewatered when flow past the evaporator (5). Finally, cooled and dewatered air will be blown into the main condenser (6), in where it will be heated again. The sludge temperature becomes high and the humidity becomes low with time, which will cause a high condensing temperature and evaporating temperature. When inlet air temperature of the drying closet above 55℃, auxiliary fan will work to enhance condensing effect. The compulsory procedure, which decreases the evaporating temperature, can ensure the stable operation of the system.

2.2. Experimental design

The heat pump system, which is charged with R134a, was configured with medium-high temperature Copeland compressor. Inlet and outlet temperatures of the main and auxiliary condenser, outlet temperature of the evaporator and ambient temperature were measured by T type thermocouples during the test. Wind speed (E + Austria, EE650) in the air inlet pipe, power consumption and water discharge mass were also concerned. Besides, the in-out pressure differences (DPL310) and air temperature and humidity (Siemens QFM3160) of the drying closet, evaporator and condenser were focused on.

The initial sludge moisture content was 60.52%, and it came to 16.73% in the end. The total operating time was 15h. Power consumption and dewater amount were recorded every 10 minutes. Other data were collected by Agilent 34970A every 30s.

3. Experimental results and analysis

3.1. Inlet and outlet temperature of the condenser and evaporator

Temperatures of main and auxiliary condenser as well as the evaporator temperature are shown in Fig.2a. One can see that the inlet and outlet temperatures of the main and auxiliary condenser as well as the outlet temperature of evaporator all have obvious temperature fluctuation; the evaporator outlet temperature shows a small range. The temperature fluctuation of condenser is related to the system control mode. When the inlet air temperature of drying closet is higher than 55℃, the auxiliary fan starts to work, which obviously increases the heat transfer capacity of the auxiliary condenser and significantly decreases the inlet temperature of the throttle valve. Thus decreases the evaporation temperature. Known from the analysis on the test, the inlet temperature of the main condenser fluctuated between 48.2-54.4℃, the average value was 51.7 ℃; the inlet temperature of the auxiliary condenser fluctuated between 46.7-53.0℃, the average value was 51.2 ℃; the outlet temperature of the auxiliary condenser fluctuated between 40.1-51.1℃, the average value was 47.3 ℃; Besides that, the outlet temperature of the evaporator showed a trend of increasing in the beginning, and then tended to stable, the values ranged from 8.6℃ to 16.2℃, and the average value was 13.7℃. Inspired by the above results, one can see that the evaporation and condensation temperatures, although there were some fluctuations, were relatively stable during the test. The whole system, which was not halted work during the experiment, is proved to be reliable and stable.

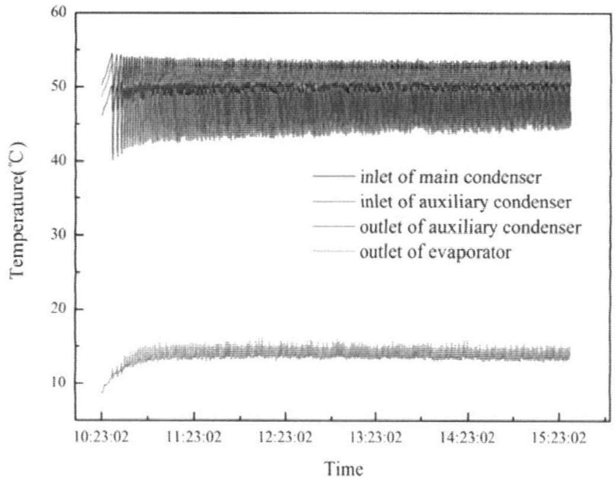

Fig2a. Variations in the inlet and outlet temperature the of the condenser and evaporator

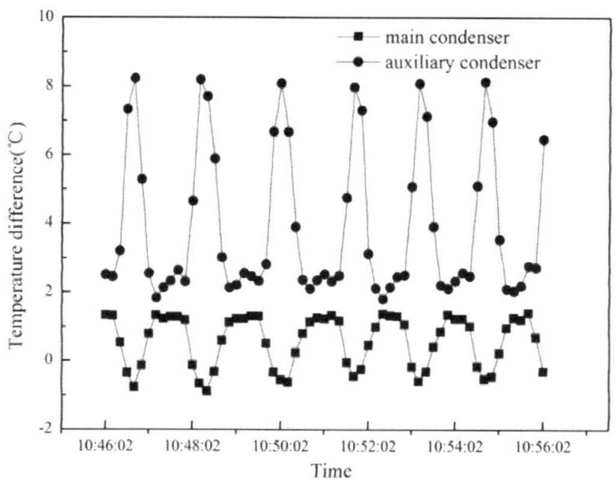

Fig2b. Variations in temperature difference in main and auxiliary condenser

In order to show the detail influence of auxiliary fan to inlet and outlet temperature of the condenser, the temperature differences between the in-out of the main and the auxiliary condenser are compared in this paper. Since the system operated stably, only 10 minutes were selected. The comparison results are shown in fig2b. From fig.2b, one can easily conclude that the temperature difference with considerable fluctuation corresponding the time of auxiliary fan working. The working of the auxiliary fan affects the inlet and outlet temperature of the main and auxiliary condenser, but delivers the opposite effect. The effect on the auxiliary condenser is more obvious. When the fan was not working, the temperature difference between the in-out of the main condenser was about 1.2 ℃, and the value was 2.5℃ for the auxiliary condenser. When the fan started to work, the temperature difference between the in-out of the auxiliary condenser increased rapidly, the value normally lay between 7.0℃-8.5℃. The evaporation temperature decreased with the outlet temperature of the auxiliary condenser decreased. It further decreased the compressor inlet temperature and outlet temperature. The changes also resulted in a decrease of inlet temperature of the main condenser, for this reason, the temperature difference between the in-out of the main condenser was a negative value, the value ranged between -0.9~-0.3℃.

3.2. Inlet and outlet temperature & humidity and moisture content of dry closet and evaporator

The outlet air temperature and humidity of the drying closet and evaporator directly affect the condensate rate during the dewatering process. The air temperature and humidity of the drying closet and evaporator during the test are presented in fig.3a. From the figure, one can see that the outlet air humidity of the drying closet and evaporator are both decrease gradually with time, it is from 89.0% to 36.7% in the former and from 99.9% to 88.4% in the later, the air humidity in the drying closet drops faster. The outlet air dry-bulb temperatures of the drying closet and evaporator are both increases gradually with time, it is from 28.0℃ to 42.1℃ in the former and from 19.7℃ to 25.0℃ in the later. The value of the evaporator outlet increases slower. The sudden drop of the air humidity in the outlet of the drying closet in the beginning of the test is mainly caused by the sludge heap. Heaped sludge rose the sludge temperature and then vaporized the free water in the sludge. The water vapor was absorbed by the temperature and humidity sensor, resulted in overvalue in the beginning of the test.

Variations in the air moisture content at the outlet of the drying closet and evaporator are shown in Fig.3b. One can notice that the outlet air moisture content of the two devices both show a trend of increase first and then decrease. The value of the air moisture content in the outlet of the drying closet reaches its maximum about 50 minutes after the test began, while the time is about 120 minutes at the evaporator. It can be seen from the figure that the air moisture content at the outlet of the drying closet gradually increases from 21.6g/kg to 23.1g/kg at the beginning, and then gradually decreases to 19.5 g/kg in the end. While the value of that of evaporator is increasing from 14.6 g/kg to 18.4 g/kg in the beginning, and then decreasing to 18.0 g/kg in the end. However, there is a very slow moisture content decreasing rate in the evaporator at the second half of the test[7].

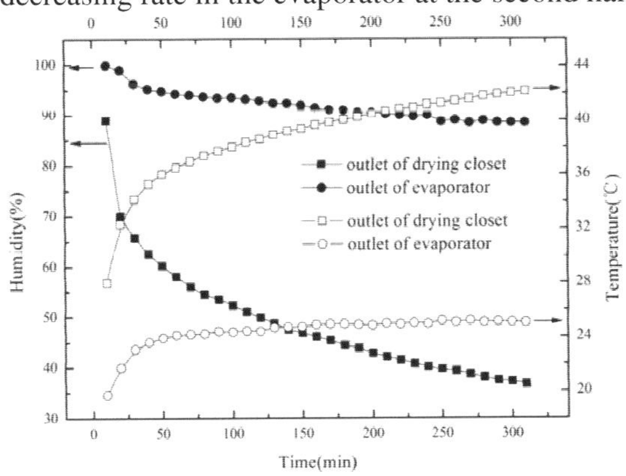

Fig. 3a Variations in outlet air temperature and humidity of drying closet and evaporator

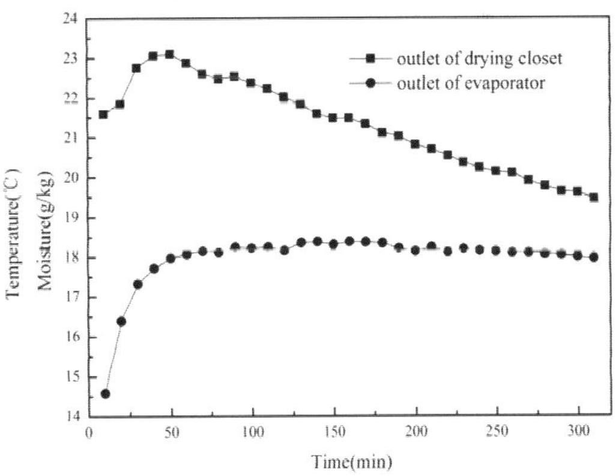

Fig.3b Variation in outlet air moisture of drying closet and evaporator

3.3. System wind resistance

In the process of sludge drying, the total wind resistance of the system directly influences the fan efficiency and determines the fan design and model-selection. The total wind resistance of the whole system consists of the resistances in the dry closet, evaporator, condenser and air cycle pipes. The wind speed and pressure differences between the in-out of the drying box, evaporator and condenser are shown in Fig. 4.

It can be observed in the figure that the pressure difference between the in-out of the condenser and the wind speed is relatively stable during the test time. The pressure differences between the in-out of the drying closet and evaporator show some certain fluctuation, but the fluctuation is relatively stable. Due to the sludge morphology change during the drying, the pressure difference between the inlet and outlet of the drying closet was the largest; the value ranged between 510-730pa and the mean value was 607Pa. There was also a big pressure difference in the evaporator due to the 90° corner between the inlet and outlet and the small fin spacing; the test data show that the pressure difference in the evaporator ranged between 350-540 Pa, the average value was 462 Pa. Thanks to the big fin spacing and air horizontal flow, the wind resistance in the condenser was relatively small, the value range between 90-96Pa, with an average value of 93Pa. In conclusion, system wind resistance was mainly determined by the drying closet and evaporator, the sum of the two devices accounting for more than 90% of the whole system resistance.

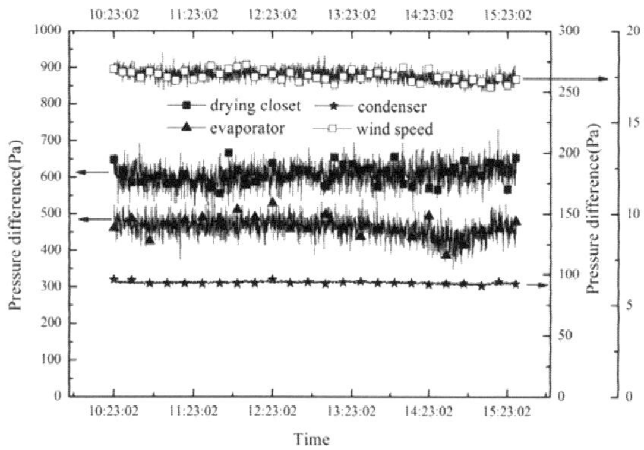

Fig4. Variation in pressure difference wind speed

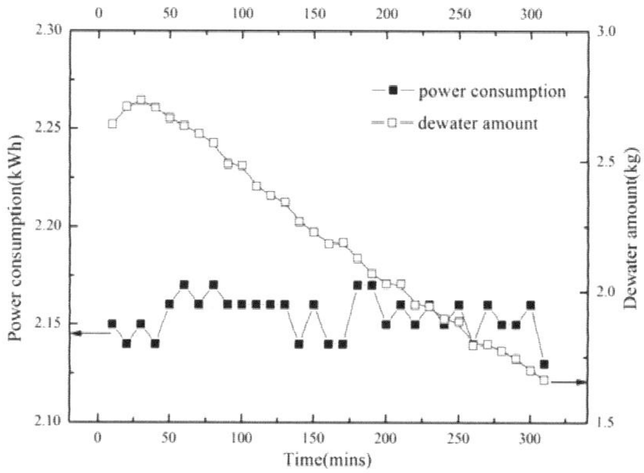

Fig. 5 Variation in power consumptionand and dewater amount

3.4. Power consumption and dewater amount analysis

Power consumption and the dewater amount of every 10 minutes are shown in Fig. 5. From the Fig.5, one can see that the dewater amount show a trend of increase first and then decrease, while the power consumption presents a relatively stable state. The variation trend of dewater amount curve agrees well with the moisture content difference between the outlets of drying closet and evaporator. According to the data, the dewater amount every 10 minutes ranged between 1.67-2.74 kg, and the average value was 2.21kg. The power consumption every 10 minutes ranged between 2.10-2.17kWh, the average value was 2.15 kWh. The system SPC (Specific Power Consumption) was 0.97 kWh/kg, and it can reach to 0.9 kWh/kg when the system is further optimized.

4. Conclusion

A small low-temperature heat pump sludge drying system was constructed in this paper, and the operation characteristics of the system were experimental studied. The main conclusions are as follows: (1)The set of auxiliary fan, although led the temperature fluctuation in the main and auxiliary condensers, guaranteed system stable operation.(2)System wind resistance was mainly determined by the drying closet and evaporator. The sum of the two devices accounting for more than 90% of the whole system resistance.(3)The SPC of the system was 0.97 kWh/kg.

References

[1] Renpeng G, XingzhongY, ZhibinW, HouW, LongbJ, Yifu L and Guangming Z 2017 *Sci Total Environ* 609(31) 1433-42

[2] Quanguo Z, Jianjun H, Duu-Jong L, Yingju C and Yu-Jen L 2017 *Bioresource Technol* **243** 1159-72

[3] Lyes B, Patricia A and Angélique L 2013 *Renew Sust Energy Rev* **28** 29-43.

[4] Fenfen Z, Huimin J, Zhaolong Z, Luyao Z, Jiawei W, Jun H and Hui Z 2012 *Procedia Environ. Sci* **16** 357-62.
[5] BartłomiejMichał C, Jacek N and Piotr K 2015 *J. Cleaner Prod* **90** 1-15.
[6] Will C, Gerry C and Zhifa S. 2011 *Energy* **36(8)** 4616-24
[7] Bennamoun L 2012 *Renew Sust Energy Rev* **16(1)** 1061 - 1073.

Acknowledgments
This work was supported by the Shanghai Colleges and Universities Young Teachers Training Funding Scheme (ZZsdl15076).

IOP Conf. Series: Earth and Environmental Science **168** (2018) 012012 doi:10.1088/1755-1315/168/1/012012

Wind Speed Prediction in Non-Monitored Areas Based on Topographic Radial Basis Neural Network (T-RBNN)

Salisu Muhammad Lawan, Wan Azlan Wan Zainal Abidin and U. Abubakar
Universiti Malaysia Sarawak, Malaysia

Email: salisumuhdlawan@gmail.com

Abstract. This paper shows an improved method for the prediction of wind speed in the areas where wind monitoring station is not available. The model has nine meteorological inputs, and one output, which is wind speed. The model was developed using Matlab/Simulink (R2016). The model was trained, tested and validated for accuracy purposes. The overall performance of the model was judged using statistical measures. It was realized that the developed model is capable of reproducing wind speed in the areas not covered by measurements. The root mean square and covariance of the developed model was 7.18 % and 0.0098 respectively.

1. Introduction

Wind power is well-recognized renewable energy producing, clean, safeguarded and foreseeable electric power. It has absolutely negligible carbon emission, and has reduced operational cost, that can be operated during the day and night times, with virtually zero greenhouse gas emission [1-2]. Before a wind energy system is installed, detailed wind resource assessment (WRA) must be conducted properly. The most important parameter of the WRA is wind speed, followed by wind direction. Because of this, wind speed must be measured or predicted with a high degree of accuracy.

According to [3], most of the early research work conducted on wind speed prediction were performed using either physical or numerical mathematical models, for example [3-4]. However, it was realized in many studies that due to the unpredictable nature of wind speed as it propagates in the atmosphere, it will be difficult to construct a reliable mathematical function, which will model the wind speed perfectly. Furthermore, according to no mathematical model either physical or numerical that will give an approximate solution of the wind speed model.

Generally, due to those weaknesses mentioned above, researchers have devoted time in finding a suitable model that will give an acceptable solution. Soft computing prediction models are found to be acceptable in terms of reliability and accuracy.

Based on the aforementioned, a lot of studies were carried out reported in scientific studies. Generally, wind speed predictions are classified based upon prediction periods/horizons, that is very short-term, short-term, medium-term and long-term. Wind speed prediction using data mining approach has successfully implemented in [5]. Instead of predicting wind power directly, authors [6] in predicting wind speed using support vector machine (SVM). Simulating annealing and SVM was proposed to overcome the use of SVM alone [7]. The fuzzy logic approach has been tested in [8] to predict wind speed values. Wind speed prediction using Radial Basic Function Neural Network (RBFNN) was suggested by [9], the model was validated using real wind speed data, and it was found to be feasible, to reduce wind speed uncertainties. One study [10] proposed the application of Artificial Neural Network (ANN) and Artificial Intelligent (AI). ANN and Markov chain model was

IOP Conf. Series: Earth and Environmental Science **168** (2018) 012012 doi:10.1088/1755-1315/168/1/012012

implemented for short-term wind speed [11], it was reported that prediction errors and wind speed uncertainties have reduced drastically. Feed forward Neural Network designed was trained with conjugate gradient algorithm and other training algorithms. The outcomes of this show an ANN approach has lees errors. Reference wind speed data was utilized to predict wind speed of four cities in turkey by means of ANN [12]. RBFNN and Empirical Mode Decomposition was applied for short-term wind speed prediction, results comparison with conventional methods indicate the suitability of the proposed method in terms of accuracy. Cluster analysis and ANNs were successfully developed to predict wind speed in Spain [13]. An improved ANN was built to predict wind speed and direction; simulation results demonstrate the accuracy of the suggested method.

Based on the short review reported here, it is acknowledged that ANNs have simple structure, fast convergence, and self-adapting, high accuracy, unlike fuzzy logic, which require a lot of fuzzy membership's rules and problems of parameter input selection. Because of this reason ANNs are widely accepted as a predictive tool in many reported studies. Furthermore, it can be noted from the review, none of the listed research works, consider the effect of terrain in developing the ANN model. Wind speed is affected by terrain variation, because of this; this study tries to bridge the gaps by introducing a terrain-based model named topographic-RBFNN (T-RBFNN) to predict the wind speed in the location with no wind station based on the nearby wind monitoring station.

2. Methodology

2.1. Study area description and data collection

The study area is the Kano University of Science and Technology, Wudil. Kano State, Nigeria (Latitude: 11° 48' 18" N and Longitude 8° 50' 43" E). The DevisPro weather Monitoring was mounted at 5m height shown in Figure 1. The Anemometer is used to measure the wind speed and wind direction at an update rate of 5 second and averages every one hour. The average hourly data were then stored in a computer system permanent memory attached to the Davis-Pro monitoring system. The data were exported into Microsoft Excel files or further analysis.

Figure 1. Position of the wind monitoring system.

2.2. Terrain data generation

Topographical maps are expensive and mostly they are restricted for security reasons. A simple, cost effective technique has been applied here to generate a terrain data. Two software was hybridized, GEPlot and Google Earth. The areas were mapped and viewed using Google Earth, While, GEPlot was utilized to point and demarcate the variations between marked points, and to locate the latitude and longitude of each location pointed out. All the data points were exported to Terrain Zonum Solutions (TZS) to view the digital elevation model (DEM). The DEM was viewed in Google Earth; finally the terrain data were generated and stored in a database system. Figure 2 shows the summary of the designed methodology.

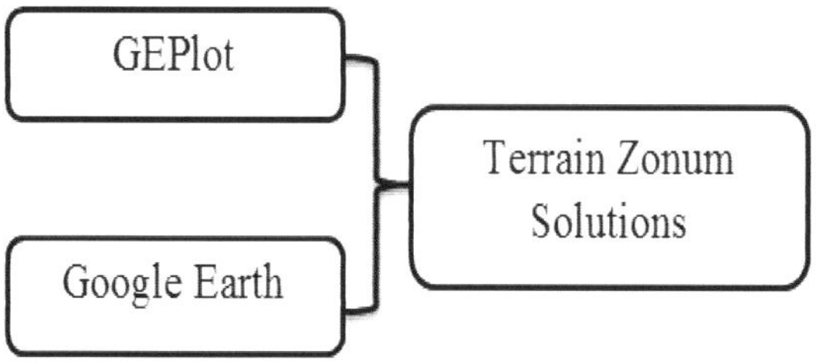

Figure 2. Terrain data generation.

2.3. T-RBFNN design

The ANN can be used without mathematical formulation. It works like the human being brain. The system learns to perform task once it is trained. ANNs have input, output and hidden layers. The neurons are connected to a large number of weights in which signal or information can pass through. Once the inputs are applied, the neurons will receive it and perform nonlinear operations. Generally, Matlab was commonly used to design and train the model. The neural network used in this work is RBFNN, among the most powerful for short-t prediction. Figure 3 shows the architecture of the ANN used in this paper.

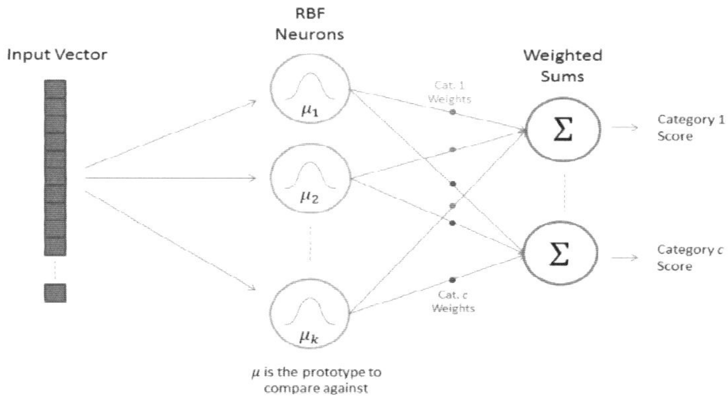

Figure 3. RBFNN model.

The input vector is the n-dimensional vector that is used to classify. The entire input vector is shown to each of the RBF neurons. Each RBF neuron stores a "prototype" vector that is only one in the vectors in the training set. Each RBF neuron measures up the input vector to its model, and produces a value in between 0 and 1 which can be a determine of likeness. If the input is the same for the model, then the output of that RBF neuron is going to be 1. As the range relating to the input and prototype increases, the response drops off tremendously to 0. The appearance of the RBF neuron's reaction is a bell curve, as shown in the network structure diagram. The transfer function used in this paper is a logistic-sigmoid function whose equation is:

$$\log istic(u) = \frac{1}{1+e^{-u}} \tag{1}$$

The ANN is trained to solve problems rather program to do so. Learning could be possible via proper training. Training is a method to teach the network on how to behave. Learning is the outcomes of the training. Learning can be supervised or unsupervised, depending on the set goals. In this paper, supervised learning is adopted, because training data are applied to infer the RBFNN model. Measured wind speed for a period of four years (4) (2013-2016) were segmented into three (3) parts, from 2013-2015 was used for the training, while 2015 and 2016 data were applied to test and validate the designed network. Matlab (R2013) is used to design and train the proposed model. The model has nine inputs (relative humidity (%), latitude (0N), longitude (0E), altitude (m), and month (days)) a monthly wind speed as the output (See Figure 4).

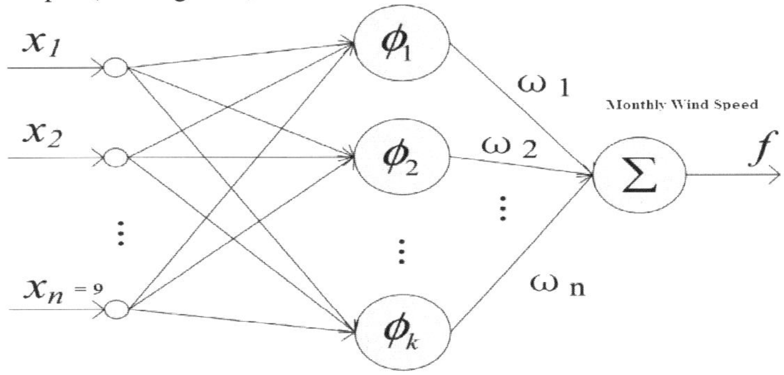

Figure 4. Developed RBFNN model.

Prior to the training, all the data were scaled to [0, 1] using minimum-maximum quotient rule, so as to avoid network confusion. To overcome the slow convergence of the network an improved algorithm, Levenberg-Marquardt algorithm (LMA) is used, in order to minimize the error between inputs and the outputs. Furthermore, to avoid under-fitting and over-fitting of the network, the number of neurons in the hidden layer carefully varied from 5-250 with a step of 5, until the optimum network is realized. Once the training, testing and validation are completed, the suitability of the model was evaluated statistically using:

$$R = \frac{\sum_{i=1}^{N}(t_i - \bar{t}) * (o_i - \bar{o})}{\sqrt{\sum_{i=1}^{N}\left(t_i - \bar{t}\right)^2} * \sqrt{\sum_{i=1}^{N}(o_i - \bar{o})^2}} \tag{2}$$

$$MAPE = \left[\frac{1}{N}\sum_{i=1}^{N}\left|\frac{t_i - o_i}{t_i}\right|\right] * 100 \tag{3}$$

N represents the number of data points, and t_i, o_i stands for the target (reference) value and the T-RBFNN predicted value for data point *i*. Bars indicate average values.

3. Results and Discussion

The third phase of the data used in this paper was predicted by means of NN models in a region with no monitoring based on the proximity of available wind stations taking into account the roughness factor and terrain variations. As discussed previously, a new t T-RBFNN approach was proposed and designed. The model consists of three layers: input, hidden, and output. The input/output vectors are the same in both structures. For better convergence, the normalized inputs, target, and sample data were chosen. It was also found that 152 neurons in a hidden layer trained L–M provides the best performances and the fewest prediction errors. Similarly, for the RBFNN design, the number of neurons in the hidden layer, type of activation, and training algorithms was carefully selected consistent with the methodology. T-RBFNN modelling starts with the training phase, and testing data sets must be designed and built. An epoch refers to a single iteration through the process of updating

IOP Conf. Series: Earth and Environmental Science **168** (2018) 012012 doi:10.1088/1755-1315/168/1/012012

the weight of each link in the network. The developed model was employed to estimate the wind speed values at two locations nearby. The time series meteorological data from 2013-2016 observed at the KUST central research lab station were used to train the forecast model and from 2011-2012 was applied to test the prediction results. Several models were designed and trained to identify the optimum network. Simulations were carried out to calculate the approximate wind speed values.

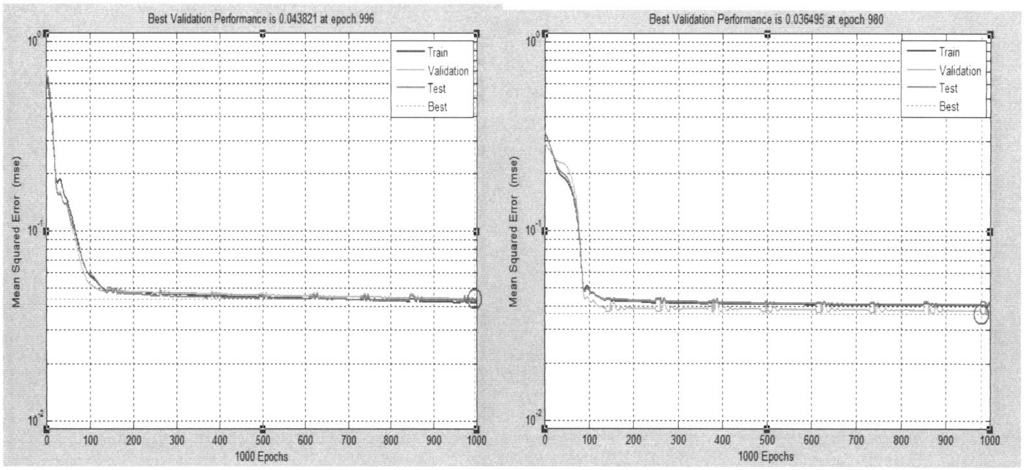

Figure 5. Developed RBFNN model ANN training (a) location 1 (b) location 2.

During the learning process, the weights were set one at a time using wind shear data between the reference area and a target position so that the desired input/output relation of the network would be achieved. The iteration is executed repeatedly until the difference between the real output of the network and the preferred output is a negligible value, which signifies that no additional advancement in the network was achieved. An example of MSE versus epoch for the learning and cross-validation is displayed in Figure 5 a and b, acquired from the reference data. It is very clear from the diagram that, right after 100-950 epochs, the MSE between the T-RBFNN predictions and test data were steadily minimized from to 0.043821 and 0.03649, at best validation points of 996 and 980 which indicates the finest validation effectiveness. Extending the training process could run to the R-BFNN to memorize the training data sets and behaves in poor generalization capability of the T-RBFNN model. The GOF of the above approach has been examined based on the correlation R; the best network has the values of 0.80276, 0.79777, and 0.81081 for lcation1, while location 2 has 0.98314, 0.9827, and 0.98292 for the training, testing, and whole data sets. The most effective training function choice was carried out by contrasting the regression results of the various training functions, such as in Figures 6 a and b), in which the match line indicates the pattern line of training. The closer the trend line is on the straight line (*Y*=*T*), the superior prediction outcomes can be found in the T-RBFNN prediction application. The marked series shows the situation in which the outcome results of the T-RBFNN model are exactly the same as all the target data; this signifies that this is the finest situation for the developed prediction tool. In all the error model graphs, small oscillations were noticed above 100, this would probably be as a result of changes occurred in the roughness length parameters between diverse classes. However, at the starting of the propagation, normally the wind is Geostrophic, afterward; there is an effect of a large-scale terrain slope. Although, the oscillation is minimal as a result of the perturbation due to wind flow.

Furthermore, a comparison was made between the observed wind speed and the predicted wind speed, the result is depicted in Figure 7. As shown in the figure, there is a high positive correlation between the measured wind speed and the predicted wind speed. It is important to note that the common metric for wind speed correlation has been reported [14], values from 0.75-1.00 could be accepted as good correlation. In order to demonstrate the suitability of the proposed method, a results comparison with other models in the literature shows values between 0.55-0.76 [15-21].

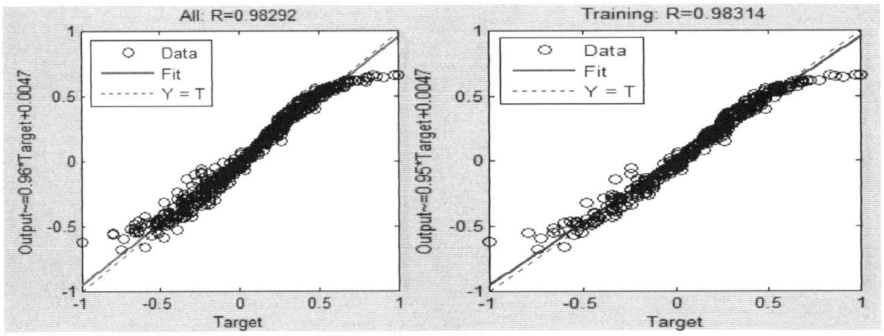

Figure 6. Regression results for (a) Location 1 (b) Location 2

The training approach for those models was thoroughly looked into. The model effectiveness was assessed according to statistical solutions correlation coefficient and mean absolute percentage error (MAPE). These values are acquired mathematically making use of Equation 2 and 3 accordingly. Table 1 shows the summary of the network based on training with most appropriate algorithms and the model with the best possible perdition precision. No results of the inappropriate neurons were provided in this paper.

Table 1: Statistical and error values of ANN

Reference Station	Target Areas	R	MAPE (%)
	Location 1	0.812	5.01
	Location 2	0.627	5.63

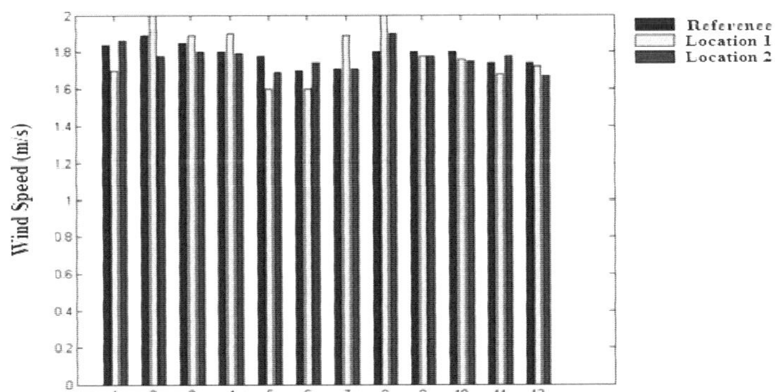

Fig. 7 Comparison between the predicted wind speed and measured reference wind speed (a) Location 1 (b) Location 2.

4. Conclusion

This paper demonstrates the potentiality of using an improved T-RBNN to predict wind speed in the areas not covered by measurements. The developed model was tested and validated using the nearby wind station. It was found that the model is suitable to reproduce wind speed with acceptable accuracy. Future works can consider using T-RBFNN with ARIMA for longer time prediction.

References

[1] M. M. Aman, G. B. Jasmon, a. Ghufran, a. H. a. Bakar, and H. Mokhlis, "Investigating possible wind energy potential to meet the power shortage in Karachi," Renew. Sustain. Energy Rev., vol. 18, pp. 528–542, Feb. 2013.
[2] I. Youm, J. Sarr, M. Sall, A. Ndiaye, and M. M. Kane, "Analysis of wind data and wind energy potential along the northern coast of Senegal," vol. 8, pp. 95–108, 2005.
[3] F. L. Ludwig, "Mass-Consistent Flow Fields From Wind Observations In Rough Terrain + I," no. 3, 1980.
[4] S. Lee, "Application of Artificial Neural Network for Mineral Potential Mapping," 2009.
[5] A. Kusiak, H. Zheng, Z. Song, and I. Engineer-, "Wind Farm Power Prediction : A Data-Mining Approach," no. September 2008, pp. 275–293, 2009.
[6] J. Zeng and W. Qiao, "Support vector machine-based short-term wind power forecasting," 2011 IEEE/PES Power Syst. Conf. Expo., pp. 1–8, Mar. 2011.
[7] R. Sensing and T. Hui, "Combining simulate anneal algorithm with support vector regression to forecast wind speed," 2010 Second IITA Int. Conf. Geosci. Remote Sens., no. 4, pp. 92–94, Aug. 2010.
[8] M. Monfared, S. K. Y. Nikravesh, and H. Rastegar, "A Novel Fuzzy Predictor for Wind Speed," pp. 840–843, 2007.
[9] W. Rutger and L. J. M. Rothkrantz, "Short-Term Wind Power Prediction with Radial Basis Networks," pp. 1–12, 2008.
[10] B. A. Sapranova, "USE OF ARTIFICIAL NEURAL NETWORKS FOR," pp. 3–5, 2012.
[11] S. a. Pourmousavi Kani and M. M. Ardehali, "Very short-term wind speed prediction: A new artificial neural network–Markov chain model," Energy Convers. Manag., vol. 52, no. 1, pp. 738–745, Jan. 2011.
[12] C. Emeksiz, "International Journal of Scientific Knowledge," vol. 1, no. 3, 2012.
[13] J. Lorenzo and M. Juan, "Short-Term Wind Power Forecast Based on Cluster Analysis and Artificial Neural Networks," pp. 191–198, 2011.
[14] K. Jin, S. Florida, W. Management, and W. P. Beach, "Estimating Missing Data of Wind Speeds Using Neural Network," pp. 343–348, 2002.
[15] K. Sreelakshmi and P. Ramakanthkumar, "Neural Networks for Short Term Wind Speed Prediction," pp. 721–725, 2008.
[16] M. Hayashi and B. Kermanshahi, "Application of Artificial Neural Network for Wind Speed Prediction and Determination of Wind Power Generation Output."
[17] S. Lawan, W. Abidin, T. Masri, W. Chai, and A. Baharun, "Wind power generation via ground wind station and topographical feedforward neural network (T-FFNN) model for small-scale applications," Journal of Cleaner Production, vol. 143, pp. 1246-1259, 2017.
[18] S. M. Lawan, W. A. W. Z. Abidin, A. M. Lawan, M. Mustapha, and S. L. Bichi, "ANN AND GIS-Assisted Methodology For Wind Resource Assessment (WRA) In Sarawak." J. Teknologi, Vol 77, No 12,2015
[19] S. M. Lawan, W. A. W. Z. Abidin, S. Lawan, and A. M. Lawan, "An Artificial Intelligence Strategy for the Prediction of Wind Speed and Direction in Sarawak for Wind Energy Mapping," in Recent Advances in Mathematical Sciences, Singapore: Springer Singapore, 2016, pp. 71–82.
[20] S. M. Lawan, W. A. W. Z. Abidin, W. Y. Chai, A. Baharun, and T. Masri, "Wind Energy Potential in Kuching Areas of Sarawak for Small-Scale Power Application," Int. J. Eng. Res. Africa, vol. 15, pp. 1–10, Apr. 2015.

[21] S. M. Lawan W. A. W. Z. Abidin, W. Y. Chai, A. Baharun, and T. Masri., "A Methodology for Wind Energy Evaluation in Complex Terrain Regions of Sarawak," Int. J. Electr. Eng. Informatics - , vol. 7, no. 2, 2015

Acknowledgments
This work was carried under postdoctoral fellowship of Universiti Malaysia Sarawak (UNIMAS). Similarly, we are grateful to the Kano University of Science and Technology, Wudil for their assistance in providing us the data used in this research work.

IOP Conf. Series: Earth and Environmental Science **168** (2018) 012013 doi:10.1088/1755-1315/168/1/012013

Preparation and Characterization of Ce-Zr/Al-MCM-41 Bimetallic Catalyst for Transesterification of Jatropha Oil to Biodiesel

Z I Jibril, A Ramli and K Jumbri

Department of Fundamental and Applied Sciences, Universiti Teknologi PETRONAS, Malaysia

E-mail: anita_ramli@utp.edu.my

Abstract. A series of bimetallic Ce-Zr heterogeneous catalysts supported on Al-MCM-41 aluminosilicate with cerium and zirconium loading ranging from 2.5 to 7.5 wt%, were prepared by co-impregnation method. The characterization of the catalysts obtained by various analytical methods; XRD, N_2 adsorption-desorption, FESEM, EDX and TPD NH3. The result shows that metals were well incorporated in the framework of the support, catalysts retained the mesoporous character and stability of structure after impregnation. The catalysts were active in one step transesterification of non-edible jatropha oil to produce fatty acid methyl esters (FAME) with methanol to oil ratio 6:1 at 90°C, after 4 h reaction time. The FAME yield was found to have the maximum activity, higher than 80% with catalyst 5 wt% Ce and 5 wt% Zr loadings supported by Al-MCM-41 with catalyst loading 5 wt%. The catalyst was recovered by simple filtration.

1. Introduction

Recently, increasing attention has been paid to mesoporous molecular sieves having ordered hexagonal arrangements, high specific surface areas, and uniform pore diameters [1]. Modification of mesoporous aluminosilicates gives rise to a platform for wider application possibilities in industry and catalysis. The structurally stable mesoporous material, Al-MCM-41 has both hydrothermal stability and acidity properties unlike the parent material, MCM-41 [2].

According to Avhad and Marchetti [3], an ideal catalyst should have superior activity and selectivity, be abundant, less expensive, durable, reusable, and be easy to separate. Although Al-MCM-41 have the requirements necessary for solid catalysts, the activity in alkyl ester conversion was found to be low and this weak performance was related to the acidity of the catalyst [4]. Active metals are incorporated into the framework of Al-MCM-41 as a support, in order to regulate the acidity. Moreover, catalyst efficiency depends on factors such as surface area, pore size and activity sites [5].

Distribution of acid strength on the surface of molecular sieves is regulated to improve their activity, selectivity as well as the stability in biodiesel formation reaction processes. Ceria-zirconia-alumina samples synthesized by sol-gel process exhibited an increase in thermal instability at high temperature and high concentration of the ceria which was regulated with an increase in zirconia concentration [6]. According to the study conducted by Neto and Schmal [7] on iso-synthesis reaction with ceria and ceria-zirconia solid catalysts, strong basic sites and high selectivity was presented by ceria-zirconia as compared to ceria only.

Heterogeneous catalyst with 20 wt % ceria, used in transesterification reaction to produce biodiesel was found to be the most active catalyst among others [8]. However, development of a stable catalyst with suitable particle size that can be reused remains inadequately addressed, as reported by Linares et al. [9]. For the purpose of this study novel heterogeneous bimetallic catalyst Ce-Zr/Al-MCM-41, was synthesized and characterized for application in transesterification of jatropha oil to biodiesel.

2. Experimental

2.1. Materials and Methods

The chemicals used for the synthesis of bimetallic heterogeneous catalysts, Ce-Zr/Al-MCM-41 were aluminosilicate, Al-MCM-41 supplied by ACS Material, the metal precursors cerium nitrate hexahydrate, $Ce(NO_3)_3.6H_2O$ and zirconium oxynitrate hydrate $Zr(NO_3)_2.xH_2O$ were supplied by Sigma Aldrich. All chemicals were of analytical grade and used without further purification. The Jatropha oil used for the purpose of this study was purchased from BIONAS®.

Al-MCM-41 support (Si/Al= 25) was activated in air at 500°C for 12 h. The bimetallic catalysts were prepared with 10 wt% of Ce and Zr metals by co-impregnation method. The catalysts were prepared by varying the weight percent of the active metals, 7.5 wt% Ce-2.5 wt% Zr (CZA7.5), 5.0 wt% Ce-5.0 wt% Zr (CZA5.0) and 2.5 wt% Ce-7.5 wt% Zr (CZA2.5) on the calcined support. Firstly, aqueous solutions containing Ce and Zr precursors were dissolved separately before they were added to the support and stirred for 10 h, then dried at 96°C for 24 h, followed by calcination in air at 500°C for 5 h.

2.2. Characterization Technique

Catalysts were characterized by X-ray diffraction (XRD) spectrometry using X'pert^3 Powder and Empyrean, the scanning was performed in the $2\theta = 1 - 10°$ scale with 0.01° step size. Surface area of the samples by N_2 adsorption-desorption isotherms at 77 K were recorded with Surface Area Analyser and Porosity System (Micromeritics ASAP 2020) N_2-adsorber, Field Emission Scanning Electron Microscopy (FESEM) was acquired on ZEISS 55 Supra VP microscope equipped with Energy Dispersive X-ray (EDX), as well as studies on acidity by TPD NH_3 (Thermo Fisher Scientific, TPDRO 1100).

The catalysts were directly applied in transesterification reaction of the Jatropha oil with high free fatty acid content (FFA %) for the initial study on the catalytic activity, without pre-treatment (esterification reaction). The feedstock was first analyzed to determine properties such as acid value, moisture content, density, viscosity, specific gravity, saponification value and flash point. Moisture content was removed from the oil at 65°C for 1 h, prior to the transesterification reaction. A lab-scale transesterification of jatropha oil was carried out in a three neck round bottom flask attached to a reflux system at temperature 90°C, reaction time 4 h, catalyst loading of 5 wt% (to the weight of oil) and methanol to oil molar ratio 6: 1.

3. Result And Discussion

3.1. Catalyst Characterization

Figure 1 shows the XRD diffractograms for the bimetallic catalysts and the calcined support. As shown in figure 1(a), the calcined support, Al-MCM-41 shows a broad peak at $2\theta = 2°$ followed by two more peaks at 3 and 4° at lower intensities. However, the subsequent peaks disappear with impregnation of active metals, which shows that the metals are successfully incorporated in the framework of the support and possess disordered hexagonal structure. The prominent peak found in all the catalysts is characteristic peak of the MCM-41 [10]. Decrease in intensity with decrease in the surface area of the support which was found from 753 ± 2.01m^2/g (see Table 1), may be related to loss of crystalline structure. Moreover, owing to the decomposition of the nitrates during calcination process, no peaks were found corresponding to the cerium and zirconium precursors, $Ce(NO_3)_3.6H_2O$ and $Zr(NO_3)_2.xH_2O$.

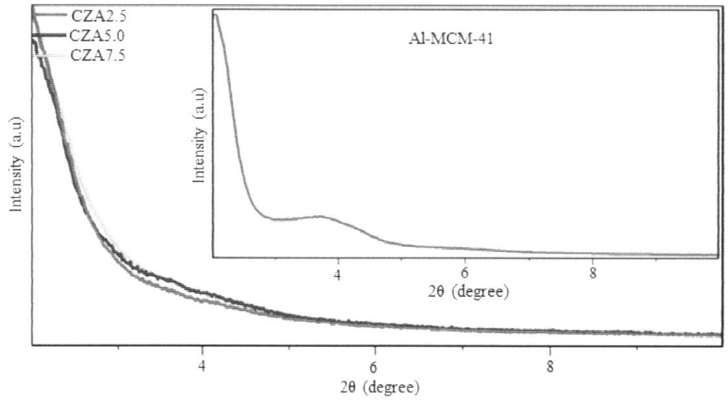

Figure 1. XRD patterns of 7.5Ce-2.5Zr/Al-MCM-41 (CZA7.5), 5.0Ce-5.0Zr/Al-MCM-41 (CZA5.0) and 2.5Ce-7.5Zr/Al-MCM-41 (CZA2.5) catalysts in comparison with the calcined support Al-MCM-41.

The N_2 adsorption–desorption isotherms for the calcined support and the synthesized catalysts are shown in Figure 2(a). It can be seen that all the samples exhibited type IV adsorption isotherms, with hysteresis loops typical of uniform mesoporous silicate materials, more prominent with a sharp inflection for Al-MCM-41 [11]. Figure 2(b) shows the pore distribution of the prepared catalysts. The pore distribution is within the mesoporous range, but a forward shift is seen from the calcined support showing a decrease in both the pore volume and diameter as the active metals are deposited on the surface of the support.

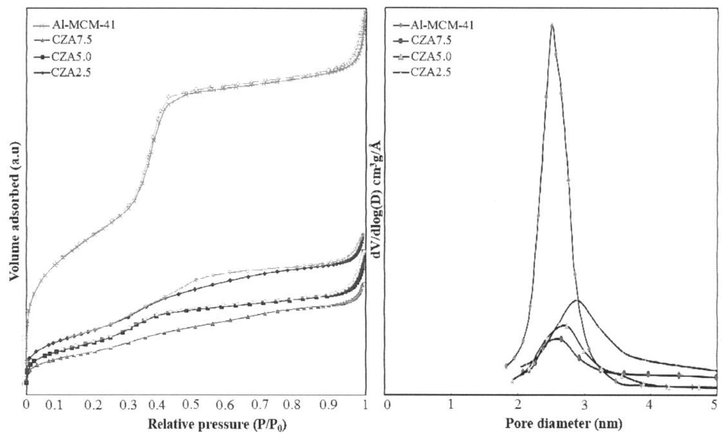

Figure 2. (a) N_2 adsorption- desorption isotherms and (b) pore distribution of Al-MCM-41, 7.5Ce-2.5Zr/Al-MCM-41 (CZA7.5), 5.0Ce-5.0Zr/Al-MCM-41 (CZA5.0) and 2.5Ce-7.5Zr/Al-MCM-41 (CZA2.5) catalysts.

Table 1 shows the textural properties of the Al-MCM-41 based catalysts. It can be seen that the surface area of the calcined support decreases with incorporation of the active metals. The surface area significantly reduced when the amount of Ce incorporated is higher as compared to Zr. The same trend is observed for the average pore volume, and pore diameter where Ce is higher than and equal to the amount of Zr loading on the support (Table 1).

Table 1. Textural parameters of the calcined support, Al-MCM-41 before and after co-impregnation with Ce and Zr

Catalyst	$d_{(100)}$	$^{*}a_0$ (nm)	$^{\#}t$ (nm)	Surface area (m^2/g)	Pore diameter (nm)	Pore volume (cm^3/g)
Al-MCM-41	4.17	4.82	0.65	753	3.50	0.60
CZA7.5	4.17	4.80	0.64	206	4.40	0.30
CZA5.0	4.19	4.84	0.65	261	4.30	0.30
CZA2.5	4.17	4.82	0.65	307	4.10	0.40

*Unit cell parameter = $2d_{100}/\sqrt{3}$
Wall thickness = $(2d_{100}/\sqrt{3} - d_{100})$

An increase in cell parameter is attributed to the incorporation of larger metals in the mesoporous network [12]. The cell parameter increases in all the catalysts as compared with the support, except for CZA2.5 which is similar to the calcined support and CZA7.5, where the size was found to have decreased (Table 1). The wall thickness as shown in table 1, the prepared catalysts have the same thickness as the support except for CZA7.5, which is also very identical to the support. The thicker the cell wall the better the acidity and stability of the catalysts [12].

FESEM micrographs of the calcined support, Al-MCM-41 was studied and presented in Figure 3 (a), with a well-defined hexagonal arrangement and a fairly uniform pore structure similar to literature [13]. However, after impregnation, the levelled structure faced some changes. The presence of the observed agglomerated particles in all the synthesized catalyst was an indication of the incorporation of cerium and zirconium metals showing reconstructed morphologies, agglomerated with no definitive shape but more close to spheroidal (Figure 3 b - d).

From the micrographs, it is observed that increase in loading of ceria and zirconia particles on the support increases the agglomeration. However, ceria (up to 4 wt % on MCM-41) loading was uniform according to literature [14]. Nevertheless, when the ratio of ceria and zirconia incorporation is 1:1, the micrographs was found to be compacted and more uniformly spread (Figure 3 c). The active metals impregnation on the support was successful as confirmed from the EDX result attached to the micrographs of the catalysts in comparison with the support, Al-MCM-41 (Figure 3 a – d).

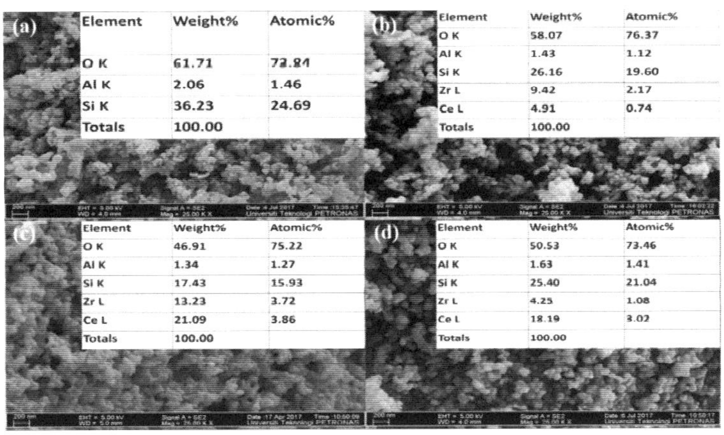

Figure 3. FESEM micrographs and EDX of (a) Al-MCM-41, (b) 7.5Ce-2.5Zr/Al-MCM-41 (CZA7.5), (c)5.0Ce-5.0Zr/Al-MCM-41 (CZA5.0) and (d) 2.5Ce-7.5Zr/Al-MCM-41 (CZA2.5) catalysts.

According to the TPD NH_3 profiles of the samples (Figure 4), sharp peaks at low temperatures (between 100- 200 °C) were detected corresponding to weak acids. Desorption peaks of the catalysts, decreases with increase in Ce-Zr incorporation, as such the highest peak was found to be that of the support. Studies on the acidity of the support reported that modified MCM-41 have two peaks, a sharp peak at low temperature and a weak peak at high temperature assigned to weak acids (NH_3 coordination with weak acid sites) and strong acids, respectively [15].

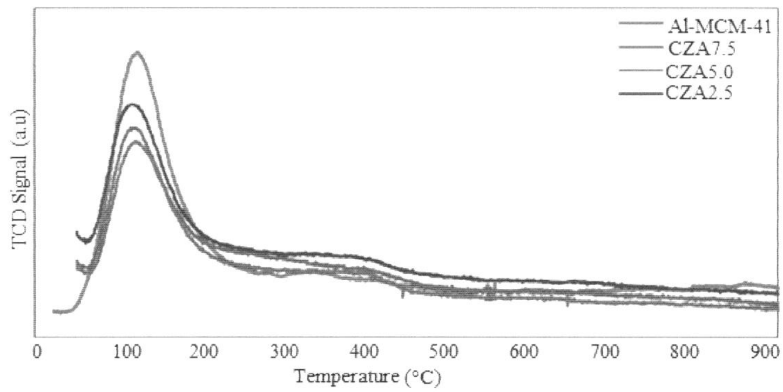

Figure 4. TPD-NH_3 profiles of Al-MCM-41, 7.5Ce-2.5Zr/Al-MCM-41 (CZA7.5), 5.0Ce-5.0Zr/Al-MCM-41 (CZA5.0) and 2.5Ce-7.5Zr/Al-MCM-41 (CZA2.5) bimetallic catalysts.

3.2. Catalytic Performance

The bimetallic catalysts were directly applied in one-step transesterification reaction process using jatropha oil and methanol to produce biodiesel. After the reaction catalyst was separated by filtration, biodiesel was washed with hot distilled water and dried with Na_2SO_4. The conversion rate of biodiesel was found to be highest for catalyst CZA5.0 at reaction temperature 90°C, reaction time 4 h, molar ratio 6: 1 and catalyst loading 5 wt% (Figure 4). The arrangement in terms of catalytic activity in decreasing order is 5.0Ce-5.0Zr/Al-MCM-41, 2.5Ce-7.5Zr/Al-MCM-41, 7.5Ce-2.5Zr/Al-MCM-41and Al-MCM-41 with the least activity.

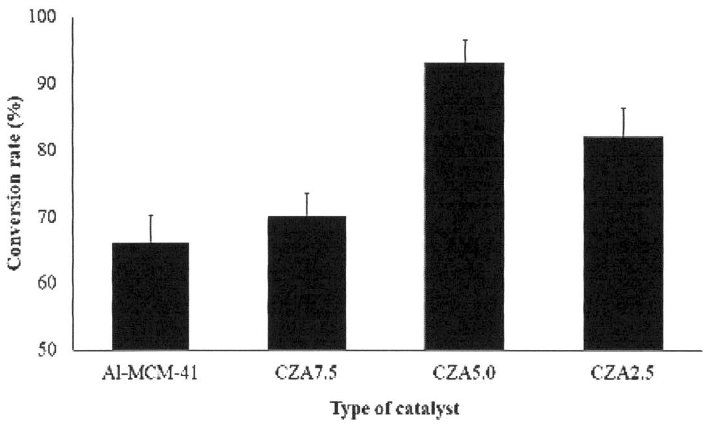

Figure 5. Biodiesel conversion rate (%) using Al-MCM-41, 7.5Ce-2.5Zr/Al-MCM-41 (CZA7.5), 5.0Ce-5.0Zr/Al-MCM-41 (CZA5.0) and 2.5Ce-7.5Zr/Al-MCM-41 (CZA2.5) catalysts.

4. Conclusion

In this study the bimetallic ceria-zirconia heterogeneous catalyst supported on mesoporous aluminosilicate, Al-MCM-41 support was successfully prepared by co-impregnation method. The active metals were successfully deposited on the surface of the support as confirmed by various characterization techniques. It was found out that the catalysts maintained the main characteristic peak of the parent hexagonal structured MCM-41 with a slight disorder as the amount of the ceria-zirconia increases. As compared with the calcined support, the catalysts showed decrease in surface area, in the same manner. The activity of the calcined support was enhanced by the impregnated active metals, 93% biodiesel yield was found for catalyst with the same amount of metals loaded 5.0Ce-5.0Zr/Al-MCM-41.

References

[1] Araujo R S, Azevedo D C, Rodríguez-Castellón E, Jiménez-López A, Cavalcante CL 2008 *J. of Mol. Catal. A: Chem.* **281** 154-63
[2] Xia Z, Fu J, Duan A, Han L, Wu H, Zhao Z, Xu C, Wang D, Wang B, Meng Q 2017 *Catalysts* **7** 141
[3] Avhad M R, Marchetti J M 2016 *Cat Rev* **58** 157-208
[4] Wang Z, Yu S 2016 *Adv. Chem. Eng. Sci.* **6** 305
[5] Stekrova M, Kumar N, Mäki-Arvela P, Ardashov OV, Volcho KP, Salakhutdinov NF, Murzin DY 2013 *Materials* **17** 2103-18
[6] Dumont M R, Nunes E H, Vasconcelos W L *Ceram. Int.* **42** 9488-95
[7] Neto R C, Schmal M 2013 *Appl. Catal. A: Gen.* **15** 131-42
[8] Farooq M, Ramli A, Subbarao D 2011 *J. Chem. Eng. Data* **5** 26-32
[9] Linares N, Silvestre-Albero A N M, Serrano E, Silvestre-Albero J, García-Martínez J 2014 *Chem. Soc. Rev.* **43** 7681-717.
[10] Eswaramoorthi I, Sundaramurthy V, Lingappan N Zirconium Containing Al-MCM-41– 2001 *Stud. Surf. Sci. Catal.* **1** 4796-802
[11] Thommes M, Kaneko K, Neimark A V, Olivier J P, Rodriguez-Reinoso F, Rouquerol J, Sing K S 2015 *Pure Appl. Chem.* **1** 1051-69
[12] Liu Y 2015 *Mesoporous Zeolites: Preparation, Characterization and Applications* (Germany: Wiley-VCH) p 115-56
[13] Boukoussa B, Hamacha R, Morsli A, Bengueddach A 2013 *Arabian J. Chem.* **10** 2160-69
[14] Babu C M, Vinodh R, Abidov A, Ravikumar R, Peng M M, Cha W S and Jang H T 2016 *Int. J. Bio-Sci. Bio-Tech.* **8** 171- 82
[15] Marakatti V S, Manjunathan P, Halgeri A B, Shanbhag G V 2016 *Catal. Sci. & Technol.* **6** 2268-79

Acknowledgment
The authors acknowledge financial support from Yayasan Universiti Teknologi PETRONAS.

IOP Conf. Series: Earth and Environmental Science **168** (2018) 012014 doi:10.1088/1755-1315/168/1/012014

Performance Analysis of Current Injection Techniques for Shunt Active Power Filter

Ikramullah[1], Muhammad Ashraf[2] and Abdur Rehman[3]

[1,2,3]Department of Electrical Engineering, Capital University of Science and Technology(CUST),Islamabad, Pakistan

[1]ikramullah_73@yahoo.com, [2]ashraf@cust.edu.pk, [3]rehmanmstel@gmail.com

Abstract. Shunt Active Power Filter (SAPF) can be used for compensation of current harmonics generated by non-linear loads. The process involves an injection of equal but opposite current to mitigate the distortion current supplied back by nonlinear load to the voltage supply. Thus, the injected current will shape the supply current to a sinusoidal. SAPF consists of signal processing part and injection part. Processing part generates reference current wave and the injection part injects the reference current parallel to the load. In this paper a comparison is made between three different control strategies for current injection in three-phase three-wire shunt active power filter. The current injection is performed by generating gate pulses pattern for the voltage source inverter using a suitable controller. In this paper three control techniques are compared including Hysteresis Current Control (HCC), Delta Modulation Control (DMC) and One Cycle Control (OCC). The performance of these techniques is evaluated by the accuracy of the injected current in parallel to the load by comparing the required current to be injected and the measured value of the injected current. The performance is also evaluated by measuring Total Harmonic Distortion (THD) of the source current before and after compensation. The results are evaluated and compared using Matlab/Simulink.

1. Introduction

With the advancement of technology, the dependency on the electrical energy has been increased greatly. The performance efficiency of different electric apparatus is directly affected by power quality. The power-quality problem is identified by a distortion in the voltage waveform of the power source from a sine wave, or in the amplitude from an established reference level [1]. Different types of disturbances that affect the power quality are voltage fluctuations, voltage sags, voltage unbalance, transients and harmonics [2].

Harmonics is known to be a major issue out of all power quality problems [3]. Harmonics in the power system are increasing rapidly due to the large scale use of power electronics equipment. Harmonics can be defined as currents or voltages which are integer multiples of fundamental frequency [4]. Harmonic currents are caused by nonlinear loads connected with the distribution system. These current harmonics will be responsible for power factor reduction, power system voltage fluctuations, decrease in efficiency and interference during communications. The Harmonic currents are increasing dramatically in all residential, commercial, and industrial installations as the nonlinear loads are increasing rapidly over the past few years. Different methods [5] like K-rated transformers, separate neutral conductors, zigzag autotransformers, phase shifting, series line inductors and harmonic filters are employed for mitigation of harmonics.

Active power filters (APF) have been used for harmonics suppression and reactive power compensation in order to enhance the power quality. Gyugi and Strycula [6] were the first to introduce the concept of Active Power Filters. Due to significant growth in power electronics, the use of APF has become the dynamic solution for reduction of harmonics because of having unique advantages as compared to passive filter devices [7]. The Shunt Active Power Filter connected in parallel with the load is used for mitigation of current harmonics. The process involves detection of harmonic component, generation of reference current and generation of the gate pulses for the power circuit to inject the reference current. The power circuit consists of AC to DC converter with a DC link capacitor and DC to AC voltage inverter. An active power filter cancels out the current harmonics introduced by the nonlinear load by injecting reference current in parallel to the load, thus making the current waveform supplied through the source, sinusoidal. The basic structure of Shunt Active Power Filter is shown in 'figure 1'. The source voltage V_s is supposed pure sinusoidal. The load current is represented by I_L. The reference current generation block detects the harmonic current waveform and generates the reference current I_R, which is anti-phase to the harmonic current. The current injection controller produces the switching pulses for voltage source inverter in order to inject I_R at the point of common coupling (PCC). The injected current with addition to load current will make the source current very near to pure sinusoid. The parameter used to evaluate the performance of APF is Total Harmonic Distortion (THD). A THD value of the source current close to zero proves a very good performance of APF in terms of harmonics mitigation.

Reference current can be extracted using either time domain techniques like Instantaneous Reactive Power (p-q) theory, Synchronous Reference Frame (d-q) theory and perfect harmonic cancellation or frequency domain techniques like Discrete Fourier Transform (DFT) [8] and sine multiplication method. DC link voltage control is one of the important issues in voltage source inverter. PI control is used to regulate the DC link voltage. Different controllers are used to inject the reference current at the point of common coupling like Hysteresis Current Controller, Deadbeat Controller, Delta Modulation Controller, One Cycle Controller Predictive Controller and Sliding Mode Controller.

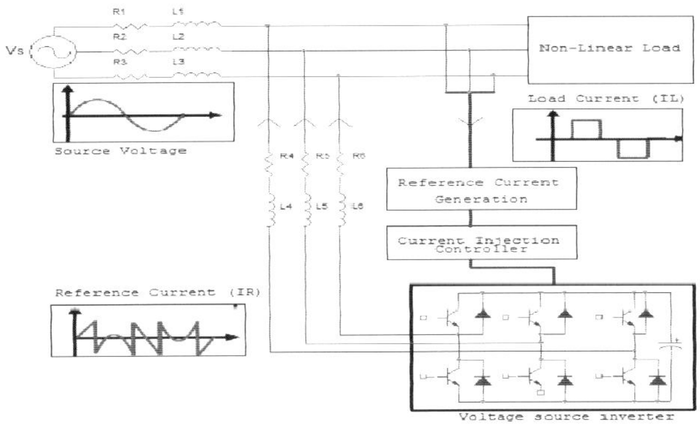

Figure 1: Basic structure of Shunt Active Power Filter

2. Reference Current Generation Techniques

Reference current generation is an important part of active power filter. Reference current is actually 180 degree out of phase with the harmonics current. The reference current added to the load current will make the source current sinusoidal. Many techniques are used for reference current generation but in this paper Instantaneous Reactive Power theory which is also known as p-q theory is used due to its simplicity and accuracy.

2.1. Instantaneous Reactive Power (p-q) Theory

This concept is very popular and useful for Active Power Filter application, and principally consists of a variable transformation from a, ***b, c,*** reference frame of the instantaneous power, voltage, and current signals to α, β reference frame [9].

Most APFs have been designed on the basis of p-q theory to calculate the desired reference currents. In this theory, instantaneous three phase current and voltages are transformed from a - b - c coordinates to α - ß - 0 co-ordinates [10] as shown in equation (1) and (2).

$$\begin{bmatrix} v_0 \\ v_\alpha \\ v_\beta \end{bmatrix} = \sqrt{\frac{2}{3}} \begin{bmatrix} 1/\sqrt{2} & 1/\sqrt{2} & 1/\sqrt{2} \\ 1 & -1/2 & 1/2 \\ 0 & \sqrt{3}/2 & -\sqrt{3}/2 \end{bmatrix} \begin{bmatrix} v_a \\ v_b \\ v_c \end{bmatrix} \quad (1)$$

$$\begin{bmatrix} i_0 \\ i_\alpha \\ i_\beta \end{bmatrix} = \sqrt{\frac{2}{3}} \begin{bmatrix} 1/\sqrt{2} & 1/\sqrt{2} & 1/\sqrt{2} \\ 1 & -1/2 & 1/2 \\ 0 & \sqrt{3}/2 & -\sqrt{3}/2 \end{bmatrix} \begin{bmatrix} i_a \\ i_b \\ i_c \end{bmatrix}$$
(2)

The source currents and phase neutral voltages are used to calculate the instantaneous real and imaginary power components as given in equation (3).

$$\begin{bmatrix} p_0 \\ p \\ q \end{bmatrix} = \begin{bmatrix} v_0 & 0 & 0 \\ 0 & v_\alpha & v_\beta \\ 0 & -v_\beta & v_\alpha \end{bmatrix} \begin{bmatrix} i_0 \\ i_\alpha \\ i_\beta \end{bmatrix} \quad (3)$$

The reference current in α- β co-ordinates is given by (4).

$$\begin{bmatrix} i_\alpha * \\ i_\beta * \end{bmatrix} = \frac{1}{v_\alpha 2 + v_\beta 2} \begin{bmatrix} v_\alpha & v_\beta \\ -v_\beta & v_\alpha \end{bmatrix} * \begin{bmatrix} p \\ q \end{bmatrix} \quad (4)$$

The reference current in a-b-c co-ordinates is given by (5).

$$\begin{bmatrix} i_a * \\ i_b * \\ i_c * \end{bmatrix} = \sqrt{\frac{2}{3}} \begin{bmatrix} 1 & 0 & 1/\sqrt{2} \\ -1/2 & \sqrt{3}/2 & 1/\sqrt{2} \\ -1/2 & -\sqrt{3}/2 & 1/\sqrt{2} \end{bmatrix} * \begin{bmatrix} i_0 * \\ i_\alpha * \\ i_\beta * \end{bmatrix} \quad (5)$$

3. Current Injection Techniques

The reference current generated must be injected in parallel to the load with high accuracy in order to make the source current sinusoidal. In this paper three different current injection techniques are implemented to evaluate the performance of active power filter. The techniques analyzed are Hysteresis Current Control (HCC), Delta Modulation Control (DMC) and One Cycle Control (OCC).

3.1. Hysteresis Current Control

In hysteresis current controller a fixed hysteresis band is set and is compared with a current error signal to generate the switching pulses of inverter. It is a closed loop system which senses the current error and generates the switching pulses of IGBT when the error exceeds an assigned band. This technique has the advantages of simplicity, good accuracy, robustness and fast dynamic response.

The switching pattern of hysteresis current control is shown in 'figure 2(a)'in which the hysteresis band is set with a lower and upper limit [11]. When the compensating current reaches the upper or lower limit, the controller increases or decreases the compensating current so that it follows the reference current[12]. The schematic diagram of HCC is shown in 'figure 2(b)'where I_{ca}, I_{cb} and I_{cc}

are compensating currents and I_{ca*}, I_{cb*} and I_{cc*} are reference currents. These currents are compared to generate the six switching pulses from g1 to g6. These pulses are used for switching of IGBTs.

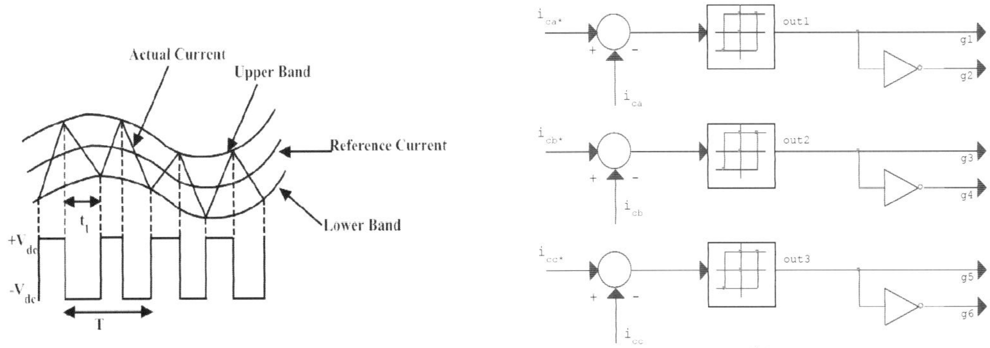

(a) (b)

Figure 2. Hysteresis Current Control (a) Switching pattern, (b) Schematic diagram.

3.2. Delta Modulation Control

The Delta Modulation Control (DMC) is also based on a nonlinear control [13] as shown in 'figure 3(a)'. The switching signals of six IGBTs are generated by applying limit comparators and D-type flip-flops. The main advantage of delta modulation current control technique is that it is simple and easy to implement. I_{ca}, I_{cb} and I_{cc} are the compensating currents and $I_{ca}*$, $I_{cb}*$ and $I_{cc}*$ are the reference currents for three phases. If I_{ca} is over than $I_{ca}*$, the output of comparator (y1) is 0. In contrast, if I_{ca} is less than $I_{ca}*$ then y1 is 1. The output of comparator is applied to D-type flip flop for producing the switching pulses. The clock signal is used to determine the output of D-type flip flop. The output Q is set equal to the output of comparator y1 upon transition of the clock signal from 0 to 1, thus generating the gate pulse g1 and output $\bar{Q}$ generates the gate pulse g2. The gate pulses g3 to g6 for the other two phases are generated in the same manner.

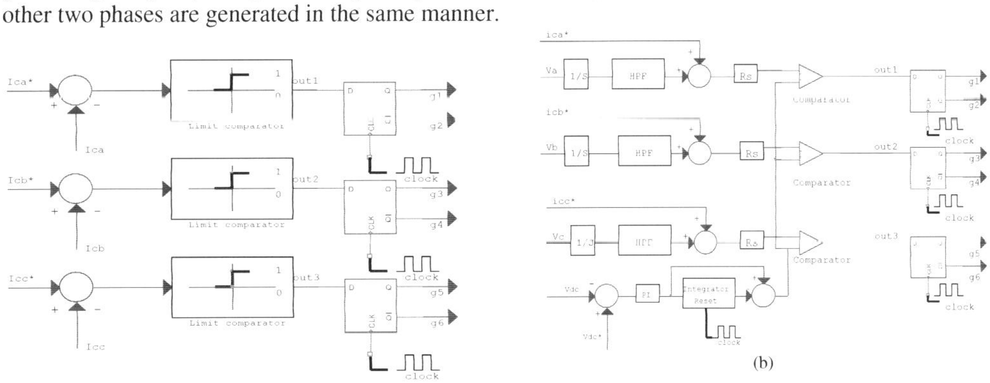

(b)

(a)

Figure 3. (a) DMC schematic, (b) OCC schematic

3.3. One Cycle Control

One-cycle control is a nonlinear control method which has fast transient response and good tracking performance [14]. This technique is implemented to control the duty ratio of the switch in real time. The input voltage perturbations are rejected by One-Cycle Control method in only one switching

IOP Conf. Series: Earth and Environmental Science **168** (2018) 012014 doi:10.1088/1755-1315/168/1/012014

cycle and follows the control reference very quickly. This new control method is very common and directly applicable to all switching converters. This method takes advantage of the pulsed and nonlinear nature of switching converters and attains instantaneous control of the average value of the chopped voltage or current [15].

The block diagram of OCC is shown in 'figure 3(b)' where i_{ca}*, i_{cb}* and i_{cc}* are the reference currents and V_a, V_b and V_c are the source voltages. R_s is the current sensing resistor. A key feature of the OCC control method is the fact that the ramp generated by the integrator circuit is reset at the end of each switching cycle and the ramp starts again from zero at the beginning of the subsequent cycle. Consequently, this method is termed, OCC, or "One Cycle Control".

4. Simulation Results

The simulation results for all the three techniques have been obtained by using Matlab/Simulink. For the purpose of simulation a three phase three wire system with balanced load has been considered. The reference current is generated using p-q theory and initially in order to check its accuracy this current is added to the load current which becomes sinusoidal giving a THD of 0.9%. In the next step this reference current is injected using three control strategies. The performance of the three techniques is analyzed by calculating the THD of the source current and computing the error between the reference current wave and injected current wave. Table1 shows the Source parameters used for simulation.

Table 1. Source parameters

Parameter	Line-line RMS voltage	Source frequency	Source resistance	Line inductance	Line resistance
Value	380 V	50 Hz	0.001Ω	1μH	0.01 Ω

4.1. Results of HCC:

The Shunt Active Power Filter is simulated first by using Hysteresis current controller for current injection. The load current waveform is shown in Fig7 and its corresponding THD is shown in Fig8 which shows that THD is 29.89%. Fig9 shows the waveform of the reference current generated using p-q theory.

(a) (b)

Figure 4. (a) Load current, (b) Reference current

The injected current waveform using HCC is shown in Fig10. The injected current waveform is subtracted from the reference current waveform thus producing an error as shown in Fig11. The minimum possible error proves the accuracy of current injection technique.

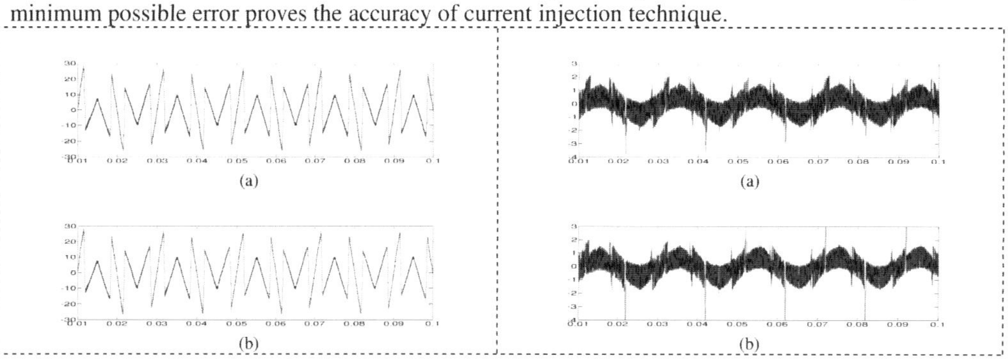

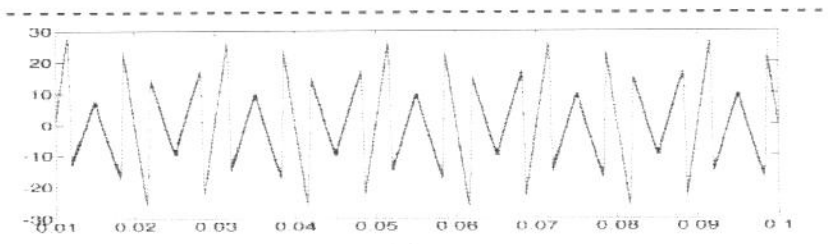

(c)

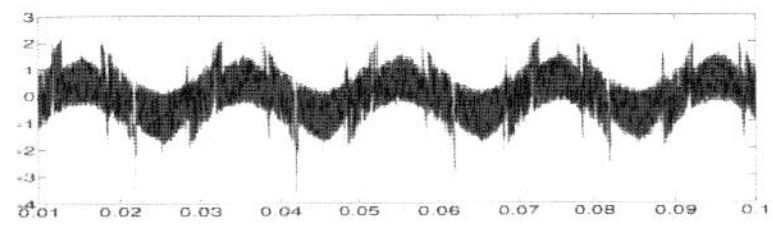

(c)

Figure 5. Injected current (a) HCC (b) DMC (c) OCC

Figure 6. Error (a) HCC (b) DMC (c) OCC

4.2. Results of DMC:

The Shunt Active Power Filter is simulated again by using Delta Modulation controller for current injection. The current injected using DMC is shown in Fig13 and its corresponding error is shown in Fig14. It can be seen that difference in error between HCC and DMC technique is very small but DMC error plot also contains some spikes at regular intervals. The source current waveform after compensation is shown in Fig15.

4.3. Results of OCC:

The last simulation of SAPF is carried out using One Cycle Control for current injection. The injected current waveform is shown in Fig16 and the error between reference current and injected current is shown in Fig17. It can be seen that this error is larger than the other techniques. Fig18 shows the compensated source current. Table 2 shows the performance of the techniques used for current injection by measuring THD of the source current before and after compensation. It also shows the amplitude of error between injected current and reference current.

Table 2. THD of source current

Control Technique	**HCC**	**DMC**	**OCC**
THD before compensation	29.89%	29.89%	29.89%
THD after compensation	1.55%	1.83%	2. 9%
Error Amplitude	0.755	0.789	1.38

5. Conclusion

In this paper three different control strategies used for current injection in shunt active power filters are analyzed. The three techniques considered are Hysteresis Current Control (HCC), Delta Modulation Control (DMC) and One Cycle Control (OCC). A three-phase three-wire system with a non-linear load is simulated to evaluate the performance of these techniques. The parameters used for performance evaluation are THD and error calculation between reference current and injected current. After comparing the results it is concluded that HCC is able to inject the reference current with good accuracy and minimum error as compared to DMC and OCC. Also THD of the source current is reduced to a lower value by using HCC.

6. References

[1] Singh, Bhim, Kamal Al-Haddad, and Ambrish Chandra. "A review of active filters for power quality improvement." *IEEE transactions on industrial electronics* 46.5 (1999): 960-971.

[2] Moravek, Jan, et al. "Power quality issues related to power flow control in systems with renewable energy micro sources." *Electric Power Engineering (EPE), 2016 17th International Scientific Conference on*. IEEE, 2016.

[3] Pandey, Shubham, Maneesh K. Tiwari, and Dileep K. Shukla. "Harmonic mitigation techniques in modern power system: A Review."*Global Journal of Multidisciplinary Studies,* Vol. 5, Issue-9, August-2016.

[4] Pejovic, Predrag. *Three-phase diode rectifiers with low harmonics: current injection methods.*

Springer Science & Business Media, 2007.

[5] Anju Jacob, Babitha T Abraham, Nisha Prakash and Riya Philip. "A review of active power filters in power system applications." *International Journal of Advanced Research in Electrical, Electronics and Instrumentation Engineering,* Vol. 3, Issue 6, June 2014.

[6] L. Gyugyi, E. C. Strycula, "Active ac power filters", *in Proc. IEEE/IAS Annu. Meeting*, 1976, pp. 81-87.

[7] Swain, Sushree Diptimayee, Pravat Kumar Ray, and Kanungo Barada Mohanty. "Improvement of power quality using a robust hybrid series active power filter." *IEEE Transactions on Power Electronics* 32.5 (2017): 3490-3498.

[8] Chauhan, Siddharthsingh K., et al. "Analysis, design and digital implementation of a shunt active power filter with different schemes of reference current generation." *IET Power Electronics* 7.3 (2013): 627-639.

[9] H. Akagi, Y. Kanazawa, and A. Nabae. "Generalized theory of the instantaneous reactive power in three-phase circuits."in *Proc. IEEJInt. Power Electronics Conf. (IPEC-Tokyo)*, 2004, pp. 1375-1386.

[10] Willems, Jacques L. "A new interpretation of the Akagi-Nabae power components for nonsinusoidal three-phase situations." *IEEE Transactions on Instrumentation and Measurement* 41.4 (1992): 523-527.

[11] Belonkar, A., and M. Salodkar. "Performance Analysis of Shunt Active Power Filter with Different Switching Signal Generation Techniques." *Presentado en Recent Advances and Applications of Electrical Engineering: Proceedings of the 9th WSEAS International Conference on Applications of Electrical Engineering (AEE'10), Penang, Malaysia*. 2013.

[12] Bhople, Satish U., and Santhosh Kumar Rayarao. "Comparison of Various Reference Current Generation Techniques for Performance Analysis of Shunt Active Power Filter using MATLAB Simulation." *International Journal of Current Engineering and Technology* 6.2 (2016): 606-611.

[13] Alargt, Farag S., Ahmed S. Ashur, and Ahmad H. Kharaz. "Novel adaptive delta modulation controller for Interleaved Boost DC-DC converters." In *Systems and Control (ICSC), 2016 5th International Conference on*, pp. 20-25. IEEE, 2016.

[14] Ramani Kannan1, Lokesh N, Khairul Nisak Md Hasan and Aravind CV. "Implementation of One cycle Controller for Single phase Bi-directional." *International Journal of Applied Engineering Research*. 12, no. 6 (2017): 804-812.

[15] Sreeraj, E. S., et al. "An active harmonic filter based on one-cycle control." *IEEE Transactions on Industrial Electronics*61.8 (2014): 3799-3809.

IOP Conf. Series: Earth and Environmental Science **168** (2018) 012015 doi:10.1088/1755-1315/168/1/012015

Smart Grid Resources Optimization Using Heuristic Approach

A Touzene[1], S Al Yahyai[2], and A Oukil[3]

1 Sultan Qaboos University, Computer Science Dept., Muscat, Oman
2 Mazoon Electricity Company, Department of IT, Muscat, Oman
3 Sultan Qaboos University, Oper. Manag. & Busi. Stat. Dept, Muscat, Oman

First author's e-mail: touzene@squ.edu.om

Abstract. In this paper, we propose a new layered architecture for Smart Grid Resource Optimization using heuristic approach (SGROH). SGROH allows Smart Grid Constituencies (SGC) such as Power Generators (PG), Power Transporters (PT), Power Distributors (PD), and Power Consumers (PC) to optimize their pay-offs using the smart grid. The proposed optimization heuristic aims to support power consumers with sustainable energy from power distributors, transparently at the best price on a real time basis (variable pricing). It will also decide automatically for the power distributors the best power generators and connecting transporter lines. The SGROH engine is based on a greedy heuristic which aims at optimizing the profits and resource utilization over the whole smart grid. Experimental study shows that SGROH heuristic gives nearly optimal solutions.

1. Introduction

Nowadays, the general trend worldwide is towards upgrading existing power grids to Smart Grids. The key motivations behind modernizing the power grid are essentially the accommodation of future demand growth, the enhancement of efficiency, reliability, and sustainability. Power plants (generators) generate electrical energy, conventional and renewable, transmitted through high voltage lines (transporters) to distribution centers (distributors) where it is stepped down to lower voltages and delivered to end-consumers. A better system utilization should strike a balance between the expected consumer's demand and the electrical energy to be generated. In many countries, the three components (generation, transportation, distribution) are owned and operated by different entities via many to many relationships. The consumer is directly connected to one among the electricity distributors operating within his/her geographical location. The consumer expects a continuous supply to his/her demand with a minimum electricity bill. In the meantime, electricity distributors, as businesses, target the maximum selling profile along a minimum electricity purchase. Similar objectives are set by electricity generators through selling profile maximization while minimizing electricity generation. Finally, electricity transporters aim at maintaining a balance between lines' capacity and the amount of electricity transported. A recent study showed that the utilization rate of the energy generated is below 55% [1], that is, almost half the production is wasted. In this paper, we address this issue by developing a decision model that supports generating only the amount of energy that is needed by the consumer. There are several other challenges on the current electricity grids, such as variable load demand [2] heterogeneity of the used technologies [3], and incorporating more renewable energy generation in response to new policies [4]. In order to meet such challenges, the need for changes in the current electricity grids, power design and operations has been recognized and discussed for several years under the Smart Grid concept [5]. The smart grid has been proposed as a new electrical grid to modernize current power grids and enhance their efficiency, reliability and sustainability [6]. Smart grids enforce the conventional electrical grids with a degree of intelligence based on tools strongly related to information and communication technologies [7]. Smart Grid concept considers different components of the electrical system, such as a smart transmission system, an energy management system, advanced distribution management, advanced metering, a demand response, and an electrical storage system [8]. Sensing, communication, decision making, and power flow capabilities are the key factors of a smart grid. Enhancing these factors will improve the efficiency and the reliability of the grid through (1) speeding up the response to consumer's demand, (2) optimizing the capacity utilization and (3) increasing the flexibility of the overall resource management of the grid system.

In this paper, we propose a Smart Grid Resources Optimization heuristic approach (SGROH) that optimizes the resource management of the smart grid. SGROH acts as a broker to facilitate the interaction between the four smart grid constituencies (SGC): power consumers (PC), power generators (PG), power transporters (PT) and power distributors (PD). SGROH's layered architecture tempts to optimize the management of the resources by decoupling the different SGCs. The decoupling approach provides more flexibility and freedom for real time connections between different SGCs. As an example, a PC can be connected to a given PD at a certain time of the day and may end up connected to another PD at another time. From the smart grid perspective, this will guaranty the best quality of service (QoS) for the PC at the best market price dynamically. In related literature, several solutions have been proposed to manage the grid resources as part of the demand side management process. A comprehensive review of the existing demand side resource management solutions is presented in [9]. Most of the proposed systems restrict the resource optimization process to only two components of the grid system (PC-PD or PD-PG). Our optimization process includes all the four SGCs (PC-PD-PT-PG). Some of the proposed solutions focus on developing scheduling algorithms to maximize the efficiency of the two components (generation, distribution) of the grid. Sergio et al. [6], proposed a dynamic energy management model with distributed energy resources. This model focuses on optimizing the resources between a PD and a PC with uncertain consumer load demand, distributed renewable energy resources and the energy storage devices owned by the consumer. The authors developed a real time pricing model to minimize the long-term average total cost to support consumer's load demand. An energy scheduling model using a multi-objective function was presented in [10] with as objectives the minimization of the operation cost alongside the maximization of the minimum reserve in a day-ahead. The scheduling model focuses on different power generation resources including conventional power plants and renewable energy resources. Due to the heterogeneous nature of the hardware devices used in the electrical grid system, [7] proposed a software layer that abstracts this hardware heterogeneity and provides homogenous-looking appearance to the application layer of the smart grid. This middleware focuses on the data exchange between the smart devices and the smart applications to simplify the integration process and avoid dealing with data of different formats. Smart grid systems involve two way communications between the consumer and the grid through an advanced metering and monitoring system. The quality and reliability of the data collected is a key factor for optimizing the operations of the smart grid. Therefore, data mining and analytics tools are essential for the effective management and utilization of the available data [11]. A comprehensive review of the state-of-the-art for the exploitation of big data tools for dynamic energy management in smart grid platforms is presented in [12]. The rest of the paper is organized as follows. Section 2 describes the Architecture and Services. The system modelling and the heuristic approach is presented on Section 3. Section 4 presents the results and section 5 concludes the paper.

2. SGROH Architecture

Please follow these instructions as carefully as possible so all articles within a conference have the same style to the title page. This paragraph follows a section title so it should not be indented. Smart grid systems are the new generation of power grid systems in terms of technology, communication, and innovations. These innovations are endless and serve the comfort of the power consumer by providing them the best Quality of Service (QoS) at the best prices. As far as innovation is concerned, we propose a realistic smart grid resource management architecture which governs all possible smart grid constituencies (components) SGCs. Our layered smart grid middleware architecture SGROH decouples the interaction between the different SGCs and focuses on the services provided by each one of them. It is worth to note that the smart grid resource allocation network is dynamically reconfigurable. As an example a PD currently connected to a given PG might be reconnected to another PG later. As mentioned earlier, our SGROH system manages different Power Generation companies (PGs), different Power Transportation companies (PTs), different Power Distribution companies (PDs), and different Power Consumers (PCs). In our architecture, we consider the possibility that some PCs may act as PGs and sell power energy to the smart grid if they fulfill a certain QoS level. In SGROH, connecting a PC to a PD, or connecting a PD to a PG using a PT must be as smart as possible, automatic, variable in time and, most importantly provide the best QoS to all SGCs within the terms of the Service Level Agreements SLAs.

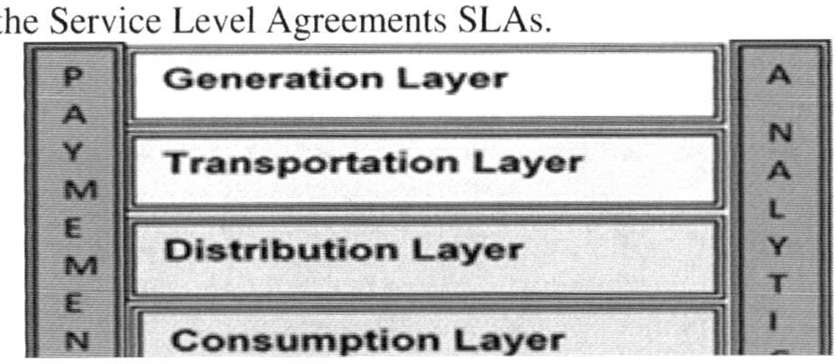

Fig. 1. SGROH Layered Architecture.

As shown in Fig. 1, the proposed SGROH consists of four horizontal power related layers: generation, transportation, distribution, and consumption. SGROH contains also two vertical layers across the four horizontal layers, that is, payment/billing and data collection and analytics. In what follows, we describe the services provided at each of the different layers along with their interactions.

Power Generation Layer: This layer deals with different sources of power energy generators. It manages different PGs taking into account the parameters of a PG such as: the type of energy (fossil fuel, hydro, wind, photo-voltaic, etc.); the cost of energy generation which depends on the type of generations for the specific PG; the period of time in which the PG is available for supplying the energy. The PG's parameters are collected periodically using the smart grid communication network.

Power Transportation Layer: In our model, we consider the power transportation lines (TL) with related power equipment, like transformers which might belong to different PTs. The parameters of a TL are its location, its capacity, and its transportation cost.

Power Distribution Layer: We assume that many PDs are competing to deliver the power to the consumers. A PD is characterized by: the distributor supplied power, which is the load purchased from different generators; the distributor power cost, which might be variable, off-pick, normal, or high pick prices; the distributor failure rate. This parameter is very important when selecting a given PD to connect to a PC, based on the QoS and the SLA; the distributor capacity which quantifies the maximum power load that can be supported by the infrastructure of the given PD. This takes into account the low voltage lines (11KVA), the sub-stations, and the transformers.

Power Consumption Layer: This layer deals with the end-users or the power consumers (PCs). PCs can be individuals or companies supplying power energy to their factories or businesses. One of the most important characteristic of a PC is its power demand. Usually the power demand is collected through smart meters connected directly to the Smart Grid System. A feed-back about the power consumption will be forwarded from SGROH to the PC about how power consumption could be reduced and its impact on the next electricity bill. Another important aspect is the possibility for a PC to sell his/her excess generated power (photo-voltaic) energy to the smart grid.

Data Analytics and Optimization Layer: Data analytics and optimization layer is one of the most important layers which implements the real smartness of the proposed SGROH architecture. It is a cross layer which collects data from different layers, analyses them and takes the optimal decisions based on the current configuration of the smart grid resource allocation. Data are collected continuously at different layers. As example, at the PC layer, smart meter readings are recorded and directly injected into the data analytics engine to reflect the PC's current demand. The data analytics engine performs some statistics on the PC's power usage at different time slots and builds the associated prediction profile to anticipate his/her future power demand. This analytical layer can alert a PC about his/her power consumption level and recommends power consumption hints for reducing the electricity bill.

Payment Layer: The payment cross-layer deals with bills and payments at different layers. At the PC layer, power billing is based on the consumed power at different prices during different time slots. We assume that, at a given time slot, a PC is connected to only one PD. In the case where PC produces power and sells it to the grid, the amount of power along with the price is recorded for each time slot. At the PD layer, the payment information is based on the amount of power offered by a given distributor to satisfy the demand of a set of PCs. At the PT layer, the payment information is related to the amount of power transported at a specific unit cost to supply a given power distributor PD during a given time slot. Similarly, at the PG layer, the payment information is recorded concerning the amount of power produced by a given generator and the associated cost, in order to service a given PD for any time slot. The overall and individual bills for all SGCs result from solving the optimization problem which models the smart grid operations.

3. SGROH System Model

We consider an electric power distribution network consisting of: a set consumers, denoted $I = \{1,2,\ldots,m\}$; a set of distributors, denoted $J = \{1,2,\ldots,n\}$; a set of generators, denoted $K = \{1,2,\ldots,p\}$; and a set of transporters, denoted $L = \{1,2,\ldots,q\}$. We consider that the smart grid system is time-slot based where resource management decisions are made in each time slot *t*. We also assume that the whole system is completely driven by the consumers' demands. Let $D_i(t)$ be consumer i's demand in time slot t. We assume $\{D_i(t)\}_{t=0}^{+\infty}$ is an independent and identically distributed (i.i.d.) non-negative stochastic process, which is deterministically bounded, i.e., $0 \le D_i(t) \le D_i^{\max}$. It is assumed that the demand for each consumer should be fully satisfied by a single distributor. Therefore, the binary decision variable $x_{ij}(t)$ is introduced

$$x_{ij} = \begin{cases} 1 & \text{if } D_i \text{ assigned dist.} j \\ 0 & \text{otherwise} \end{cases} \tag{1}$$

We also assume that each distributor *j*, registered with the smart grid system, should serve at least one consumer:

$$\sum_i x_{ij}(t) \ge 1 \text{, for all } j \in J. \tag{2}$$

With regards to the relationship between distributors and generators, we introduce the positive decision variable y_{jk}^l which denotes the amount of power that distributor *j* receives through line *l* from generator *k*. Therefore, the total amount of power received by distributor *j* through all lines from all generators should be equal to the demand of all the consumers (flow conservation) assigned to distributor *j*.

$$\sum_k \sum_l y_{kj}^l(t) = \sum_i x_{ij}(t) D_i(t) \text{, for all } j \in J. \tag{3}$$

IOP Conf. Series: Earth and Environmental Science **168** (2018) 012015 doi:10.1088/1755-1315/168/1/012015

On another hand, the total power produced by all generators must satisfy the demand of all distributers:

$$\sum_k \sum_l \sum_j y_{kj}^l(t) \le \sum_k G_k(t) \quad \text{for all } k \in K. \tag{4}$$

Where G_k is the amount of power produced by generator k in time slot t. Moreover, during each time slot t, the smart grid system ensures that each generator k is not idle.

$$\sum_l \sum_j y_{kj}^l(t) \ge G^{\min} \quad \text{for all } k \in K. \tag{5}$$

Where $G^{\min}$ is the minimum power level required to operate the generator economically. Similarly, each transporter l must participate to supplying the distributors:

$$\sum_k \sum_j y_{kj}^l(t) \ge T^{\min} \quad \text{for all } l \in L. \tag{6}$$

Where $T^{\min}$ is the minimum power level required to operate the transporter economically. Finally, each transporter has its line capacity. Hence, the amount of power transported in time slot t through transporter l is bounded by its line capacity:

$$y_{kj}^l(t) \le T^{\max}, \text{ for all } l \in L. \tag{7}$$

Recall that the objective of this system is to optimize the smart grid resource utilization through the minimization of the overall cost, including generation, transportation, distribution, and consumption costs.

3.1. Consumption Cost Function

Let c_j be the unit selling price offered by distributor j to the grid. In fact, the smart grid middleware is looking not only to minimize the total consumers' bills but also to encourage consumers to reduce their energy consumption through a discount scheme, as it will be shown in the billing functions section. Furthermore, such a strategy will contribute to reducing the total energy generation which will have a positive impact on the Carbone footprint. Accordingly, a higher discount price will be given to those consumers with lower power demand: The total consumers cost is:

$$\text{ConsumCost} = \sum_i \sum_j D_i(t) x_{ij}(t) c_j(t)$$

3.2. Distribution Cost Function

The use ofLet us assume that each distributor can buy the power from different generators through different transportation lines. We denote by $C_k(t)$ the selling price applied by generator k in time slot t and $C_{kj}^l(t)$ is the transportation price offered by transporter l to transport one unit of power from generator k to distributor j. Therefore, the total distribution cost function is:

$$\text{DistCost} = \sum_k \sum_l \sum_j (c_k(t) + c_{kj}^l(t)) y_{kj}^l(t)$$

The overall cost function for the smart grid includes both the consumption cost and the distribution cost. Our objective is to minimize the overall smart grid cost (GridCost):

$$\min \left(\begin{array}{l} \sum_j \sum_i x_{ij}(t) D_i(t) c_j(t) + \\ \qquad \sum_k \sum_l \sum_j (c_k(t) + c_{kj}^l(t)) y_{kj}^l(t) \end{array} \right) \tag{8}$$

The SGROH optimization model is solved using a greedy heuristic algorithm which takes into account the defined constraints (1) to (7) and the objective function (8). This optimization problem could be modeled as a Mixed Integer Linear Program and solved using CPLEX software. Real life smart grids are often designed to serve millions of consumers using large numbers of distributors, generators and transporters. To solve a MILP of such a scale, CPLEX requires more computational time than entailed for quick decisions within time slots as short as one hour. As an alternative, we develop a heuristic approach to enable SGROH to provide an approximate solution for resource management when needed and, hence, save computational time and memory requirement.

3.3. SGROH Heuristic Description

The main idea of the SGROH heuristic is to assign lower demand consumers to cheaper cost distributors. This strategy privileges low demand consumers by servicing them at the best market prices prior to high demand consumers. The heuristic operates as follows. First, consumers are sorted according to their respective load demands in an ascending order, while distributors are sorted by price charge in an ascending order. Next, the heuristic starts by assigning low demand consumers to the first low price distributor until its load capacity is reached. Then, the following batches of consumers are assigned to the next distributor, and so forth, until all consumers are assigned. At the distribution layer, a different marketing strategy is adopted: higher distributor demand is assigned to lower price generator and transporter. Here, generators and transportation lines are sorted based on their price charges in an ascending order. Then, assign the most loaded distributor to the lower price generator through the lower price transportation line, until either the transportation line is saturated or the generator's capacity is reached. If the current transportation line is saturated, assign the remaining load of the distributor to the next transportation line in the corresponding sorted list. In case the generator capacity is reached, assign the remaining load to the next generator in the generators' sorted list. Once the distributor's load is fulfilled, proceed with the next distributor in the distributors' list until all distributors have been assigned.

SGROH Heuristic

```
1. Sort consumers according to their demand (ascending)
2. Sort distributors according to their distribution costs (ascending)
3. for all consumers
  {if(Distributor_Assigned_Load[j]+Consumer_Load[i]<=
      Distributor_capacity[j])
    { Assign consumer i to distributor j
      Distributor_Assigned_Load[j]+= Consumer_Load[i]
    }
    else
    { Go to next distributor (j++); Assign consumer i to new distributor j;
      Distributor_Assigned_Load[j]+= Consumer_Load[i]
    }
  }
4. Sort distributors according to their assigned load (descending)
5. Sort generators according to their generation costs(descending)
6. for all distributors
  {remaining_Generator_Load = distributor_load[j]
   while more generators
   {if ( Generator_Can_Take_More[k])
         {if ( remaining_Generator_Load - generator_capacity[k] > 0)
          {remaining_Generator_Load-=generator_capacity[k];
           remaining_Line_Load=generator_capacity[k];
           Generator_Can_Take_More[k]=false;
          }
          else
          {generator_capacity[k]-=remaining_Generator_Load;
           remaining_Line_Load=remaining_Generator_Load;
           remaining_Generator_Load=0;
          }
     6.1 for all transportation line
                   {if(remaining_Line_Load - line_Capacity[j][k][l] >0)
                    { remaining_Line_Load-=line_Capacity[j][k][l]
                      transport line_capacity[j][k][l] trough the line
                    }
                    else
                    { transport remaining_Line_Load trough the line
                      remaining_Line_Load=0
```

```
                }
              }
          Go to next generator (k++)
        }
        else
          Go to next generator (k++)
    }
  }
```

4. Experiments and discussions

In this section, an illustrative example of SGROH is presented to give an overview of the quality of our greedy heuristic. A system of ten customers, three distributors, three generators and two transporters was used to simulate the grid system. The system was tested under different load conditions namely low load, medium load and heavy load. An average of 15Kw, 20Kw, 30Kw consumer demands was considered for low, medium and heavy load conditions, respectively. To simulate the variable rate scheme, one distributor charge and one generator charge was increased by 5% compared to the low load scenario. During the heavy load scenario, the charges of all distributors and all generators have been increased by 5%. During the evaluation process, the distribution cost and overall system cost (distribution, generation, and transportation) were recorded for a period of 24 hours. The grid system is solved by both SGROH heuristic algorithm written in C++ and using CPLEX 12.6.2. [13]. Note that CPLEX software provide us with optimal solution when the heuristic gives near optimal solution. Table 1, shows the system total cost for the grid system with both MILP's (CPLEX) and the heuristic algorithm under different load conditions. It can be seen that under a low load condition, both algorithms behave similarly. For medium and high loads, MILP's solution is 2% better than the heuristic.

Table 1. Total Cost.

Load	MILP	Heuristic
Low Load	5064	5064
Medium Load	7296	7344
Heavy Load	11952	12216

In this example, the heuristic algorithm produces a relatively good resource allocation solution (close to the optimum). The advantage of the heuristic approach is its potential to solve large scale problems faster which may not be the case for MILP solver using CPLEX. As far as our experiment are concerned we could not run CPLEX for more than 50000 customers. As we mentioned earlier, real grid system consists of millions of users and large number of distributor, transporters, and generators.

5. Conclusion

We proposed a new layered architecture for resource management optimization SGROH. Each layer is responsible for collecting data. The SGROH engine selects the best resource allocation for all the smart grid constituencies and ensures that only the needed power energy is generated in each time slot. A heuristic approach has been developed and tested successfully on a prototype SGROH. The advantage of the heuristic algorithm is that it can be used as an efficient fast decision tool with nearly optimal solution when dealing with large scale smart grid systems. It is worthwhile noting that, in the ongoing research work, we are investigating the implementation of SGROH on real-life power grids.

6. References

[1] Strbac G 2008 Demand side management: Benefits and challenges *Energy Policy* **36** 12 pp 4419-4426.

[2] Jayanta M, Yigzaw Y, Brian N 2009 Optimizing the economic viability of grid-connected photovoltaic systems *Applied Energy* **86** 1 pp 985-999
[3] AK D, Ismail M 2012 Design of isolated hybrid systems minimizing costs and pollutant emissions *Renewable Energy* **44** 1 pp 215-224.
[4] Ippolito M, Silvestre M, Sanseverino E, Zizzo G 2014 Multi-objective optimized management of electrical energy storage systems in an islanded network with renewable energy sources under different design scenarios *Energy* **64** 4 pp 648-662.
[5] Aleena G, Sivraj P, Sasi K 2015 Resource Management on Smart Micro Grid by Embedded Networking *Procedia Technology* **21** 2 pp 468–473.
[6] Sergio S, Ming L, Pan L, Yong F 2013 Dynamic energy Management for Smart Grid with Distributed Energy Resources *IEEE Transaction on Smart Grid*, **4** 4 pp 2139-2151.
[7] Diego R, Martinez J, Rodriguez-Molina J, Cuerva A 2014 A Semantic Middleware Architecture Focused on Data and Heterogeneity Management within the Smart Grid *Energies* **7** 3 pp 5953-5994.
[8] Boukettay G, Krichen L 2014 A dynamic power management strategy of a grid connected hybrid generation system using wind, photovoltaic and Flywheel Energy Storage System in residential applications *Energy* **71** 1 pp148-159.
[9] RazaKhan A, Mahmood A, Safdar A, Khan Z, Khan N 2016 Load forecasting, dynamic pricing and DSM in smart grid: A review *Renewable and Sustainable Energy Reviews* **54** 1 pp 1311–1322.
[10] Soares J, Ghazvini M, Vale Z, Oliveira M 2016 A multi-objective model for the day-ahead energy resource scheduling of a smart grid with high penetration of sensitive loads *Applied Energy* **162** 1 pp 1074-1088.
[11] Fan Z, Chen Q, Kalogridis G, Tan S, Kaleshi D 2012 The power of data: data analytics for M2M and smart grid *In Proc. 3rd IEEE PES ISGT Europe* pp 1-8.
[12] Diamantoulakis P, Kapinas V, Karagiannidis G 2015 Big Data Analytics for Dynamic Energy Management in Smart Grids *Big Data Research* **2** 1 pp 94–101.
[13] ILOG CPLEX 2006 High-performance software for mathematical programming and optimization.

Acknowledgments
The authors gratefully acknowledge the support of Internal Grant IG/SCI/COMP/18/01, Sultan Qaboos University, Oman

IOP Conf. Series: Earth and Environmental Science **168** (2018) 012016 doi:10.1088/1755-1315/168/1/012016

Analysis of CO-Emission from Domestic Gas Cooker with Variation of Natural Gas Composition

Zhiguang Chen, Chaokui Qin, Pengfei Duan

School of Mechanical Engineering,Tongji University,Shanghai, China
e-mail: chenzhiguang@tongji.edu.cn

Abstract. For domestic gas cookers, the operating safety is the most important factor for end-users. The increases of CO emission with the variable natural gas constituents will seriously influence indoor pollution and affect human health. Response of CO emission of domestic gas cooker when interchanged by variable natural gas constituents was experimentally investigated from 8 kinds of Chinese domestic gas cookers while using 11 kinds of gas sources. The applicability of Weaver and AGA index on the CO emission prediction was also analyzed. It was found that well-adjusted gas cookers were flexible enough to give satisfactory performance without materially increasing CO emission. CO emission was found to be increasing with increasing Wobbe index of gas and Yellow-tipping. when considering the change of gas components, it can use the change of gas Wobbe Index to predict the CO performance of domestic gas cookers; Some limits should be imposed on Wobbe index of gas to be delivered nationally, so as to ensure no CO-related issues will be encountered.

1. Introduction

Natural gas has been recognized as one of the most important fuels for domestic gas cookers in many countries. With the development of global natural gas trade, liquefied natural gas (LNG) begins to play an increasingly important role in those countries where gas demand steadily increases and indigenous gas production gradually declines [1]. The constituents of LNG are different from those of pipeline natural gas (PNG), particularly the heavier hydrocarbons in LNG, may lead to combustion-related difficulties. For domestic gas cookers, yellow-tip, lift, flash back and incomplete combustion (excessively high concentration of CO) have been reported to occur [2].

Due to emphasis on environmental protection, indoor air pollution (particularly CO emission) has become of great concern. CO is considered as a strong greenhouse gas and very hazardous to human health, and emissions from natural gas burners have always been strictly controlled [3]. Domestic gas cookers are individually small, but so numerous that, they could be significant influenced on CO emission. Such small-scale devices are expected to have different emission factors compared to the open, large-scale combustion. Experiences have shown that slight changes in combustion conditions can have large impacts on emission factors [4].

In China, conventional Bunsen type burner is the most popular in domestic gas cookers market, and natural gas is commonly used though its constituents vary greatly from a city to another. For example, the natural gases supplied into the distribution network of Guangdong province is from different sources, including 3 PNGs, 2 offshore natural gases (OSGs) and 6 LNGs. The variable Wobbe index may strongly affects gas cookers' performance, including the combustion characteristics (such as yellow-tips, lifting and flash back etc.), thermal efficiency, heat input rating and pollutant emissions (especially CO emission). In order to evaluate impacts of gas constituent variations upon gas cookers performance, an experimental research was carried out. 8 sets of gas cookers covering three types of ports, round, square and ribbon, were selected as test samples. All sampled gas cookers incorporated atmospheric burners and injectors made of cast-iron, diffusion/distribution head made of aluminum or cast-iron, together with burner head made of casting copper alloys (as shown in Figure 1). Port intensity of all the selected gas cookers falls between 7.0 and 9.0 W/mm^2. The results obtained in this work would help clarify the most crucial parameter influencing CO emission for variable natural gas constituents. The applicability of well-established AGA and Weaver indexes was also discussed. The experimental results are of great importance as to control the quality of natural gas to be introduced into the local gas network without causing CO emission-related problems.

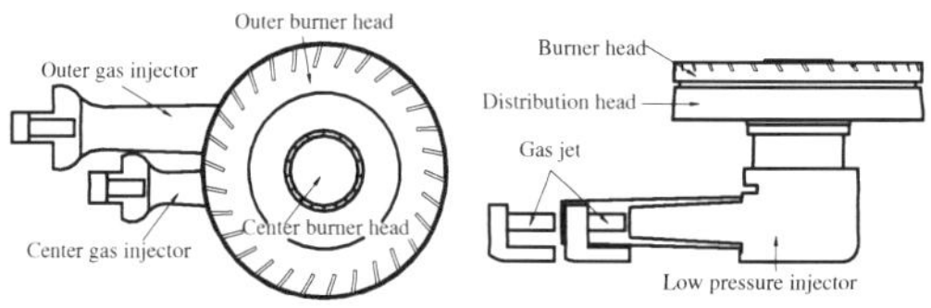

Figure 1. Schematic diagram of the atmospheric burners.

2. Experiment Setup

2.1. Experimental Facilities

Test rig included two sub-systems: gas-blending and CO emission measurement system, as illustrated in Figure 2. The 11 test gases were provided by gas-blending sub-system through which methane, ethane, propane, butane, nitrogen and carbon dioxide, were blended to give exactly the same constituents as test gases. Also the same Wobbe index and combustion potential were achieved. The gas-blending sub-system includes a 5m^3 storage tank. During gas-blending operation, natural gas available in the laboratory (a PNG) was used to purge the tank. Pure components such as methane, ethane, propane etc. were serially piped through a gas meter which has full scale of 10m^3/h and ±0.2% FS precision to monitor gas flow into the storage tank. Gas pressure gauges and a heating circuit were incorporated to improve precision. A propelling mixer located in the top of storage tank helped mix uniformly the fed constituents. The purities of individual components involved were as follows: methane 99%, ethane 99.5%, propane 99.95%, butane 99.95%, nitrogen 99.999% and carbon dioxide 99.6%. After all the individual constituents were fed into the storage tank, the mixture remained for 3-5 hours while propeller was working. Then gas was sampled and analyzed by means of gas chromatography. When the constituents of blended gas fell within permissible limitations compared with test gases (listed in Table I) [5], the blended gases could be regarded as identical to test gases.

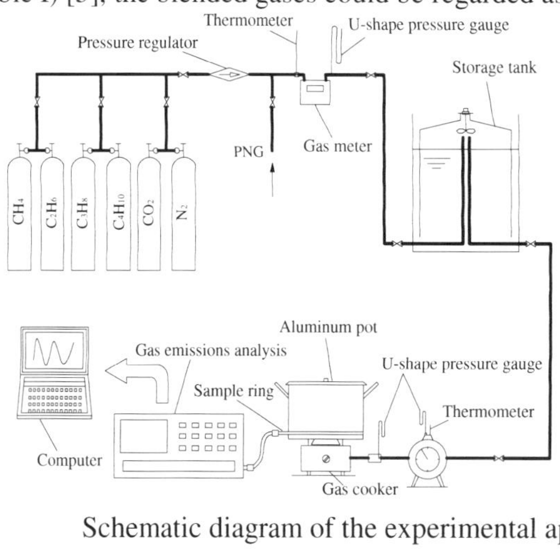

Figure 2. Schematic diagram of the experimental apparatus.

IOP Conf. Series: Earth and Environmental Science **168** (2018) 012016 doi:10.1088/1755-1315/168/1/012016

TABLE I. PRECISION AND REPRODUCIBILITY OF GAS CHROMATOGRAPHY

Component range (mole %)	Reproducibility (%)	Precision (%)
0~0.1	0.01	0.02
0.1~1.0	0.04	0.07
1.0~5.0	0.07	0.10
5.0~10	0.08	0.12
>10	0.20	0.30

Gas cooker CO emission measurement sub-system was designed according to Chinese National Standard GB 16410-2007[6]. As illustrated in Figure 2, the concentrations of the combustion product components were measured continuously by exhaust analyzer (KM9106, the accuracy of CO, CO_2 and O_2 is ±10ppm), and the sampled data can be input into computer per 10 seconds. It was reported [7]-[10] that CO emission from domestic gas cooker was closely related with distance between burner head and pot bottom. CO emission would decrease with increasing distance and gradually stabilized toward a specific distance. This means impingement of flame upon cold bottom of pot can quench combustion process to some extent. The location of sampling ring, corresponding to the time required for unburned intermediate to be completely combusted, can have significant influence upon CO emission readings. In Chinese National Standard it is specified that CO emission and thermal efficiency of sampled burner should be measured simultaneously. In order to achieve reliable readings of CO emission, sampling ring should be located as close to the bottom of aluminum pot as possible, usually 20~40 mm. Meanwhile O_2 reading of exhaust should not exceed 14%. Moreover, the radial clearance between sampling ring and the pot should be 0.5~1.0 mm, so that no excessive air will be entrained during sampling process [6]. It should be pointed out that only those CO readings strictly according to this step can be comparable. Despite the thermal efficiency was measured at the same time, discussions about its changing pattern with gas constituents were not included in this paper.

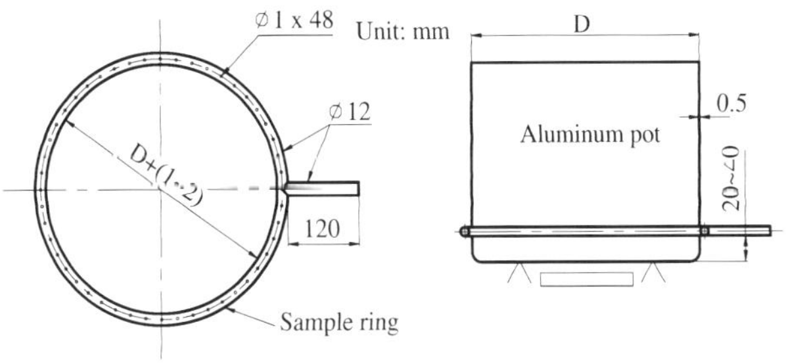

Figure 3. Schematic diagram of the sample ring and test position.

2.2. Experimental Procedure

First of all, initial adjustments were made for 8 sets of gas cookers, which were the most commonly used cookers in China, under adjustment gas. The shape and hardness of an atmospheric flame depend mainly upon primary air given a fixed burner head. Inner cones will change from "soft" to "distinct" when the primary air rate increases and the color of flame will change from yellow to clear blue. The primary air shutters of 8 sets of domestic gas cookers were so adjusted that resulting flames fell between class -2 to +2 acorrding to AGA flame code. Such flames were considered to be most flexible according to previous researches. 8 sets of gas cookers were observed to give satisfactory performance under LNG4, namely no lift, no flash-back, no yellow-tipping occurred and CO emission remained below specified limit. Secondly, the air shutters remained unchanged, CO emission under the other 10 gases were measured in sequence. The measurement procedure of CO emission was carried out according to Chinese national standard GB16410-2007 "Domestic gas stove" [6]. The CO sampling procedure was as follows: maintaining 2kPa pressure at inlet of gas cooker, to ignite the main burner and after 15 min of ignition, taking sample of exhaust by means of sample ring where the concentration of oxygen must be lower than 14% (as shown in Figure 3).

3. Results and Discussion

For domestic gas cookers, CO emission are the most significant representatives. There exists clearly defined limits on CO emission for the sake of safety, changes in CO emission are known to be directly related to combustion chemistry. During the experiment, the most difficult aspect of CO measurement is that some of sampled cookers burned with excessively high air or that the sampled exhaust contained too high percentage of air. In order to compare values for samples with different dilution percentages of air, the CO values are corrected to a standard value (air free) The units for CO is ppm (parts per million) in the sample.

In this paper, the normalized CO (a ratio relative to CO under adjustment gas) was used for discussion in order to reveal the effect of gas constituent variations. The main advantage of "ratio" approach is that the measured CO data are normalized and more appropriate for comparison of different burner types or test conditions. Another reason is that CO measurement for domestic gas cooker depends upon ambient conditions and can be very sensitive to sample location. During experiment, several of the appliances were found to give different CO emission values for the same gas tested on different days. However, the general trend for changes in the emission was the same. In order to further investigate CO emission resulting from gas constituent variations, the measured CO and normalized CO were plotted as functions of several indexes, as shown in Figure 5~8.

3.1. Thermal Efficiency

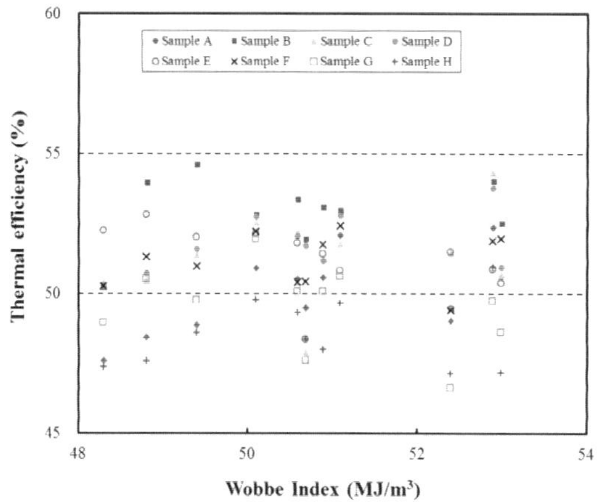

Figure 4. Thermal efficiency changing with Wobbe index.

The thermal efficiency of gas cooker is directly affected by burner structure, secondary air supplying and some other factors. It was prescribedthat thermal efficiency must not be lower than 55% (for on-top cooker) or 50% (for embedded cooker) [6]. The experiment results are shown in Figure4. It can conclude that with the change of gas composition the thermal efficiency performance of 3 samples, Sample A, G and H, are not very satisfactory. For all 3*11=33 operation points, 21 operation points about 63% can be considered as "unqualified" in terms of efficiency which are under 50%. Meanwhile the other 5 samples are seem like to be qualified in efficiency when gas compositions were changed.

3.2. Effect of Wobbe index

Figure 5 illustrates the CO emission of all samples under different natural gas constituents. The change of CO emission with the gas constituents doesn't follow any regular pattern, and most of samples were smaller than 500ppm. It was prescribed that CO emission (air-free) must not be higher than 500ppm[6]. CO emission measurements show that CO emission of a domestic gas cooker can be related with its structure and initial adjustment. Given a suitable structure, those well-adjusted cookers, e.g. Sample A, D and H, can give quite low emission. The normalized CO is plotted as a function of Wobbe index (J_H), as shown in Figure5b. It is obvious that the magnitude of CO peaks increases with Wobbe index, implying there exist authentic impact of gas quality upon CO emission of domestic gas cooker. And it was found that 5 samples can keep lower carbon monoxide emission on all natural gases, accounting for 62.5%; for all 88 operation points, 74 operation points about 84% can be considered as "qualified" in terms of CO emission. So it can be concluded that well-adjusted initial state of domestic gas cookers can be very helpful when fuelled by variable-constituent natural gases, without materially increasing CO emission.

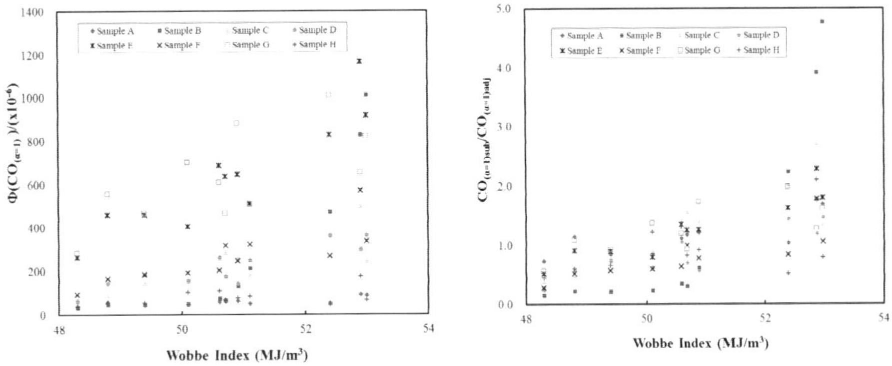

(a) CO (air-free) (b) Normalized CO

Figure 5. CO emission changing with Wobbe index

3.3. *Influence of Incomplete Combustion Index and Primary Air Ratio Index*

Figure6 shows the tendency of normalized CO for 8 sets of cookers changing with Weaver incomplete combustion index JI and primary air ratio index J_A which is fairly similar with the tendency of Wobbe index (J_H). Because LNG5&6 have the similar constituents and properties, it can be found from Figure6 that the highest emission in CO corresponds to the highest absolute value for J_I and J_A. Meanwhile when J_I is close to the limit value, CO emission problem tends to occur for gas cookers. So with the change of gas constituents, CO emission closely corresponds to the Wobbe index. It is very necessary to specify the range of Wobbe index to make sure that there will not be CO emission problems when gases interchange.

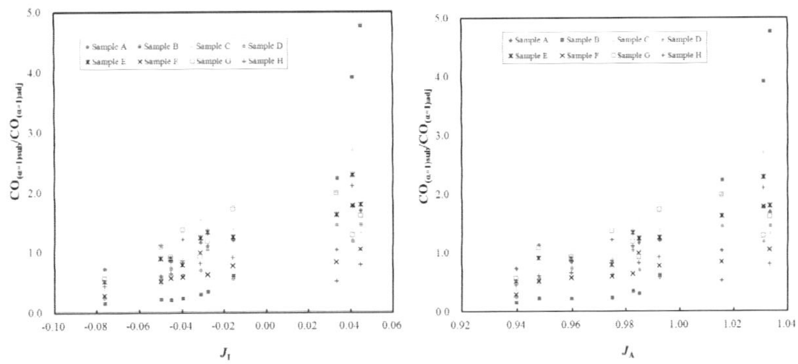

Figure 6. Normalized CO ratio data vs. J_I and J_A.

3.4. Effect of Yellow-tipping Index and Lift Index

Figure 7 and Figure 8 illustrate the tendency of normalized CO for 8 sets of gas cookers changing with yellow-tipping index (I_Y and J_Y) and lift index (I_L and J_L). The trend is complete converse since the formulae are different between AGA and Weaver. It was found that when the yellow-tipping index I_Y and J_Y are close to their own limit values, CO emission of gas cookers are observed to increase obviously. But the tendency for normalized CO and lift index I_L and J_L is absolutely different, the lowest CO emission appear in the area which lift index I_L and J_L are close to the limit values.

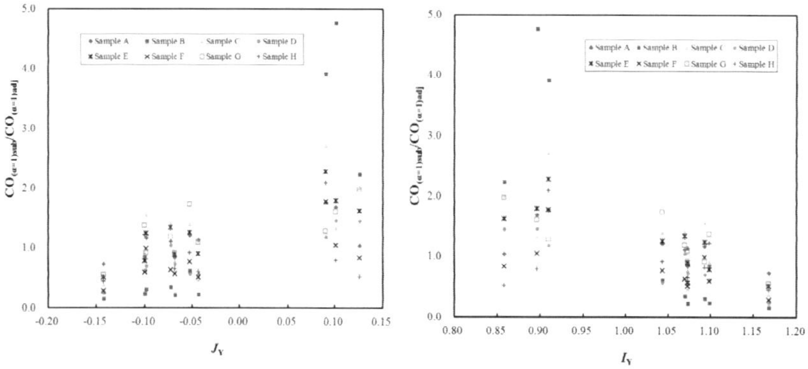

Figure 7. Normalized CO data vs. J_Y and I_Y.

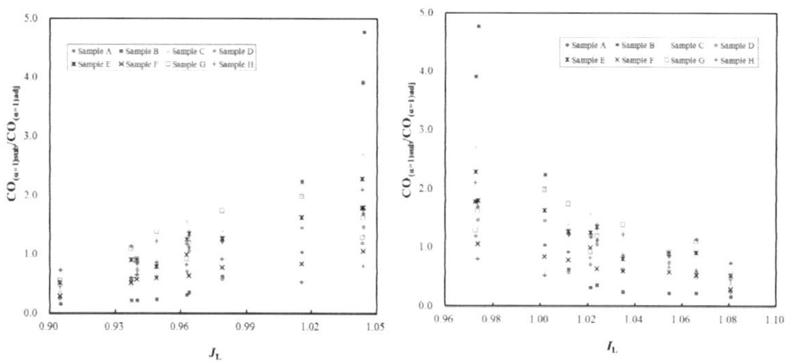

Figure 8. Normalized CO data vs. J_L and I_L.

Seemingly yellow-tipping tends to lead to CO issues more likely than lift. Furthermore, the highest emission in CO does not correspond to the highest value for J_Y or the lowest value for I_Y. It means that yellow-tipping is largely affected by gas constituents and it can't be predicted only by Wobbe index or CO emission value, and vice versa.

4. Conclusion

Through the experimental study on CO emission of 8 sets of domestic gas cookers under 11 kinds of natural gas constituents, it can be conclude:

(1) Well-adjusted domestic gas cookers are flexible enough to give satisfactory performance under gas constituents discussed without materially increasing CO emission.

(2) With gas constituent variation, CO emission closely corresponds to the Wobbe index and it appears that CO emission increases with increasing Wobbe Index .It is very necessary to specify the range of Wobbe index to make sure that there will not be CO emission problems.

(3) When substitute gas takes place of adjustment gas, yellow-tipping is more likely than lift to cause the CO emission problem. But yellow-tipping is largely affected by gas constituent, so it can't be predicted only by Wobbe index or CO emission value, and vice versa.

For domestic gas cookers, the operating safety is the most important factor for end-users. The increases of CO emission with the variable natural gas constituents will seriously influence indoor pollution and affect human health. So it is very necessary to specify the fluctuation range of gas Wobbe index and define qualified adjustment gas to make sure that there will not be no CO emission problems when gas interchange. On the other hand, it is helpful for the manufacturers to adjust the initial combustion condition of domestic gas cooker, and make the gas cookers will not be easily impacted by changing gas constituent.

5. References

[1] National Bureau of statistics. China Statistical Yearbook 2015[J].. Beijing: China Statistics Press, 2016.
[2] Qi WANG, Wenxue GAO, Zijun ZHAO, etc. Research Progress in Interchangeability of Gases in China.[J] GAS&HEAT, 2013(2),B14-B20.
[3] Zhuang Yong-mao, Shi Hui-bang edited. Combustion and pollution control. 1st ed. Shanghai: Tongji University Press; 1998, p. 78-80. (in Chinese)
[4] Keeman Lee, Jong-Min Kim, Byeonghun Yu, Chang-Eon Lee, Seungro Lee. Effect of various gas compositions on gas interchangeability and combustion characteristics for domestic appliances [J]. Mechanical Science and Technology, 2013, 27(4): 1191-1201.
[5] General Administration of Quality Supervision, Inspection and Quarantine of the People's Republic of China (AQSIQ). GBT13610-2014 Gas composition analysis by gas chromatography. Beijing: China Standard Press; 2014. (in Chinese).
[6] General Administration of Quality Supervision, Inspection and Quarantine of the People's Republic of China (AQSIQ).GB 16410-2007 Domestic gas stove. Beijing: China Standard Press; 2007. (in Chinese)
[7] H.B. Li, T.T. Wong, C.W. Leung, S.D.Probert. Thermal performances and CO emissions of gas-fired cooker-top burners. Applied Energy 2006; 83:1326–1338.
[8] Shuhn-Shyurng Hou, Chien-Ying Lee, Ta-Hui Lin. Effciency and emissions of a new domestic gas burner with a swirling flame. Energy Conversion and Management 2007; 48:1401–1410.
[9] H.S. Zhen, C.W. Leung, T.T. Wong, Improvement of domestic cooking flames by utilizing swirling flows[J].Fuel, 2016(3), 153-156
[10] Yangjun Zhang, Wenxue Gao, Yanyan Yu, etc. Primary air ratio change and gas interchangeability index correct for domestic gas cooker burning multi-source natural gases[J]. Journal of Natural Gas Science and Engineering, 2016(9), 276-282.

IOP Conf. Series: Earth and Environmental Science **168** (2018) 012017 doi:10.1088/1755-1315/168/1/012017

Web-Based Design and Implementation of Smart Home Appliances Control System

Samuel Bimenyimana*, Aimable Ishimwe, Godwin Norense Osarumwense Asemota, Cecilia Messa Kemunto, Lingling Li

School of Electrical Engineering, Hebei University of Technology, Tianjin 300401,China

emails: s0785213122@gmail.com; ishaim2010@gmail.com, asemotaegno@gmail.com; cessmesa83@gmail.com; lilinglinglaoshi@126.com

Abstract. Minimising energy wastage is vital for economic development. This paper deployed a web-based design and implementation of smart electric home appliances control, which enable owners to remotely use web access to efficiently control their home appliances energy usage. Node MCU, breadboard, light emitting diodes (LED), relays, resistors, electric wires, Wi-Fi device (smart phone, tablets or computer), jumper wires, transistor, direct current (DC) motor, universal serial board (USB) cable, printed circuit board (PCB), C programming language and Arduino IDE (Integrated Development Environment) were used, for circuit design and implementation. C programming codes were written and compiled in Arduino IDE device, then uploaded into Node MCU through USB cable to control energy usage of electrical appliance (DC motor) connected to the circuit. The control was achieved by turning ON the DC motor when needed and OFF, when not in use, remotely via web. Whenever appliance was OFF, LED showed red color and green LED lights up when appliance was ON. Developed control system performed excellently and efficiently by significantly minimizing energy wastage, when appliances are not in use. Future research should consider smart home control systems able to record quantities of energy consumed at any period of the day.

1. Introduction

Technological advancements especially in information and communications technology (ICT), equally has a significant impact on control systems including Electrical Engineering. Control systems engineering [1] consist of 'sets of equipment and software able to manage, command, direct or regulate the behaviour of other devices or systems for desired results' [2]. Today, people use electrical appliances to perform domestic chores like cleaning, dusting, sweeping, vacuuming, washing dishes, laundry, preparing meals and cleanse bathrooms [3, 4]. This study used design and programming to implement an accurate and efficient control system of electric home appliance (DC motor as Sample), to either switch ON whenever it is needed, or OFF, if not needed, using web access of node multipoint control unit (MCU) [5]. It is an internet of things (IoT) endpoint device that uses local area network (LAN) to provide three or more terminals and gateways for multipoint conference participation [5]. MCU is ('an open-source, interactive, programmable, low cost, simple, smart, WI-FI enabled device' [6]) connected with other equipment for energy loss reduction. Efficient energy usage is achieved in this study, because only connected smart electric home appliances will consume energy, when commanded. This paper is organised into: introduction, related works, materials and methods, design and implementation, and conclusion.

IOP Conf. Series: Earth and Environmental Science **168** (2018) 012017 doi:10.1088/1755-1315/168/1/012017

2. Related Works

Jabbar and Kawaitkar [7], implemented smart home control using low cost Arduino & Android designs, [8, 9] developed lightweight wireless and traditional home appliances control systems using Global Systems Mobile (GSM) network and PIC 16F877A. IoT cloud data storage was used to implement smart power monitoring and control [10], [11] used mobile telephony for home appliances remote control, and [12] deployed electricity load management system in smart home conttrol. Similarly, Labview was used to design and implement a smart house control system [13]. Furthermore, [14-21], have worked on other aspects of smart home electrical appliances energy control. The major contribution of this paper, is that there are not many people who have worked on smart home equipment control systems using Node MCU and web access. Further refinements can be made to extend electric home appliances control to include recording quantities of energy consumed at certain time instants and/or intervals.

3. Materials and Methods

Node MCU, breadboard, LED, relay, resistors, WI-FI devices (smart phone, tablets or computer), Jumper and electric wires, Transistor, DC motor (appliance controlled), USB cable, Printed Circuit Board (PCB), were used. Electric circuit designed on breadboard, wrote and compiled C program codes in Arduino IDE, uploaded into node MCU by USB cable for implementation and proper control.

3.1. NODE MCU

'NodeMCU is a single-board microcontroller, an eLua based firmware for the ESP8266 Wi-Fi SOC from Espressif. The NodeMCU *firmware* is a companion project to the popular NodeMCU device kits, ready-made open source development boards with ESP8266-12E chips' (Fig. 1) [22, 23].

It is 'an open- source effort that takes prototyping of internet of things applications to the next level by providing a development kit, which includes hardware module with robust firmware' [24].

3.2. RELAY

Relays are designed to open and close circuits electromechanically and electronically (Fig. 2) [25]. They have two contacts (NO: normally open and NC: normally closed contact). Once a relay is energized, 'the coil activates the armature which operates either to close normally open contacts or to open the normally closed contacts' [26, 27].

3.3. ARDUINO IDE

'Arduino Integrated Development Environment (Arduino Software (IDE)) contains text editor for writing code, message area, text console, toolbar with buttons for common functionalities and a host of menus. Thus, uploaded programs connected and communicated with the Arduino and Genuino hardware for needed control implementation' [28].

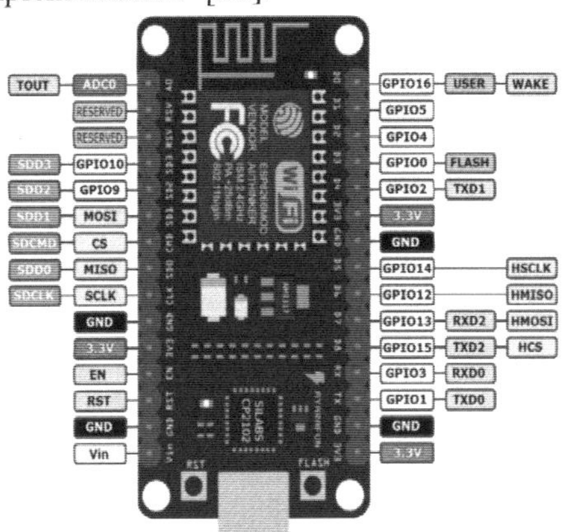

Figure 1. Node MCU pins layout [22]

IOP Conf. Series: Earth and Environmental Science **168** (2018) 012017 doi:10.1088/1755-1315/168/1/012017

Figure 2. SRD-03VDC-SL-C relay used [25]

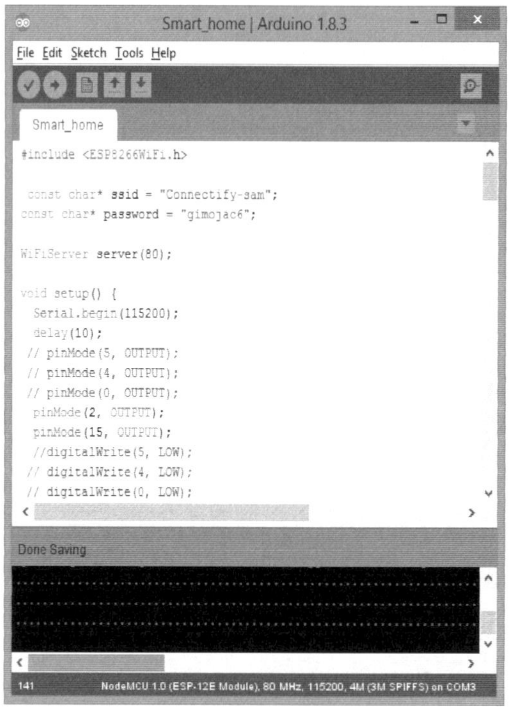

Figure 3. Arduino IDE software homepage

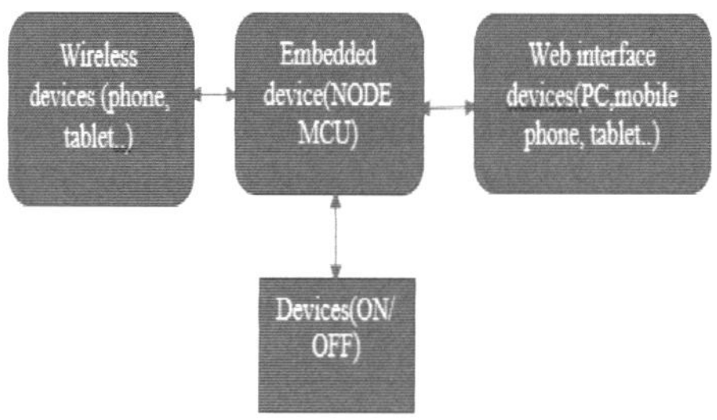

Figure 4. Block diagram of the proposed electric appliance home control system

This study uses web interface to link NODE MCU with computer or other WI-FI device, which contains in-built browser. Another function of web is to display the status of device, when ON/OFF. The computer sends signals to microcontroller (NODE MCU), which process those signals and dispatch them to relay or power transistor to actuate the power device being controlled (DC motor in our case: RF-300CA-11440).

4. Project Design and Implementation

The control system shows the status of home appliance whether ON or OFF and the user can take decision to turn it ON or OFF remotely by web interface (Fig. 5). The web interface links NODE MCU to computer or other device built into browser and shows whether device was ON or OFF. There is computer communication with microcontroller (NODE MCU), relay and power transistor to activate the DC motor power device being monitored. Relays were used to switch external voltage to power the load. Transistors were used to switch ON 3.3 V from Node MCU to power the relay, and shunt resistors were used to avoid short-circuit, between Collector and Emitter terminals of the transistor. Red LED lights if DC motor is OFF and green LED shines, when DC motor is ON. Jumper wires were used to connect devices on breadboard, while electric wires were used to connect SRD-03VDC-SL-C relay from PCB to breadboard. USB cable was used to transfer and upload the C programming codes written and compiled in the Arduino IDE software in Node MCU. Upon circuit design and C programming codes compilation, Internet Protocol address (IP address) appears in the Arduino IDE software and this IP address can be used to control the electrical appliance remotely via web access, hence facilitating the reduction of energy wastage because the electrical appliance will only be ON when needed and OFF, when not in use.

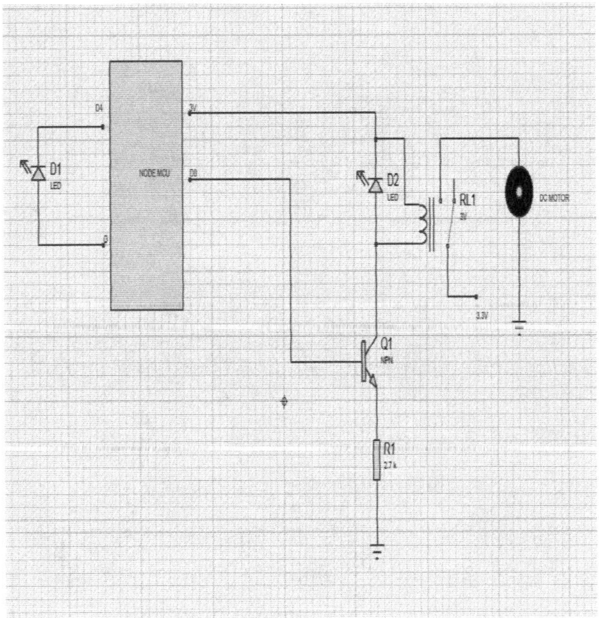

Figure 5. Circuit design outlook of the proposed control system

IOP Conf. Series: Earth and Environmental Science **168** (2018) 012017 doi:10.1088/1755-1315/168/1/012017

Figure 6. Implementation layout of control system

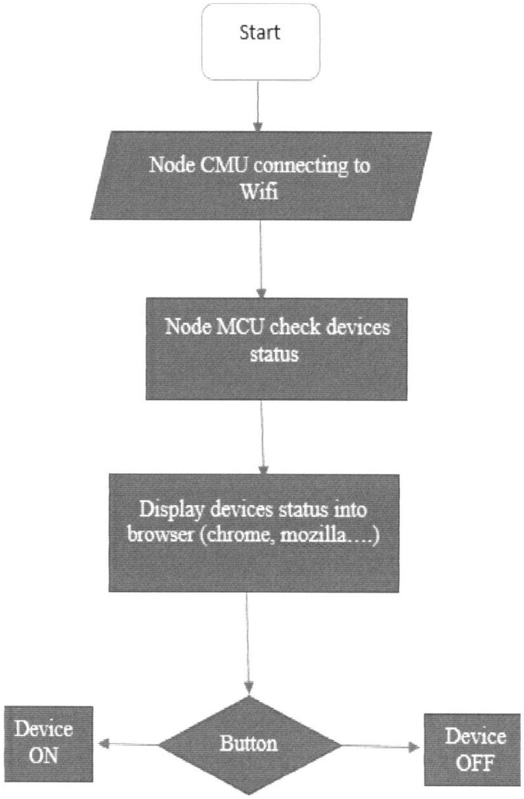

Figure 7. Flowchart of the proposed control system

Compiling Codes in Arduino IDE

```
#include <ESP8266WiFi.h>
 const char* ssid = "Connectify-sam";
const char* password = "gimojac6";

WiFiServer server(80);
 void setup() {
```

IOP Conf. Series: Earth and Environmental Science 168 (2018) 012017 doi:10.1088/1755-1315/168/1/012017

```
 Serial.begin(115200);
 delay(10);
// pinMode(5, OUTPUT);
// pinMode(4, OUTPUT);
// pinMode(0, OUTPUT);
 pinMode(2, OUTPUT);
 pinMode(15, OUTPUT);
 //digitalWrite(5, LOW);
// digitalWrite(4, LOW);
// digitalWrite(0, LOW);
 digitalWrite(15, LOW);
 digitalWrite(2, LOW);
// Connect to WiFi network
 Serial.println();
 Serial.println();
 Serial.print("Connecting to ");
 Serial.println(ssid);
WiFi.begin(ssid, password);
while (WiFi.status() != WL_CONNECTED) {
  delay(500);
  Serial.print(".");
 }
 Serial.println("");
 Serial.println("WiFi connected");
// Start the server
 server.begin();
 Serial.println("Server started");
// Print the IP address
 Serial.print("Use this URL to connect: ");
 Serial.print("http://");
 Serial.print(WiFi.localIP());
 Serial.println("/");
}
 void loop() {
  // Check if a client has connected
  WiFiClient client = server.available();
  if (!client) {
    return;
  }
  // Wait until the client sends some data
  Serial.println("new client");
  while(!client.available()){
    delay(1);
  }
  // Read the first line of the request
  String request = client.readStringUntil('\r');
  Serial.println(request);
  client.flush();
  // Match the request
```

IOP Conf. Series: Earth and Environmental Science **168** (2018) 012017 doi:10.1088/1755-1315/168/1/012017

```
  if (request.indexOf("/devon") > 0)  {
    digitalWrite(15, HIGH);
    digitalWrite(2, LOW);
  }
   if (request.indexOf("/devoff") > 0)  {
    digitalWrite(15, LOW);
    digitalWrite(2, HIGH);
   }
// Set ledPin according to the request
// Return the response
  client.println("HTTP/1.1 200 OK");
  client.println("Content-Type: text/html");
  client.println(""); //  do not forget this one
  client.println("<!DOCTYPE HTML>");
  client.println("<html>");
  client.println("<head>");
  client.println("<meta name='apple-mobile-web-app-capable' content='yes' />");
  client.println("<meta name='apple-mobile-web-app-status-bar-style' content='black-translucent' />");
 client.println("</head>");
  client.println("<body bgcolor = \"#f7a6ec\">");
  client.println("<hr/><hr>");
  client.println("<h1><center>  Smart home electic appliance control system by web  </center></h1>");
  client.println("<hr/><hr>");
  client.println("<br><br>");
  client.println("<br><br>");
  client.println("<center>");
    client.println("Sample Device(DC MOTOR)");
  client.println("<a href=\"/devon\"\"><button>Turn On </button></a>");
  client.println("<a href=\"/devoff\"\"><button>Turn Off </button></a><br />");
client.println("</center>");
  client.println("<br><br>");
  client.println("<center>");
  client.println("<table border=\"5\">");
 client.println("<tr>");
   if (digitalRead(2))
        {
client.print("<td bgcolor = \"#FF0000\"><h1>DC MOTOR is OFF</h1></td>");
        }
       else
        {
 client.print("<td bgcolor = \"#00FF00\"><h1>DC MOTOR is ON</h1></td>");
        }
 client.println("</tr>");
client.println("</table>");
client.println("</center>");
  client.println("</html>");
  delay(1);
  Serial.println("Client disonnected");
  Serial.println("");

}
```

5. Conclusion

This web-based design was used to monitor the energy usage characteristics of a sample DC motor instrumented into a model smart electric home appliances control system, where user can monitor and control it remotely from any place. The control system showed to be reliable and efficient in facilitating energy wastage and loss reduction. Further work should consider electrical appliances control systems able to both control and record the quantities of energy consumed at any time instants, using web interface and Node MCU.

6. Acknowledgment

This work was funded by the National Natural Science Foundation of China (program No.51475136), Natural Science Foundation of Hebei Province, China (Program No.E2014202230), Universities in Hebei Province, China Science and Technology Research Youth Fund (Program No.QN20151002), University Innovation Team Leader Program in Hebei Province (LJRC003).

7. References

[1] Wikipedia (2017). Control engineering. Available at: https://en.wikipedia.org/wiki/Control_engineering (Accessed 20 July 20, 2017).
[2] Electrical4u (2017). Control system: Closed Loop opene loop control system. Available at : https://www.electrical4u.com/control-system-closed-loop-open-loop-control-system/ (Accessed 20 July 20, 2017)
[3] Wikipedia (2017) Home Appliance. Available at : https://en.wikipedia.org/wiki/Home_appliance (Accessed 20 July 2017)
[4] Roberts. K (n.d) Household Chore List. Available at: http://cleaning.lovetoknow.com/Household_Chore_List (Accessed 20 July 2017)
[5] Wikipedia (2017) Multipoint control unit. Available at: https://en.wikipedia.org/wiki/Multipoint_control_unit (Accessed 21 July 2017).
[6] NODEMCU (2014). NodeMCU connect things easy. Available at : http://nodemcu.com/index_en.html (Accessed 20 July 2017).
[7] Jabbar, Z.A. and Kawitkar, R.S., 2016. Implementation of Smart Home Control by Using Low Cost Arduino & Android Design. *International Journal of Advanced Research in Computer and Communication Engineering*, *5*(2).
[8] Umelo, N. and Adekiigbe, A., A Lightweight Wireless Home Appliances Control.
[9] Lafta, H.A.R., 2014. Design and Implementation of Appliance Controller using Traditional or Smart Phone. *International Journal of Computer Applications*, *101*(9).
[10] Sindhuja, P. and Balamurugan, M.S., 2015. Smart power monitoring and control system through internet of things using cloud data storage. *Indian Journal of Science and Technology*, *8*(19).
[11] Shahriyar, R., Hoque, E., Sohan, S.M., Naim, I., Akbar, M.M. and Khan, M.K., 2008. Remote controlling of home appliances using mobile telephony. *International Journal of Smart Home*, 2(3), pp.37-54.
[12] Vázquez, F.I., Kastner, W., Gaceo, S.C. and Reinisch, C., 2011. Electricity load management in smart home control. In *12th Conference of International Building Performance Simulation Association* (pp. 957-964).
[13] Hamed, B., 2012. Design & implementation of smart house control using LabVIEW. *International Journal of Soft Computing and Engineering (IJSCE)*, *1*(6), pp.98-106.
[14] Robles, R.J. and Kim, T.H., 2010. Applications, Systems and Methods in Smart Home Technology: A. *Int. Journal of Advanced Science And Technology*, *15*.
[15] Adetiba, E., Matthews, V.O., Awelewa, A.A., Samuel, I.A. and Badejo, J.A., 2011. Automatic Electrical Appliances Control Panel Based on Infrared and Wi-Fi: A Framework for Electrical Energy Conservation. *International Journal of Scientific & Engineering Research*, *2*(7), pp.1-7.
[16] Davidoff, S., Lee, M.K., Yiu, C., Zimmerman, J. and Dey, A.K., 2006, September. Principles of smart home control. In *International Conference on Ubiquitous Computing* (pp. 19-34). Springer,

Berlin, Heidelberg.
[17] Das, N.K., Alamgir, M.Z., Das, C.K., Hasan, M.M., Rahman, M.M. and Alam, M.F., 2011. Developing A Smart Control System of Electrical Appliances By Computer Interfacing. *MIST Journal: GALAXY (DHAKA), 3*.
[18] Lobaccaro, G., Carlucci, S. and Löfström, E., 2016. A review of systems and technologies for smart homes and smart grids. *Energies*, *9*(5), p.348.
[19] Gunge, V.S. and Yalagi, P.S., 2016, March. Smart Home Automation: A Literature Review. In International Journal of Computer Applications (0975–8887) National Seminar on Recent Trends in Data Mining (RTDM 2016).
[20] Piyare, R. and Lee, S.R., 2013. Smart home-control and monitoring system using smart phone. *ICCA 2013, ASTL*, *24*, pp.83-86.
[21] Istvan, A., Fleck, R., Budd, R., Stark, K. and Tanumihardja, M., Vulcan Inc., 2005. *Smart home control of electronic devices*. U.S. Patent Application 11/120,653.
[22] Github (2017) NodeMCU devkit. Available at : https://github.com/nodemcu/nodemcu-devkit (Accessed 20 July 2017)
[23] Wikipedia (2017) NodeMCU. Available at : https://en.wikipedia.org/wiki/NodeMCU (Accessed 20 July 2017)
[24] Farazmemon (2017). A brief Introduction to NodeMCU. Available at : http://farazmemon.com/a-brief-introduction-to-nodemcu/ (Accessed 20 July 2017)
[25] Aliexpress (n.d) free shipping 10pcs/lot 3v dc 10a songle voltage relay srd-03vdc-sl-c t73 pcb dip 5pin type new & original . Available at: https://www.aliexpress.com/item-img/Free-Shipping-10pcs-lot-3V-DC-10A-SONGLE-voltage-relay-SRD-03VDC-SL-C-T73-PCB/32387116973.html (Accessed 20 july 2017)
[26] Galco(2017). How relays work. Available at : http://www.galco.com/comp/prod/relay.htm (Accessed 20 July 2017)
[27] PC-Control (2008). Relays. Available at : https://www.pc-control.co.uk/relays.htm (Accessed 20 July 2017)
[28] SM (2017). Arduino Software (IIDE). Available at: https://www.arduino.cc/en/Guide/Environment. (Accessed 20 Julay 2017)

An Economic Evaluation of Electric Vehicle Charging Infrastructure in Public Places in China

Xueqi Huang[1], Xiao Li[1], and Xinmei Yuan[1], Bo He[2], Juan Li[3]

[1]State Key Lab. of Automotive Simulation and Control, Jilin University, Changchun, China
[2]Department of Power Grid Development, State Grid Energy Research Institute, Beijing, China
[3]State Grid Tianjin Electric Power Economic Technology Research Institute, State Grid Tianjin Electric Power Company, Tianjin, China
e-mail: yuan@jlu.edu.cn, hebo@sgeri.sgcc.com.cn, zbblj@126.com

Abstract. The electric vehicle market has witnessed a big growth spurt as a result of the promotion of the transportation electrification in the global. Among electric vehicles, electric passenger cars are the most promising fleet to achieve the global climate target in the next decades, which will lead to a vast potential public charging market. Therefore, evaluating the economy of operating the charging infrastructure from the operator's perspective is especially required. The paper aims at providing an economic evaluation of charging infrastructure in public places in China. In the proposed method, the initial investment cost, the subsidy, the operating revenue and the operating cost are all modeled in detail. Using the proposed model, two scenarios representing the electric vehicle charging in the work place and in the fast charging station are assumed, and the cost and revenue of charging operators are calculated and compared. It is shown that under our assumption the payback period of the charging infrastructure in the work place is much shorter than that of fast charging station. Because the calculated payback period of fast charging station is more than ten years, current fast charging stations are unlikely to profit, and the promotion of the fast charging stations requires further policy support and smart grid technology development.

1. Introduction

Electric vehicle provides a crucial opportunity to promote the low-carbon energy transition of the transportation sector [1], [2] and has had a good run as the substitution of conventional vehicle, but it still faces a lot of barriers to widespread adoption. Among several main obstacles, from user's perspective, poor public charging infrastructures is the largest barrier after the technological barriers [3], [4]. With the development of electric vehicles, charging infrastructure is the obvious substitute for petrol station accompanied by the transition from conventional vehicle to electric vehicle. Additionally, the possibility of charging at home, work place or other public places [5] is becoming another superiority for the energy refilling of electric vehicles. Destination charging could save the unnecessary time used for driving to the petrol station and waiting for refilling the fuel tank of conventional vehicles, which even could solve technical problems, for example range anxiety [6].

On the basis of those, charging infrastructures are heavily invested by national governments [7] as well as enterprises so as to be networked and widely used. In China, the development of electric vehicles, the construction of charging infrastructures is given high priority, and policies for charging infrastructure constructions have been issued by governments. Constructing charging infrastructures in public places is one of great urgency to promote the adoption of electric cars.

The economy of constructing and operating charging infrastructures is significant for charging operators; therefore, an economic evaluation of electric vehicle charging infrastructures is given in this paper. A profit model is proposed in terms of several major costs and revenues to evaluate the annual profit and the payback period of various charging ways. Two scenarios representing the electric vehicle charging in the work place and in fast charging station are assumed, and the cost and revenue of charging operators are calculated, compared and discussed. The results show that because of very low initial investments, slow charging infrastructures are easier to profit, however, fast charging stations are unlikely to profit currently [7], and further policy and technical support are required.

The paper is arranged as follows. After introduction, the method of the charging infrastructure economic evaluation is proposed. Based on that, two charging scenarios are calculated and compared. Conclusions are discussed in the last section.

2. Method Proposed

The economy of charging infrastructures from the perspective of charging service operators includes the initial investment cost, the subsidy, the operating revenue, the operating cost. The net investment cost in the initial stage is calculated by the initial investment and the subsidy; the annual profit is the difference between the annual operating revenue and the annual operating cost. Payback period is the time when the accumulated annual profit equals to the net investment and operating cost.

2.1. Initial Investment Cost Model

For the charging infrastructure operator, the initial investment cost (*IIC*) includes land purchasing cost (*LPC*), building construction cost (BCC), power-supply system cost (*PSC*), monitoring system cost (*MSC*) and charging infrastructure cost (*CIC*), which is expressed as

$$IIC = LPC + BCC + PSC + MSC + CIC(n) \tag{1}$$

where, *LPC* is the total cost of the land where the charging infrastructure sits. The area of the land is related to the area of each section of the charging infrastructure including charging spaces, carriageways, power-supply room, monitoring room and other land for example a certain amount of reserve land which will be used in the future. LPC represents a significant portion of the initial investment cost and highly depends on the location in the city. *BCC* is the total construction cost of the related building and ground. PSC and MSC represent the costs of purchasing equipment, installing equipment and hiring labor in initial stage respectively, which is related to the scale of the charging infrastructure. Land purchase cost, building construction cost, powered-supply system cost and monitoring system cost will be saved if the charging infrastructure is constructed based on the existing public parking lots. *CIC* is the function of the number of the charging systems and the price of the charging equipment related to the charging demand. Fast charging is preferred for EV users when the battery is low and the users only stay a short time; and slow charging is a better selection for extending the battery lifespan and further incorporate within the control of smart grid. The charging infrastructure cost could be calculated by

$$CIC(n) = n_p \gamma_c (\gamma_f P_f + \gamma_s P_s) \tag{2}$$

where, n_p is the number of parking spaces in the public, γ_c is the ratio of the number of charging spaces and the number of parking spaces, γ_f is the ratio of the number of fast charging spaces and the number of charging spaces, γ_s is the ratio of the number of slow charging spaces and the number of charging spaces, P_f and P_s are the allocation prices of the total purchase cost of charging systems per fast charging space and per slow charging space respectively, including the purchasing cost, the freight cost, the installation cost, the material cost, the test cost, and the labor cost according to the charging infrastructure supplier.

2.2. Subsidies

The relationship between the electric vehicle and the charging infrastructure pose a chicken-or-egg conundrum. Hence, in the early phase of development of electric vehicles, it is difficult to benefit from operating charging infrastructures [8]. To promote electric vehicles, it is necessary to provide subsidies to support the construction and the operation of the charging infrastructure. And the way of the subsidy is various and depends on the national and local policies from the government. The total subsidies are expressed as

$$SUB = S_c + S_l + S_o \tag{3}$$

where S_c is the subsidy for purchasing the charging infrastructure, S_l is the subsidy for purchasing the land of the charging infrastructure, S_o is the subsidy by other forms such as according to the number of the new registered electric vehicles locally and the tax reduction.

2.3. Operating Revenue Model

Operating revenue (OPR) in the i th year is divided into several parts: electricity revenue (ER), service revenue (SR), other revenue (OR), expressed as

$$OPR(i) = ER(i) + SR(i) + OR(i) \tag{4}$$

EV batteries are assumed to be charged during the parking time. And electricity revenue in the i th year from the EV users is calculated by

$$ER(i) = 365 \times 24 \times n_p \gamma_c (\gamma_f \beta_u^f \beta_p^f p^f + \gamma_s \beta_u^s \beta_p^s p^s) P_e \tag{5}$$

where β_u is the utilization rate of time, β_p is the utilization rate of charging spaces, p is the rated output power of the charging equipment, P_e is the electricity selling price per kWh, including the electricity cost and the extra charging service profit.

Service revenue is gained by providing extra service except charging for example parking, management and scheduling of EVs providing by parking lot operator. If the charging operator also plays the role of parking lot operator, service revenue in the i th year is evaluated by

$$SR(i) = 365 \times 24 \times n_p \gamma_c (\gamma_f \beta_u^f \beta_p^f + \gamma_s \beta_u^s \beta_p^s) I_s \tag{6}$$

where I_s is the service fees per hour.

Other revenue such as the advertisement revenue from business cooperation is a large extra revenue for charging operators.

2.4. Operating Cost Model

The operating cost (*OC*) includes electricity purchasing cost (EPC), charging infrastructure maintenance cost (*CMC*), management cost (*MC*) and other operating cost (*OOC*). And the operating cost in the i th year is given by

$$OC(i) = EPC(i) + CMC(i,n) + MC(i) + OOC(i) \tag{7}$$

The charging infrastructure operator is a medium between the gird and EV users exactly. The electricity is purchased from the power gird and sold to EV users. And it is assumed the electric energy is transmitted from the power gird to the EV users directly without energy storage systems. Then the electricity purchasing cost is calculated by

$$EPC(i) = 365 \times 24 \times n_p \gamma_c \left(\frac{\frac{\gamma_f \beta_u^f \beta_p^f p_f}{\eta_c^f} + \frac{\gamma_s \beta_u^s \beta_p^s p_s}{\eta_c^s}}{\eta_P} \right) P_p \tag{8}$$

where η_c is the charging equipment efficiency, η_p is the power-supply system efficiency, P_p is the electricity purchasing price per kWh.

The charging infrastructure maintenance cost per year including the labor cost, related equipment repairing and replacement cost, assumed to be 10% of initial cost of the charging infrastructure.

$$CMC(i,n) = 10\% CIC(n) \tag{9}$$

Management cost is required including the labor cost for guidance and scheduling, the energy cost for lighting and cleaning and other related cost. But for the charging infrastructure based on the existing parking lot, the cost of this part will be included in the business cooperator cost within OOC.

Other operating cost per year includes the business cooperation cost paid for the service from other operators to manage the charging infrastructure, the maintenance cost for power-supply system and monitoring system, and other labor cost.

2.5. Profit Model

The revenue model is established by the initial investment cost, subsidy, operating revenue and operating cost. And yearly profit (YP) and payback period is expressed by

$$YP = OPR(i) - OC(i) \tag{10}$$

$$m = \frac{IIC - SUB}{YP} \tag{11}$$

where m is the payback period when the accumulated annual profit of the charging infrastructure in the past m years has covered the total cost to construct and operate the charging infrastructure.

3. Case Studies

Fast charging and slow charging are two typical charging ways at present, which is suitable for different charging requirements. Based on the different travel habits, two scenarios representing the two charging ways respectively are analyzed.

3.1. Work Place

According to the travel habits of working people, their electric cars usually be parked at the work place during their 8-hour work time. Therefore, the parking lot of the work place is a promising choice for charging electric vehicle, and slow charging is preferable due to the long-time parking behavior. In this analysis, it is assumed that the charging operator in the work place constructs the charging infrastructure based on the existing parking lot and cooperates with the parking lot operator; therefore, the land purchase cost and the building construction cost is saved, besides, the power-supply system and the monitoring system rely on the existing system. Because the parking and charging behaviors are approximately constant, the number of charging systems can be accurately planned according to the number of electric car owners in the work place. The initial investment cost is only the purchase cost of the slow charging systems. Business cooperation cost is considered within the other operating cost. Since the charging operator does not own the parking lot, it has no gain in service revenue and only makes revenue from electricity selling and advertising.

3.2. Fast Charging Station

Fast charging infrastructures are necessary to complete the charging network for emergency charging, where most EVs could be fully charged within one hour or less. Charging station is an independent charging infrastructure, and the land purchase cost and building construction cost are needed and take a great proportion in the initial vestment cost. Power-supply system and monitoring system are also necessary. The planning of the number of charging spaces is a complex optimization problem, relating to different parameters of EV owners' driving behavior, driving route planning, charging stations distribution in the neighborhood, etc. We simply use 10 charging spaces in this analysis according to the size of the current charging stations in China. The management cost is used for hiring employees to carry out the daily management of the charging station and the maintenance of the equipment. Electricity revenue and service revenue are gained and the advertising revenue will increases compared with charging infrastructure in the work place due to the larger service flow.

3.3. Results and Discussions

The results of the two scenarios are shown in Fig. 1 and Fig. 2 respectively. The payback period is calculated by the initial investment cost model, subsidy, operating cost model and operating revenue model. Due to the absolute value differences between the two types of charging infrastructures, the results are represented using normalized costs. The 15-year total cost of each scenario is used to normalize all the calculation results of each scenario respectively. The 15-year total cost in the work place is approximately 820,000, however, it is as high as 18,000,000 for the fast charging station. In Fig. 1 and Fig. 2, y-axis represents the normalized cost or revenue and x-axis represents the duration that the charging infrastructure is operated. The projection of the difference between the normalized revenue line and the normalized cost line on the y-axis in the given year is the net normalized annual profit that year, and the projection of the intersection of the normalized accumulated cost line and the normalized accumulated revenue line on the x-axis is the payback period for the charging infrastructure.

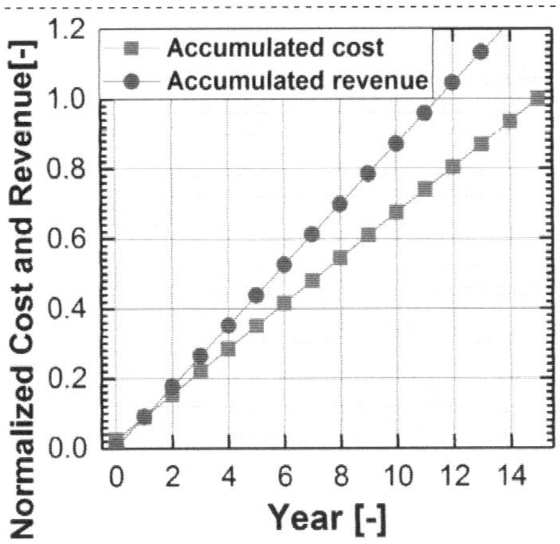

Figure 1. Normalized yearly accumulated cost and revenue (charging infrastructure in the work place).

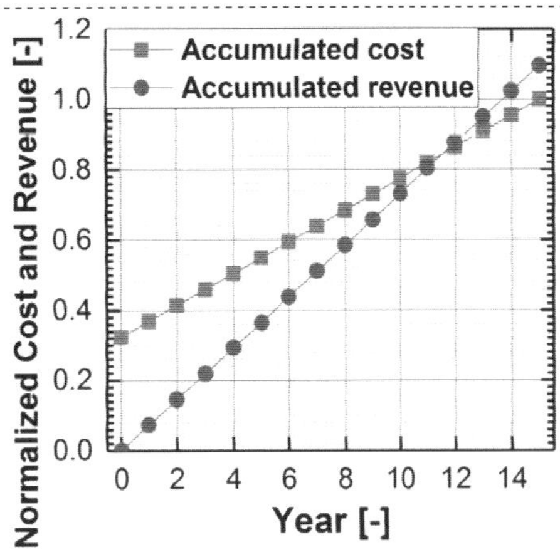

Figure 2. Normalized yearly accumulated cost and revenue (fast charging station).

According to the results shown in the figures, the net initial investment cost of the fast charging station is much higher than that of the charging infrastructure in the work place because of the costly related equipment, the building and the land. The payback period for the charging infrastructure in the work place is 1 year, which is far sooner than the payback period of 12 years for the fast charging station. However, the lifetime of the charging systems is not considered here.

Due to the characteristic of the charging way, fast charging could provide more effective service and gain higher electricity revenue and service revenue, but with the limitation of battery technology it faces the anxiety of battery capacity degeneration. On the contrast, slow charging provides an economical charging choice which is easier to be adopted. Although the charging process is time-consuming, slow charging is feasible to be promoted vigorously with inexpensive initial investment cost and wins on volume.

With the chosen assumption, the scale of the charging infrastructure, the utilization of the time and the number of the charging spaces is constant, as a result, the operating revenue and the operating cost is constant every year. As the market penetration of the electric car, the scale of the charging infrastructure, the utilization of the time, charging spaces as well as service efficiency will increase and the operating revenue including the business cooperation revenue will rise rapidly; then the payback period will be shortened. In addition, intangible resources are not calculated in the economic evaluation for example the increasing value of the land, gaining the market share and the brand cognitive, which will help operator benefit more.

4. **Conclusion**

A model to evaluate the economy of the electric vehicle charging infrastructure in the public place from the operator's perspective is built, including the initial investment cost model, subsidy, operating revenue model and operating cost model. Then two typical scenarios, the charging infrastructure based on the existing parking lot in the work place with slow charging way and the charging station with fast charging way, are calculated and compared. As a result, the payback period is 1 years for the charging infrastructure in the work place, but it is 12 years for the fast charging station. The results show that the initial cost of slow charging infrastructure is low and the corresponding investment is easy to recover in the short term, with further advantages such as very low impact on the power grid. Slow charging infrastructure is an economical choice for the large-scale promotion of electric vehicles. The initial cost of fast charging system is very high, the payback period is too long, therefore, the investment of fast charging station seems difficult to profit currently. Additionally, the planning of the fast charging station is also very complex [9], and global planning with the smart grid is needed [10]. However, fast charging station is necessary for the promotion of electric vehicles, because it is especially important to overcome the range anxiety of EV users. Therefore, on the one hand, vehicle and grid inter-connection techniques, energy storage techniques, novel business models should be applied to increase the profitability of the fast charging station; on the other hand, more favorable policy supporting is required for the operation of fast charging stations.

5. References

[1] X. Yuan, L. Li, H. Gou, and T. Dong, "Energy and environmental impact of battery electric vehicle range in China," Applied Energy, vol. 157, pp. 75-84, 2015.
[2] K. Nansai, S. Tohno, M. Kono, M. Kasahara, and Y. Moriguchi, "Life-cycle analysis of charging infrastructure for electric vehicles," Applied energy, vol. 70, no. 3, pp. 251-265, 2001.
[3] O. Egbue and S. Long, "Barriers to widespread adoption of electric vehicles: An analysis of consumer attitudes and perceptions," Energy policy, vol. 48, pp. 717-729, 2012.
[4] C. Madina, I. Zamora, and E. Zabala, "Methodology for assessing electric vehicle charging infrastructure business models," Energy Policy, vol. 89, pp. 284-293, 2016.
[5] T. G. San Román, I. Momber, M. R. Abbad, and A. S. Miralles, "Regulatory framework and business models for charging plug-in electric vehicles: Infrastructure, agents, and commercial relationships," Energy policy, vol. 39, pp. 6360-6375, 2011.
[6] J. Chynoweth, C.-Y. Chung, C. Qiu, P. Chu, and R. Gadh, "Smart electric vehicle charging infrastructure overview," in Innovative Smart Grid Technologies Conference (ISGT), 2014 IEEE PES, 2014, pp. 1-5: IEEE.

[7] L. Zhang, T. Brown, and S. Samuelsen, "Evaluation of charging infrastructure requirements and operating costs for plug-in electric vehicles," Journal of Power Sources, vol. 240, pp. 515-524, 2013.
[8] A. Schroeder and T. Traber, "The economics of fast charging infrastructure for electric vehicles," Energy Policy, vol. 43, pp. 136-144, 2012.
[9] S. Davidov and M. Pantoš, "Planning of electric vehicle infrastructure based on charging reliability and quality of service," Energy, vol. 118, pp. 1156-1167, 2017.
[10] K. Mets, T. Verschueren, W. Haerick, C. Develder, and F. De Turck, "Optimizing smart energy control strategies for plug-in hybrid electric vehicle charging," in Network Operations and Management Symposium Workshops (NOMS Wksps), 2010 IEEE/IFIP, 2010, pp. 293-299: Ieee.

Acknowledgment
This work was supported by the State Grid Tianjin Electric Power Economic Technology Research Institute for the project of research on multi-scene coordinated development strategy and planning technology of distribution network and charging infrastructures [SGTJJY00GHJS1600274]. Any opinions, findings, and conclusions or recommendations expressed herein are those of the authors and do not necessarily inflect the views of State Grid Tianjin Electric Power Economic Technology Research Institute.

IOP Conf. Series: Earth and Environmental Science **168** (2018) 012019 doi:10.1088/1755-1315/168/1/012019

Carbon additive structure influence on electrochemical double-layer capacitor electrode behavior

E.A. Kiseleva[1,*], S.A. Kochanova[1], O.V. Uryupina[2], L. Puntusova[3], A.B. Tarasenko[1]

[1] Joint Institute for High Temperatures of the Russian Academy of Sciences, Izhorskaya 13 Bldg 2, Moscow 125412, Russia
[2] "Trust for Energy and Electromobile projects" LLC, Botanicheskaya 14 Office 21, Moscow 127276, Russia

Email: kanna787@mail.ru

Abstract. Electrochemical double-layer capacitors (EDLC) are widely used as power sources for peak load or peak generation leveling in transport and stationary applications sometimes in combination with accumulator battery. For such applications EDLC need low sensitivity (from lifetime and capacity point of view) to high current mode operation, which is usually crucial for traditional accumulators' types. Capacity loss in high current mode operation due to both transport and resistive limitations is observed both for accumulators and supercapacitors but much less for latter due to initial low internal resistance and electrochemical reactions absence. Carbon nanotubes (CNT) and nanofibers (CNF) as alternative to carbon blacks as conductive additive for electrodes are good possibility to improve high current operation both for accumulators and EDLC. This paper is devoted to investigation of carbon nanotubes and nanofibers comparison with traditional carbon black additive. CNF structure usually contains more defects and has lower aspect ratio than single-walled nanotubes but this kind of materials is less expensive than CNT, that is also important for final customer. Electrode samples with Vulcan XC72R carbon black were prepared for reference measurements. All electrode samples were tested in galvanostatic charge-discharge mode at different current levels in laboratory-scale EDLC cell, characterized by impedance spectroscopy. Obtained results showed improved high-current operation for CNT-containing electrodes. Economic estimation for different kinds of additives has also been carried out.

1. Introduction

Supercapacitors are special sort of energy storage devices, operating on electrochemical double electric layer charge and discharge. Double electric layer occurs on the interface of porous carbon electrode and liquid electrolyte, penetrating into porous structure and forming large interface surface area leading to high electric capacity of such devices [1]. Faradic processes on electrodes are absent, contrary to electrochemical accumulators [2]. So high lifetime and depth of discharge, close to 100% are available for supercapacitors, but specific energy is very low. So usual applications of electrochemical double-layer supercapacitors (EDLC) are concerned with high electric power generation or consumption in short time periods due to low energy value stored. Usually transportation [3], large or micro-grid parameters regulation [4] and different ways of energy recuperation are reviewed as application niches for supercapacitors. Different ways to increase stored energy (without significant losses in power rate) are under investigation – organic electrolytes with

high voltage limit [5], new carbonaceous materials [6] with high surface area. It's worth mentioning that high surface area of carbonaceous materials usually leads to their poor electron conductivity [7], so different conductive additives are used in supercapacitor electrode to conserve power rate of the whole device in these conditions. Carbon blacks are usual product for this purpose, especially wide use in electrochemical devices has Vulcan XC72R from Cabot Corp., USA [8].
Carbon nanotubes and nanofibers, graphene-based materials are under investigation too, due to their high stability and electron conductivity [9].

Purpose of this research is to compare two kinds of industrial produced carbon nanostructures – Tuball™ single-walled carbon nanotubes (SWCNT) from OCSiAl LLC (Novosibirsk, Russia), carbon nanofibers (CNF) Taunit™ (NanoTechCenter, Tambov, Russia) with larger diameter than Tuball, but less expensive, then high-quality Tuball, with traditional additive Vulcan XC72R. Comparison includes technical and economic issues – optimal additive mustn't decrease capacity, lifetime or power rate and increase cost of electrode.

2. Experimental details

To evaluate electrochemical parameters of electrodes with different carbon additives, laboratory-scale test pouch cells were assembled. First slurry from activated carbon, conductive additive and polytetrafluorethylene (PTFE) emulsion was prepared, containing PTFE, carbon conductive additive and activated carbon. Slurry was mixed mechanically using Ultra Turrax T10 during 10 minutes, after that dried at 250 °C in Binder™ vacuum chamber for 4 h, then frictioned through mesh with 2 mm cell diameter to obtain non-agglomerated carbon material, relatively uniform covered with PTFE. As-received mass was fed to rolling press MSK-HRP-MR100DC (MTI Corp., USA) and rolled several times to obtain composite PTFE-bound carbon sheet with thickness 100-140 μm. More detailed routine for electrode preparation is given in [10]. These sheets were annealed at 250 °C in Binder™ vacuum chamber for 4 hours to finally remove residual water and solvent from electrode. In all experiments activated carbon produced in Joint Institute for High Temperatures from pyrolyzed birch by alkaline activation technique based on described in [11] was used as active material of electrode.

Table 1. Electrode specimens' composition (dry weight).

	Conductive additive type	Activted carbon, mass, %	Conductive additive, mass, %	PTFE, mass,%
1	Tuball Batt	91.09	0.01	8
2	Vulcan XC72R	82	10	8
3	Taunit	82	10	8

As-prepared electrode series with composition, given in Table 1, were tested in lab-scale laminated pouch cell with 30*30 mm active area. 1 M tetraethylammonium tetrafluoroborate (TEA-BF_4) in electrochemical grade (water content <50 ppm) dissolved in acetonitrile (AN) purchased from BASF, USA, using cellulose-based separator. Cells lamination to prevent moisture and impurities penetration from the environment was also conducted in dry box with Ar atmosphere. Galvanostatic cycling was the main approach for cell characterization. All measurements were carried out in range from 1.25 up to 2.5 V of cell voltage. Specific gravimetric capacitance during discharging was calculated via (1):

$$C_{sp} = \frac{4}{m} * \int_{to}^{T} \frac{idt}{U} \tag{1}$$

with specific gravimetric capacitance C_{sp}, time t (t_0- starting time of discharge (charge), T- end of discharging (charging) time), i – charging (discharging) current, U - cell voltage, m - total mass of the electrodes (i.e., considering carbon materials and the binder). For every type of electrode composition, three electrode pares were prepared and tested individually to calculate a mean value.Testing routine included long-term charge-discharge cycling at current density 25 mA/cm^2, short periods (about 100 cycles) at 2 mA/cm^2, then about 900 cycles at 167 mA/cm^2. After that test cells were cycled at 25mA/cm^2 again to observe capacity loss after cycling mode change (first of all, after fast cycling at 167 mA/cm^2). Results were registered for last cycles. ASK2.5.10.8 (YAROSTANMASH LLC, Russia, [12]) chemical current source analyser was used to perform cycling at given current densities. Results for long-term cycling are given in Fig. 1, for different current densities – in Fig. 2, calculated specific capacity loss for all cells at different cycling modes – in table 2. Specific capacity loss was calculated as

$$L_c = \frac{C_0 - C_{fin}}{C_0} * 100, \quad (2)$$

C_0 is initial specific capacity (before series of tests) and C_{fin} is specific capacity after series of tests. In this paper two series of tests are considered – 40000 cycles at 25mA/cm^2 and 100 cycles at 167 mA/cm^2, followed by 100 cycles at 25mA/cm^2. In the first mode $C_0$is C_{sp} after 100-th cycle and C_{fin} is C_{sp} after 40100-th cycle (100 first cycles are needed for cell polarization and training, after which parameters are more stable). In the second mode is C_{sp} after 40100–th cycle at 25 mA/cm^{2}and C_{fin} is C_{sp} after 100 cycles at 167 mA/cm^2 and 10 cycles at 25 mA/cm^2. Additional

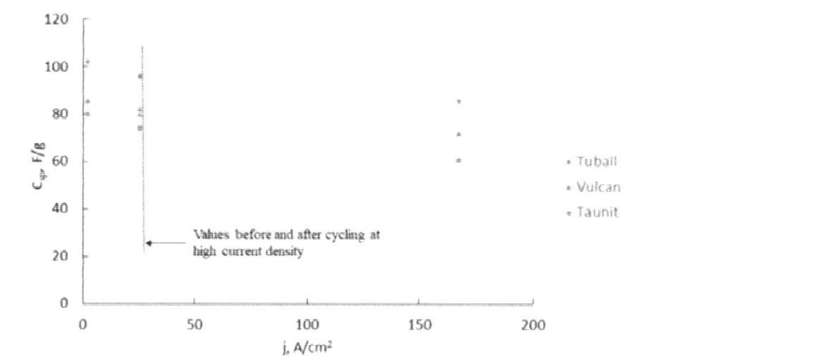

Figure 1. Long-term stability tests. Electrode parameters according to Table 1.

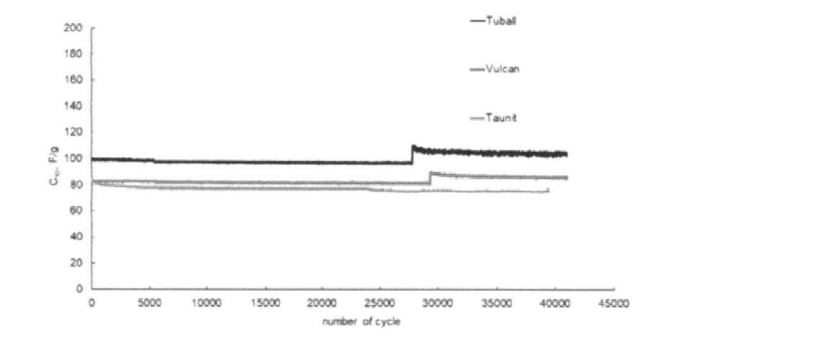

Figure 2. Cells specific capacity for different cycling current density and different conductive additives. Electrode parameters according to Table 1.

Impedance spectra were also registered for specimens with different conductive additives to evaluate electrode transport properties. Electrochemical analyzerVoltaLabPGZ 301 was used for impedance measurements. Nyqist plots for all specimens are shown in Fig. 3.

Table 2. Specific capacity loss at different modes. Electrode parameters according to Table 1.

	Lc, %	
Conductive additive type	After 40000 cycles at 25mA/cm2	After 100 cycles at 167 mA/cm2
Tuball Batt	7	1
Vulcan XC72R	3	3
Taunit	8	1

Impedance spectra were also registered for specimens with different conductive additives to evaluate electrode transport properties. Electrochemical analyser VoltaLabPGZ 301 was used for impedance measurements. Nyqist plots for all specimens are shown in Fig.3.

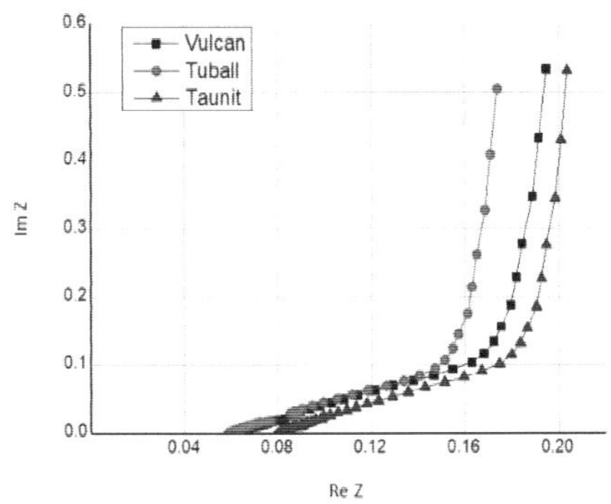

Figure 3. Nyquist plot for electrode specimens with Vulcan, Tuball and Taunit conductive additives. Electrode parameters according to Table 1.

3. Economic estimations

It is necessary to estimate different conductive additives influence on supercapacitor costs, because cost is one of the most important parameter from consumer's point of view. For this purpose, current prices for all components were reviewed. According to authors purchase experience, carbon black Vulcan XC72 R price can be estimated as 4.3 USD/kg [13], Taunit – 2000 USD/kg [14], and 8000 USD/kg for Tuball Batt (for dry weight, because price for deionized water suspension with 0,2 mass.% of CNT is about 16 USD/kg, according to consultations with OcSiAl specialists). Activated carbon, synthesised in RAS JIHT, can have specific cost as high as 15 USD/kg in case of industrial scale production. According to table 1, costs for electrode materials can be estimated as (for 1 kg of electrode active mass):

$$E_{el} = (E_c * (1 - \phi) + E_{add} * \phi) * m, \quad (3)$$

where ϕ is mass share of conductive additive in electrode, E_c – specific cost for activated carbon (USD/kg), m – electrode active mass (1 kg) and E_{add} - specific cost for conductive additive (USD/kg). Taking into account different electrode specific capacity values for different additives, we consider test results for specimens after 40000 cycles at 25 mA/cm^2: 96 F/g for Tuball-doped electrode, 85 F/g for Vulcan-doped electrode. For Taunit 10% are not optimal, so, considering 5 mA/cm^2 results for 3 mass.%, specific capacity for Taunit-doped electrode can be estimated as 94 F/g at 25 mA/cm^2.

Further specific capacity cost (USD/F) for 1 kg of active mass with different additives is calculated as:

$$E_{cap} = \frac{C_{sp} * m}{E_{el}}, \quad (4)$$

where C_{sp} – specific capacity of electrodes after 40000 cycles at 25 mA/cm^2, obtained in experiment. Calculation results are summarized in Table 3.

Table 3. Specific capacity cost estimation for electrodes with different conductive additives.

Conductive additive type	Tuball, 0,012 mass.%	Vulcan XC 72 R, 10 mass.%	Taunit, 3 mass.%
Ecap, USD/F	$0.17*10^{-3}$	$0.16*10^{-3}$	$0.79*10^{-3}$

Additional research has been carried out to evaluate optimum load for Taunit CNFs. Cells based on the same technology and the same activated carbon as described above with different Taunit load were prepared and tested. Testing protocol included charge-discharge cycles at 3 mA/cm^2. Specimens with 1, 3 and 10 wt% of CNF were prepared and tested. The results are shown at Fig. 4.

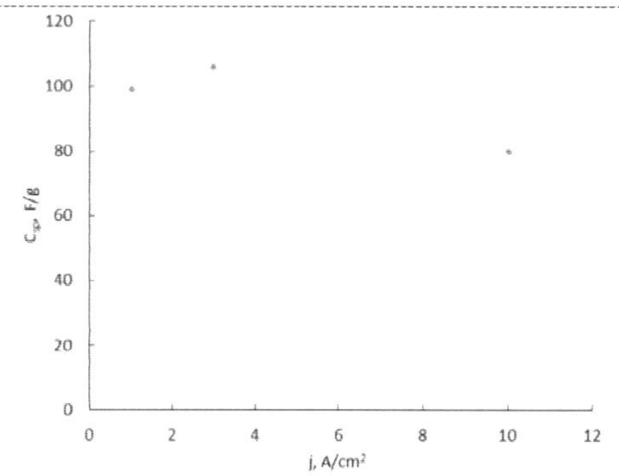

Figure 4. Load optimization for Taunit conductive additive. Current density 5 mA/cm2

4. Analysis and discussion

Carried out experiments shows that structures with high aspect ratio (nanotubes and nanofibers) show much less amount of conductive additive to obtain the same electrochemical parameters than carbon black. Moreover, SWCNT with small diameter demonstrate the smallest load of additive to provide good electrochemical performance due to their ability to form percolation conducting cluster in electrode in less concentrations, that is needed for carbon blacks, forming long micelles at large concentrations. It is possible that percolation cluster in case of long nanostructures is formed not only for electron conductivity, but also for electrolyte transfer channels along the additive – Fig. 3 shows smaller diffusion (close to horizontal) part of the impedance curve for nanotubes and nanofibers, than for carbon black. According to Fig. 2, SWCNT-based electrode lost only 10% of its capacity during transition from 25 mA/cm^2 to 167 mA/cm^2 cycling current density, compared to 12% for Vulcan and 18% for Taunit. So SWCNT in electrode structure makes electrodes less sensitive to power rate increase, than in case of carbon blacks.

As a whole, specific capacity of nanotubes and nanofibers (Figs 2,4) increased in case of their optimal concentration because of the fact, that less mass of conductive additive (with low surface area) allowed to increase portion of activated carbon having larger surface area and providing additional capacity.

The results obtained during long-term stability tests shows that capacity losses are smaller for carbon black than for nanotubes and nanofibers. But during short-term high current tests situation is reverse (Table 2). The possible cause is that accumulating of gas phase during electrolyte or active carbon degradation during cycling can occur in larger volumes for nanostructure-based electrodes due to higher amount of activated carbon – possible catalyst for electrolyte electrolysis. Additional problem for Taunit can be residual transition-metall based carbon fiber synthesis procatalyst. which also accelerates electrolyte degradation. Supplier data shows that CNFs contains about 95 mass.% of carbon and SWCNT – about 99 mass.%, though capacitive (upside) part of impedance curve on Fig 3 doesn't demonstrate larger tilt angle for CNF, than for SWCNT or carbon black.

Economic estimation shows that despite lower cost, CNFs are less competitive with carbon black than SWCNT – much higher load of CNFs demanded. In addition, Tuball is supplied as uniform slurry of SWCNT in deionized water and CNFs – as powder-like material that can influence uniformity of conductive additive distribution in electrode structure during its formation. High-quality SWCNT can be competitive with traditional carbon blacks, especially in high-power applications.

5. Conclusions

High-quality SWCNTs are competitive with carbon blacks as conductive additives to supercapacitors electrodes due to larger amount of activated carbon which substitutes carbon black, increasing electrode active area. SWCNTs also can provide additional ion transport channels along themselves, which is necessary for better operation at high power rates. CNFs provide worse performance than SWCNT due to their lower aspect ratio. Higher optimal load of CNF worsens economic competitiveness of such electrodes compared both with SWCNT and carbon black. Larger share of activated carbon in SWCNT and CNF – based electrodes can increase degradation rate in long-term stability due to larger amount degradation centers on carbon surface.

References

[1] B.E. Conway, *Electrochemical Supercapacitors—Scientific Fundamentals and Technological Applications* (Kluwer, New York, 1999).

[2] X. Andrieu, Energy Storage Syst. Electron.: New Trends Electrochem. Technol., **1** (2000) 521.

[3] M. Carignano, R. Costa-Castelló, N. Nigro, S. Junco, IFAC-Papers On Line, **50**, 1 (2017), 10052-10057.

[4] G. L. Lopez, R. S. Rodriguez, V. M. Alvarado, J.F. Gomez-Aguilar, C. Sandoval, Appl. Energy, **205** (2017), 1478-1494.

[5] C.W.Liew, S.Ramesh, A.K.Arof, Mater. Des. **92,** (2016), 829–835.

[6] I. Shakir, Electroch. Acta, **129,** (2014), 396–400.

[7] B. Xu, F. Wu, Y. Su, G. Cao et al, Electroch. Acta, **53,** (2008), 7730–7735.

[8] M. Hwang, J. Oh, J. Kang, K.Seong, Y.Piao, Electroch. Acta, **221,** (2016), 23–30.

[9] I. Shakir, Electroch. Acta, **129**, (2014), 396–400.

[10] I. V. Yanilkin, A.A. Sametov, I. N. J. Prikladn. Ch., **88**, 7, (2015). 36.

[11] A. Volperts, G. Dobele, A. Zhurinsh, Z. Zalane et al, Bulgar. Chem. Comm., **48**, Sp. Is. E (2016), 337 - 341.

[12] YAROSTANMASH LLC Home page: yarst.org

[13]http://www.shipmentsdata.com/latest-india-import-data/black-carbon/pol-united-states.php, 2017/04/17

[14] Tambov Nanotechnology Center Official website, http://www.nanotc.ru/producrions/87-cnm-taunit, 2018/02/09

Thin-Film And Crystalline Photovoltaic Modules – Outdoor Performance And Economic Estimation

E.A. Kiseleva[1,*], O.S. Popel[1], A.B. Tarasenko[1]

[1] Joint Institute for High Temperatures of the Russian Academy of Sciences, Izhorskaya 13 Bldg 2, Moscow 125412, Russia

Email: kanna787@mail.ru

Abstract. Thin film photovoltaic (PV) technologies were quite popular in 2005-2010 due to relatively low modules costs and low need in silicon or other materials. Amorphous silicon (a-Si) thin films are well known as PV active material with wider light absorption bandgap than for crystalline silicon which allows better module operation in case of diffuse radiation. Low PV energy conversion efficiency (7-11% against 16-19% for crystalline) is a main drawback of such modules. CuInGaSe2 (CIGS) thin films have no such features for absorption bandgap, but have large efficiency (13-14% for PV module and 20,4% for laboratory scale cell). It is also known, that thin film modules have much lower temperature power coefficient, which makes their operation more reliable at hot weather. Nowadays fast progress in crystalline silicon PV technologies led to lower cost for crystalline modules and made thin film technologies less competitive. In Joint Institute for High Temperatures comparative field tests of several thin film and crystalline modules were conducted in 2015-16, using small autonomous power units (PV module, charge controller, small accumulator for charge controller feed and resistive load) as a test bed. Specific energy production for each module throughout different months and the whole test period (10.2015 10.2016) was used as a main criteria for comparison. Economic efficiency (energy production to capital costs ratio for each module type) was also estimated.

1.Introduction

Fast photovoltaic growth in the beginning of the XXI century faced lack of the most widespread raw material for PV cells – solar grade polycrystalline silicon. Raw materials problem gave rise to many alternative technologies for solar cells production without polycrystalline silicon at all or with its decreased consumption. Solar cells based on thin films of amorphous Si, CdTe, $CuInGaSe_2$ (CIGS) were very popular that time despite their smaller efficiency that for crystalline silicon cells [1]. For 2008-2011 due to fast polycrystalline silicon growth, mainly due to Chinese suppliers, competitiveness of thin film technologies began to decrease – huge production capacities for polycrystalline silicon allowed not only to solve material lack problems, but also to decrease prices for efficient crystalline solar cells significantly [2]. Such global changes in the market made thin film technologies developers and producers to make efforts to increase efficiency of their modules. Oerlikon Solar and Applied Materials Inc introduced into the market tandem amorphous/microcrystalline technology for their amorphous silicon modules (a-Si), increasing module efficiency up to 8-9% (instead of pure amorphous silicon used before) and with wider light absorption band due to microcrystalline layer (mc-Si) [3]. This layer also allowed to decrease Stoebler-Wronsky effect – light-induced degradation

of hydrogenated amorphous silicon layer because of hydrogen loss. Plasma-enhanced chemical vapour deposition (PE-CVD) is used as basic technology for such modules production.
USA-based company Stion in 2010 introduced its new CIGS modules and later created 100 MW/year production factory in Taiwan, licensing technology to Taiwan Semiconductor Manufacturing Corporation [4].Stion's CIGS module has efficiency of 12-13% and is produced by magnetron sputtering of active material.
More popular multicrystalline silicon modules had efficiency 13-14% that time and up to 16% in 2017 and up to 17% for monocrystalline silicon modules [Goodrich]. For advanced monocrystalline silicon modulesbased on such technologies as Sun Power Interdigitated Back Contacts (IBC) and Panasonic HIT (Heterojunction with intrinsic thin layer) module efficiency can achieve 19-20%, though prices for such modules are higher due to quality of silicon wafers. The main problem for thin-film photovoltaic developers and producers was not efficiency but drastic price decrease for multi- and monocrystalline silicon PV modules (μc-Si and mono-Si) – thin film modules based on sophisticated functional layer technologies had very low raw material share in cost structure. For μc-Si cells and modules with their simple and approved technology (main advances made in XXI century, were concerned with passivated emitter and rear contact technology (PERC) and its modifications introduction) the price for silicon wafer and polycrystalline silicon for its production was about 60% of PV cell cost [5]. In early 2000-s in Netherlands comparative field tests for several modules types, including crystalline and amorphous silicon thin-film based were conducted [6]. The main result was that amorphous silicon thin film modules can achieve energy productivity as high as crystalline ones though nominal efficiency for thin film technologies is approximately twice lower. This result was explained by better operation of amorphous silicon thin film modules at low and diffuse insolation conditions due to wider light absorption bandgap. In this paper this research is conducted in Moscow weather conditions for CIGS, tandem a-Si||mc-Si, mono-PERC and μc-Si PV modules to evaluate specific productivity (kWh/kW) of modern PV modules.

2. Experimental

Test bed scheme for comparative test included independent small autonomous power unit for each module. Unit included PV module, MPPT (maximum power point tracking) charge controller, lead-acid accumulator battery mainly for controller feed, and resistive load to dissipate into heat all produced energy. Otherwise there was possibility for module to utilize only part of available solar radiation. Test bed scheme is given in fig. 1.
SunSaver MPPT charge controller SS-MPPT-15L from MorningStar Corp was used for module output control. This charge controller also has information channel for operation data collection and transfer to personal computer or laptop using RS 232 interface. Data on PV module, battery and load voltage and current can be collected on PC. Data acquisition was performed once in 5 seconds. Then PV module power was calculated as

$$W_{PV} = U_{PV} * I_{PV} * \Delta t \qquad (1)$$

Here U_{PV} stands for PV module voltage, I_{PV} – PV module current, Δt – measurement time step. Parameters for tested modules are given in Table 1.PV module from Canadian Solar is made using MWT-technology (metal-wrap-through, where front contacts are

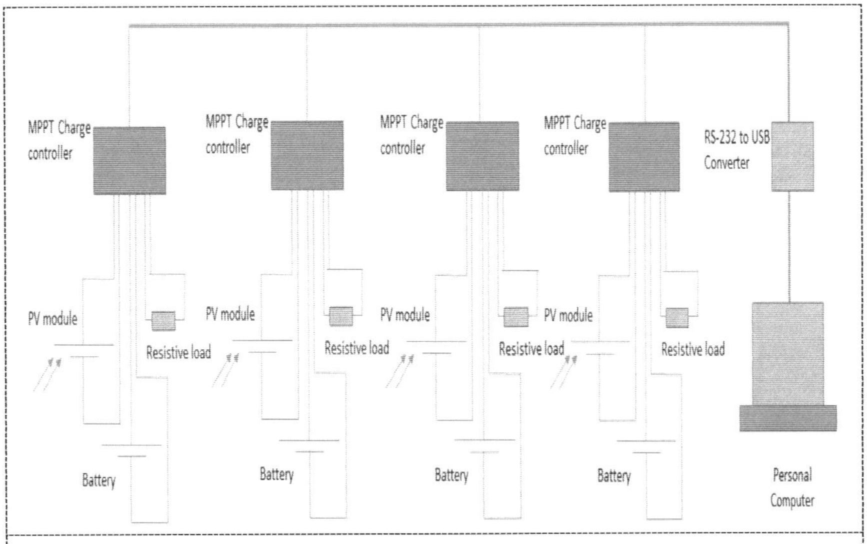

Figure 1. Test bed scheme for PV modules comparative tests.

Table 1. Tested modules parameters.

Parameters	Canadian Solar CS5A 210 MM	Green Energy Technology GET AT2	Taiwan Semiconductor Manufacturing Corporation TSMC 145A	Telecom-STV JSC TSM 200 V
Technology	Mono-Si, PERC (MWT)	Tandem a-Si‖mc-Si	CIGS	μc-Si
Open circuit voltage, V	45.5	130	65	42
MPPT voltage, V	37.9	90	0.600	34
Peak power, W (at standard test conditions)	210	125	145	200
Purchase cost, USD (for 2013)	366	214	380	360

Insolation (W/m^2) has been measured by QMS101 Pyranometer from Kipp&Zonnen, situated on the same tilt angle as PV modules. Tilt angle for all PV modules was the same and equal to 70° because of energy production in winter maximization (which is desirable for autonomous PV power systems due to load graph specific features) and winter snow coverage minimization. Specific energy production (kWh/kW) for different periods of time was estimated as

$$w_m = \frac{\sum_{\Delta t} W_{PV}}{1000 * P_{peak}} \quad (2)$$

PV energy production for each module m was calculated for the considered period of time (usually for particular month or whole year), P_{peak} stands for peak power of module at standard test conditions (STC). View of experimental test bed is given in Fig. 2. Battery voltage was used to ensure that electric load is operational, and all produced by PV modules energy is utilized (battery voltage not more than 25 V for each 24 V battery). Insolation measurements allowed to characterize weather conditions by means of daily or monthly average insolation A_d (A_m)

$$A_{d(m)} = \frac{\sum_{\Delta t} A * \Delta t}{\sum \Delta t} \quad (3)$$

Here A is insolation value, measured at Δt period of time, summing has been carried out for all Δt of considering day (A_d) or month (A_m). Comparative tests were carried out from October 2015 till December 2016

To compare economic efficiency of tested types of module, energy cost for each module type was estimated as (assuming that for 20 years energy production of each module type will be close to production from October 2015 to December 2016 in every year)

$$C_m^{sp} = \frac{c_m}{20 * W_{PV}^Y} \tag{4}$$

Where c_m is module m cost from table 1 and $W^Y{}_{PV}$ is experimentally measured energy production for module m from October 2015 till December 2016. Degradation issues are not taken into account for this estimation.

Figure 2. Experimental test bed for comparative PV modules test in Russian Academy of Sciences Joint Institute for High Temperatures, Moscow, Russia.

3.Results and discussion

Specific daily energy production at high and low average daily insolation value for selected types of PV modules is given in Fig.3. This data is a result of module energy production averaging for several days with close to each other insolation.

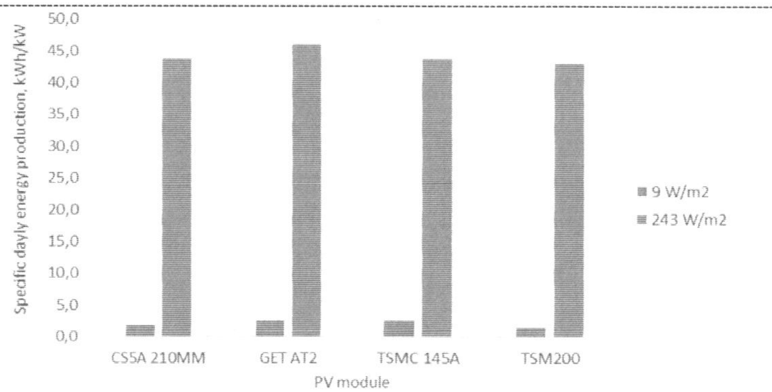

Figure 3. Specific daily energy production at high and low average daily insolation value.

From Fig.3 one can see, that a-Si||mc-Si module shows the best performance at high insolation conditions. Moreover, despite different module nominal efficiency (16.5% for CS5A 210 MM, 8.9% for GET AT2, 12.5% for TSMC145A and 14.5% for TSM 200) energy production for them at high insolation is very close to each other. The main cause can be lower power temperature coefficients for thin film modules (about 0.29-0.33%/K instead of 0.4-0.46 %/K for crystalline ones). Another specific feature is low operational current of thin film modules (1-3 A against 8-9 A for crystalline). Low operational current in this case is compensated by high voltage to conserve power level. Low currents for thin film modules are specified by contact structure – usage of transparent conductive oxides (TCO) in thin modules leads to higher resistance than in case of silver printed fingers and busbars of crystalline modules. But as a whole lower operation current means lower Ohmic losses and lower PV module operation temperature. So these 2 factors can explain high thin-film modules energy productivity at high average daily insolation value. At low insolation value a-Si layers in tandem module allows to broaden absorption bandgap. And nominal efficiency of CIGS module isn't much lower than for crystalline ones with additional advantage of low Ohmic losses due to low operation current. The worst performance at low insolation showed μc-Si module, which hasn't any additional features concerned with light absorption bandgap or construction improvements.
Fig.4 shows specific energy production for the whole measurements period.

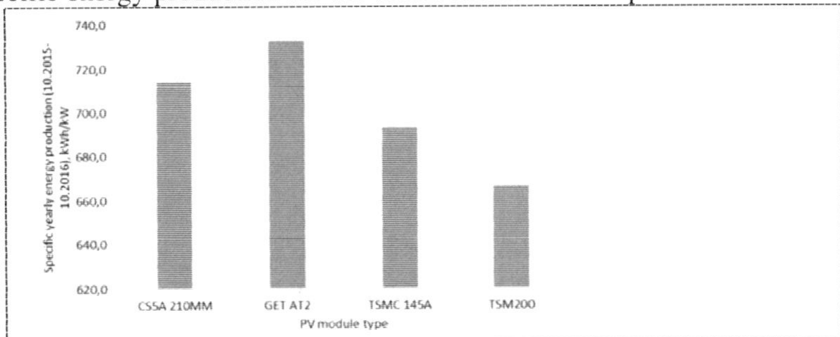

Figure 4. Specific yearly energy production for selected modules types (10.2015-10.2016).

Tandem module showed the best performance. Possible reasons were discussed above and combination of weather factors (temperature, insolation) throughout the year is an additional issue. According to these values, data from Table 1 and (4) specific energy cost for 20-year period was estimated. Results are given in Fig 5.

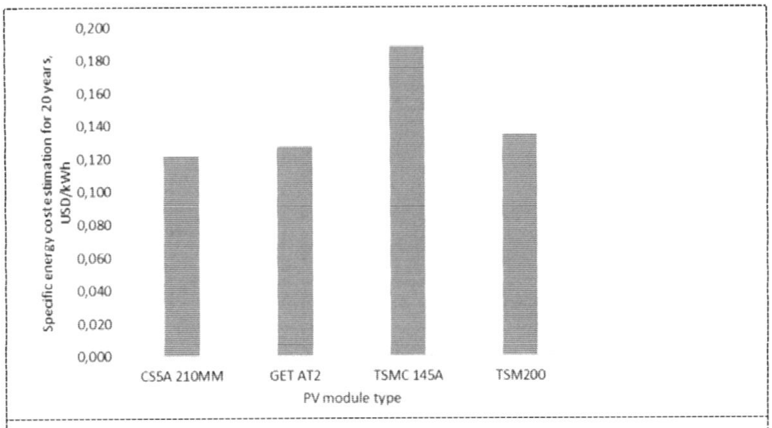

Figure 5. Specific energy cost estimation results for 20 year period.

These results showed that module price is very important factor, so important as energy production, and, may be even more. Additional capital costs, concerned with inverters, string boxes and mounting poles are not taken into account in this estimation. According to estimations from [7], high-efficient modules like monocrystalline ones will have additional cost advantages in case of real PV plants construction due to lower amount of mounting poles and string boxes.

4. Conclusions

Improved efficiency of mono- and multicrystalline PV modules and their decreased price made them more competitive to thin film PV modules. Advantage of high efficiency, specific for crystalline modules allows to decrease area of PV plant and construction cost. Despite this modern thin film modules, show good performance, comparable and in some cases higher energy productivity than crystalline ones, due to lower operation current and power temperature coefficient. Additional advantage for based on a-Si||mc-Si films modules is their wider light absorption bandgap improving operation at high share of diffuse radiation. So in some projects these modules can be competitive with crystalline, especially in case of further cost decrease for thin films technologies.

5. References

[1]V.V. Tyagi, Nurul A. A.Rahim, N.A.Rahim, Jeyraj A./L. Selvaraj, Renewable and Sustainable Energy Reviews 20 (2013), pp. 443–461
[2] E. Płaczek-Popko, Opto-Electronics Rev. 25 (2017), pp. 55–64
[3] Vikrant Sharma, O.S. Sastry ,Arun Kumar, Birinchi Bora, S.S. Chandel, Energy 72 (2014), pp. 536-546
[4]TSMC and Stion sign technology and supply agreement. https://www.printedelectronicsworld.com/articles/2363/tsmc-and-stion-sign-technology-and-supply-agreement
[5] A. Goodrich, P. Hacke, Q. Wang, B.Sopori, R.Margolis, T.L. James, M.Woodhouse, Sol. Energ. Mat&Sol.Cells, 114, 110-135 (2013).
[6] K.L. Chopra, P.D. Paulson, V. Dutta, Progress In Photovoltaics: Research And Applications Prog. (2004)
[7] S.V. Kiseleva, O.S. Popel, A.B. Tarasenko, R.R. Avezov, 2017, published in Geliotekhnika, 2017, No. 4, pp. 24–29

Acknowledgements

This research is supported by Russian Science Foundation (project № 14-50-00124).

Partial oxidation of aluminum powder for obtaining a controlled amount of aluminum oxide on the surface of aluminum

Grayr Ambaryan[1,*], Georgy Valyano[1], Andrey Zhuk[1], Eugeny Shkolnikov[1], Alexander Gromov[2], Sergey Zmanovsky[3] and Mikhail Vlaskin[1]

[1]Joint Institute for High Temperatures of the Russian Academy of Sciences, Izhorskaya 13 Bldg 2, Moscow 125412, Russia
[2]MISIS National University of Science and Technology, Moscow, 119049, Russia
[3]RUSAL Co., Moscow, Russia

*E-mail: ambaryan1991@gmail.com

Abstract. This work is devoted to the study of the process of partial oxidation of aluminum powder with hydrogen generation and the production of a metal-ceramic product that can be used in the production of various manufacturers such as additive manufacturing (AM) engineering technologies, machine building and so on. The released hydrogen can be effectively used in the form of an energy carrier that will contribute to reducing the cost of production and reduce the cost of the final product.

1. Introduction

In the world, additive manufacturing (AM) for the production of metal parts [1-5] is developing rapidly and will be a priority technology for the aviation industry, the space industry, and machine building. The processing parameters, such as laser power, scanning speed and powder feed rate determine the characteristics of AM components.

Technologies of growing metal parts have opened the possibility of producing new structural metal composites based on aluminum [6-7]. However, there are difficulties in using AM based on metals and their introduction into production. These include the quality of initial metal and parts, residual stresses, internal porosity.

The alloy containing aluminum is ideal for selective laser melting to manufacture complex components with high strength requirements. At present, a number of metal composite materials on aluminum matrixes reinforced with aluminum oxide have been developed. There is a problem of aluminum oxide just proportion in the metal, its wettability and technological yield of Al_2O_3. Aluminum powders with a grain size distribution of 20 to 120 μm for additive printing are used. The porosity is formed due to the powder size in the wide range. If using spherical aluminum particles with a small size range, it can obtain high quality parts by AM. This facilitates a more compact placement of particles in a certain volume and a good "flow" of the powder composition in the feed systems with a minimum resistance. The amount of aluminum powder, the ratio of Al/Al_2O_3 is important by growing metal parts.

The studying the process of aluminum oxidation in different ways and obtaining an aluminum oxide of 10 to 30 % is an important goal. Given the idea of certain content of aluminum oxide, these results suggest that there may be broader viability for additively manufactured parts. The studies of the change dynamics of the spherical aluminum particles are presented. For the analysis of the obtained samples, transmission electron microscopy and X-ray phase analysis were used [8-14].

2. Experimental

2.1. Description of the installation

The key step in the process of obtaining a partially oxidized aluminum powder is the autoclave oxidation of the aluminum powder. The properties of the product obtained depend on the oxidation parameters (reagents ratio, temperature, interaction time).

The scheme of the autoclave is shown in Fig.1. Inside the reactor (2), the reaction medium is stirred by an anchor 2-blade agitator. The speed of rotation of the stirrer is about 100 rpm. The mixer is made of SS316 steel with Teflon coating. The following devices are provided on the reactor cover: a manometer (M); gas outlet connector; water injection valve (V2) and a ball valve for injection of a suspension (V1), a resistance thermometer (T).

Heating of the reactor-autoclave is controlled by the control panel (controller). During the operation of the autoclave, the pressure, temperature and speed of rotation of the mixer are measured. Depending on the readings of pressure and temperature sensors, water is supplied to the reactor or the speed of rotation of the mixer is changed. The system of separation of water vapor is designed to condense water vapor from a mixture of steam and hydrogen leaving the reactor, return condensed water and supply hydrogen to the consumer (hydrogen-utilizing device).

The reactor vessel has a volume of 5000 ml. The vessel is vertical, cylindrical, made without welding and other connections made of SS316 stainless steel. All parts in contact with the reaction medium are made from SS316. The reactor is heated from an electric ceramic belt heater with thermal insulation and a temperature controller. The drive of the mixing device is an electric motor with alternating current with a frequency converter and a digital speed display Engine power of 0.18 kW The sealing of the stirrer shaft is carried out using magnetic coupling. Magnetic seal - with zero leakage (declared by the manufacturer), maintenance-free. The maximum torque is 40 kg*cm (3.92 N*m).

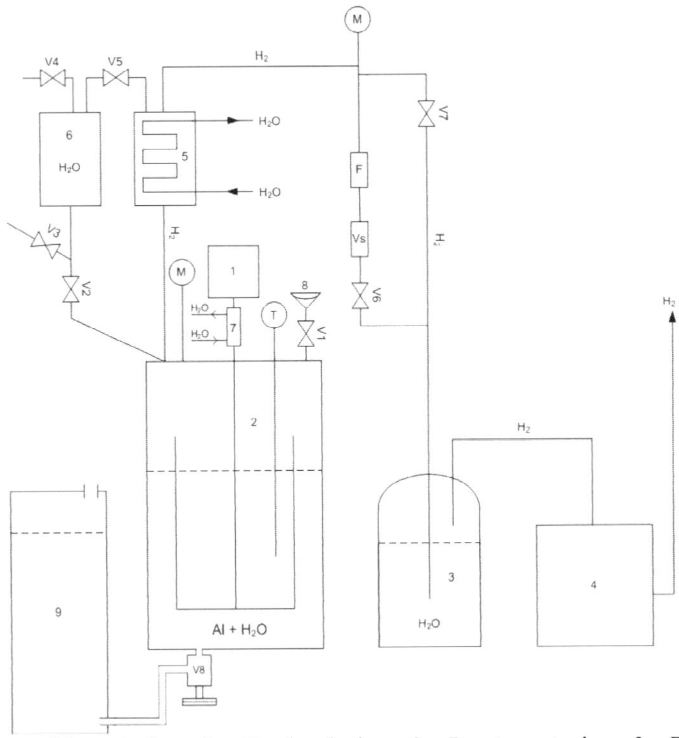

Figure 1. Scheme of the autoclave: 1 – Overhead stirrer, 2 – Reactor autoclave, 3 – Bubbler, 4 – The gas drum counter, 5 – Heat exchanger, 6 – Distilled water storage tank, 7 – Magnetic coupling with cooling circuit, 8 – Funnel for aluminum backfilling and for water filling, 9 – Product acceptance tank, M – Manometer, F – Filter, Vs – Adjustable safety valve, T – Thermocouple, V1-8 – Valves.

A mixture of steam and hydrogen comes from the autoclave reactor to the heat exchanger (5), where the water condenses and flows back into the reactor. A pressure sensor (M) is installed on the line between the heat exchanger and the pressure regulator, which measures the pressure to the regulator corresponding to the pressure inside the reactor. The heat of steam condensation leaving together with hydrogen from the reactor is transferred to the water of the second circuit. The pressure regulator (V6) maintains a certain pressure in the reactor (pressure "to itself"). After the pressure regulator hydrogen with the remaining water enters the bubbler (3), where a final separation of liquid and gaseous phases is carried out. From the bubbler hydrogen is directed to the gas drum counter (4).

2.2. Properties of the initial aluminium powder

The distribution of aluminum particles in size is shown in Fig.2. The average diameter of the particles is 40 microns.

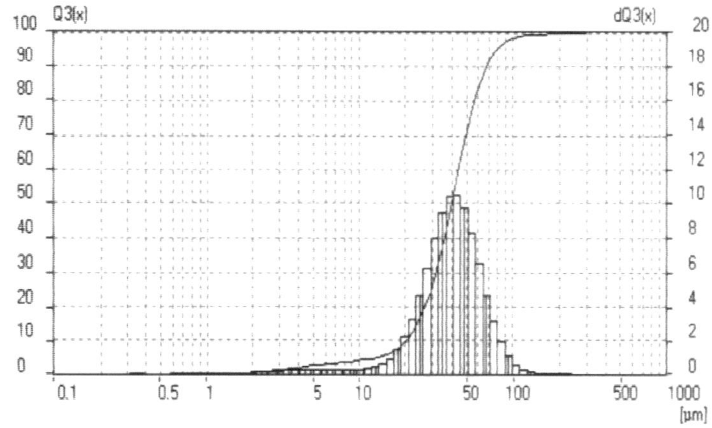

Figure 2. A histogram of particle size distribution for a sample of the initial aluminum powder.

The initial aluminum powder contains 770 μg/g of iron (0.077%). Also, based on the obtained elemental composition (analysis of the sample of aluminum was determined by mass spectrometry with inductively coupled plasma using a mass spectrometer with inductively coupled plasma ThermoScientific X-2), it follows that the total aluminum content in the powder was 99.89%.

A micrograph of the powder obtained with the JEOL JSM-7407F scanning electron microscope is shown in Fig.3. The image show that the initial aluminum powder consists of agglomerates. The size of the structural units of these agglomerates is in 1-10 microns. The particles have a relatively smooth surface, on which there are various defects (Fig.3).

Figure 3. Initial aluminum powder obtained. Scanning electron microscope image at approx. 300 μm magnification.

2.3. Description of the process of obtaining partially oxidized aluminum powder

The production of partially oxidized aluminum powder takes place according to the scheme shown in Fig.4.

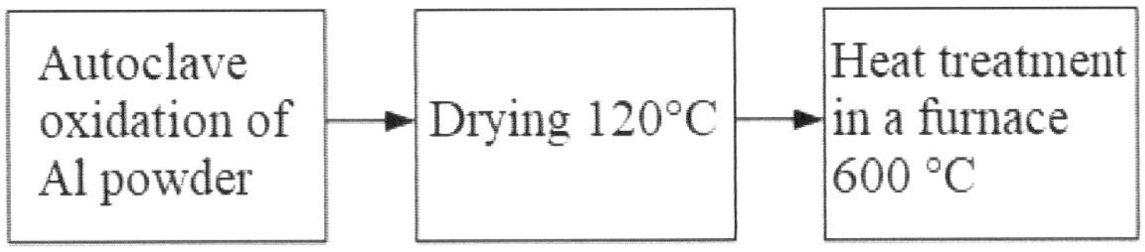

Figure 4. A block diagram of technology for partially oxidized aluminum powder production.

As can be seen from the scheme, the process of obtaining partially oxidized aluminum powder consists of three main stages: oxidation, drying and heat treatment.

Before the oxidation stage the weighing of initial aluminum powder and water is preceded. The mass ratio of water to aluminum in the suspension is selected depending on the required degree of oxidation. The distilled water was used. The prepared suspension is poured into the autoclave through the ball valve. Further, the ball valve closes. The reactor heating and stirring of the reaction space are activated. The reactor is heated by an electric heater at a given speed to a certain temperature. Throughout the experiment, the readings of pressure and temperature sensors inside the reactor, as well as the volume of hydrogen released, are monitored.

If the temperature inside the reactor exceeds a certain level, water from the distilled water storage tank (6) is added to the reactor. During the oxidation process, the hydrogen formed together with the water vapor generated inside the reactor by the thermal energy of the exothermic aluminum oxidation reaction enters the heat exchanger. In the heat exchanger, the water condenses and returns back to the reactor. When the pressure inside the reactor reaches a certain value, the pressure regulator is activated, and hydrogen from the heat exchanger enters the bubbler. Solid reaction products are removed with water from the bottom of the reactor through a shut-off valve (8) into the tank receiving the oxidation product. For more complete removal of reaction products from the reactor and its washing and cooling, additional water is supplied to the reactor from the distilled water storage tank. After that, stirring is switched off.

The resulting oxidation of the slurry is partially oxidized aluminum powder is divided into stages of deposition. As a result of sedimentation is the deposition of aluminum particles on the bottom of the tank. The phase separation takes place by decantation.

The next technological stage is drying of the oxidized aluminum powder obtained in the autoclave. To do this, the wet aluminum powder is poured into pre-suspended stainless steel tanks, weighed and placed in the Binder VD drying oven at a temperature of 120 °C.

The final stage of the process of obtaining partially oxidized aluminum powder is heat treatment. Heating of the product was carried out in the muffle furnace Nabertherm LHT 08/16 at a temperature of 600°C for 5 hours.

The result is a product that is an aluminum core coated with a porous layer of γ-Al_2O_3.

Two products with different oxidation degree were produced in this work. Those products were produced in different regimes. In the first regime, the oxidation of aluminum powder was carried out when at a temperature in the autoclave did not exceed 120 °C, the holding time in the reactor – 60 minutes In the second regime, the temperature of aluminum powder in the autoclave did not exceed 180 °C, the holding time in the reactor – 120 minutes. In first experiment the reactor was loaded by 900 g of aluminum and after 60 minutes 80 liters of hydrogen were produced. In second experiment the reactor was loaded by 1500 g of aluminum and after 120 minutes 300 liters of hydrogen were produced.

3. Results and discussion

3.1. Particle size distribution

Fig.5 shows the particle size distribution of partially oxidized aluminium. The particle diameter is in the range of 2 – 200 µm. The average diameter of the particles is 41 microns, which corresponds to the average size of the particles in the initial aluminum powder

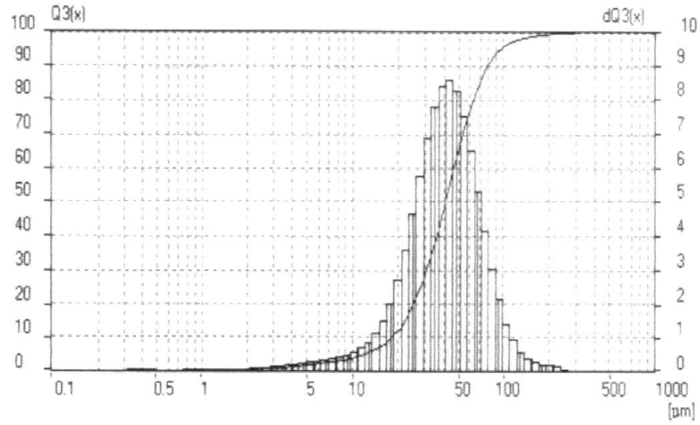

Figure 5. The histogram of distribution of particle size for the partially oxidized sample.

3.2. Microscopy and the generation of hydrogen in experiments

Fig.6a, 6b shows the product obtained in experiment with the oxidation of aluminum powder at a temperature of 120 °C, the holding time in the reactor of 60 minutes, and hydrogen yield of 80 liters.

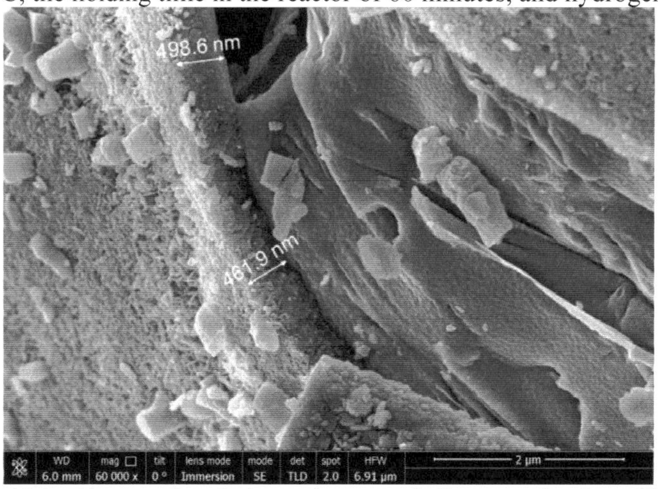

Figure 6a. Image obtained by scanning electron microscopy in the approximation of 2 microns for a sample with a hydrogen output of 80 liters.

The images show that the relatively smooth surface of the aluminum powder was coated with a layer of aluminum oxide with a thickness of up to 500 nm (Fig.6a, 6b).

The image for product of aluminum powder oxidation at a temperature of 180 °C, the holding time in the reactor of 120 minutes, and hydrogen yield of 300 liters, is shown in Fig.7. It shows that the oxidation product is covered by oxide film with the thickness of approximately 1 micron.

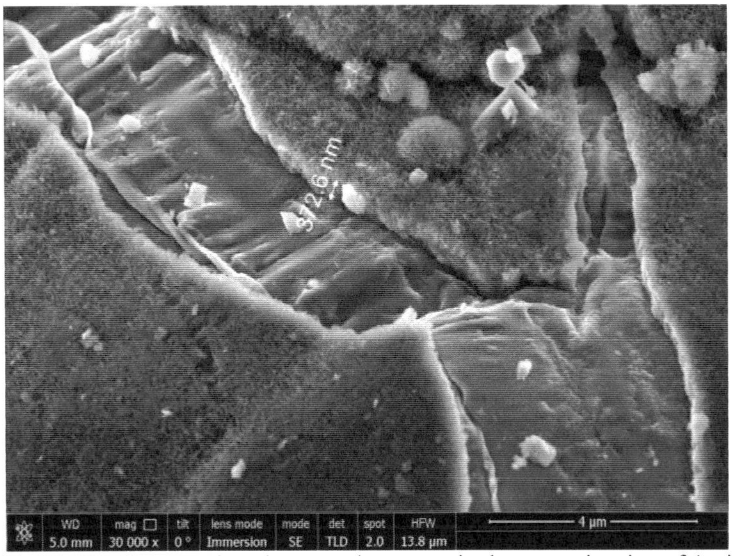

Figure 6b. Image obtained by scanning electron microscopy in the approximation of 4 microns for a sample with a hydrogen output of 80 liters.

Figure 7. Image obtained by scanning electron microscopy in the approximation of 4 microns for a sample with a hydrogen output of 300 liters.

3.3. The results of x-ray analysis

The main objective of the x-ray analysis is to identify the different phases in their mixture based on the analysis of the diffraction pattern formed by the sample under study. Determination of the substance in the mixture is carried out by a set of its interplane distances and relative intensities of the corresponding lines on the radiograph. The x-ray phase analysis was carried out on the x-ray diffractometer DRON-2 (made in USSR). As a result of the analysis of the obtained patterns, it was found that, in addition to metal aluminum, traces of the presence of fine γ-Al_2O_3 were registered in the sample. The rest of the material is in amorphous (or x-ray amorphous) state.

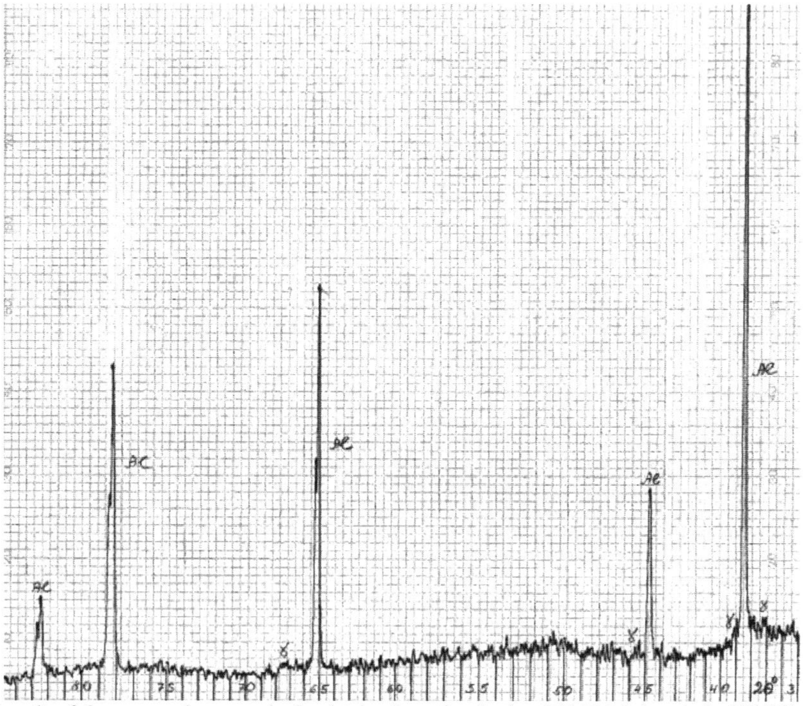

Figure 8. Result of the x-ray phase analysis obtained on the device DRON-2 (made in USSR), where the axis «X» is 2Θ for CuKα radiation, and the axis «Y» is radiation intensity.

4. Conclusion

Two samples of partially oxidized aluminum powder were produced in reactor-autoclave in distilled water. The first sample was oxidized with hydrogen output of 80 liters obtained from oxidation of 900 g of aluminum and the second sample was oxidized with hydrogen output of 300 liters obtained from oxidation of 1500 g of aluminum. According to the microscopic analysis it was shown that the products keep the spherical shape, and the thin layer of aluminium oxide (after heat treatment at 600 °C) is formed on the surface of aluminium. According to the results of x-ray analysis it was shown that products contain two crystallic forms: aluminum and -Al_2O_3 that is predominantly in amorphous state.

Acknowledgements

The study was supported through a grant of the Russian Science Foundation (project No. 14-50-00124).

The authors are grateful to Tatyana Borodina for performing X-ray phase analysis.

References

[1] Daniel R.Eyers, Andrew T.Potter, Computers in Industry, Volumes 92-93, Industrial Additive Manufacturing: A manufacturing systems perspective, 11 (2017)

[2] Erwin Rauch, Marco Unterhofer, Patrick Dallasega, Manufacturing Letters, Industry sector analysis for the application of additive manufacturing in smart and distributed manufacturing systems, 12 (2017)

[3] Y.KokX.P.Tan,P.Wang,M.L.S.Nai, N.H.Loh, E.Liua, S.B.Tor, Materials & Design, Anisotropy and heterogeneity of microstructure and mechanical properties in metal additive manufacturing: A critical review, 02 (2018).

[4] Dirk Herzog, Vanessa Seyda, Eric Wycisk, Claus Emmelmann, Acta Materialia, Additive manufacturing of metals, 9 (2016), p: 371-392

[5] Stefan Szemkus, Bernd Kempf, Simon Jahn, Journal of Materials Processing Technology, Laser additive manufacturing of contact materials, 2 (2018), p: 612-617
[6] M.S. Vlaskin, A.V. Grigorenko, A.Z. Zhuk, A.V. Lisitsyn, A.E. Sheindlin., E.I. Shko'lnikov, High Temperature 54, 3 (2016)
[7] Sergey Vorozhtsov, Alexander Vorozhtsov, Olga Kudryashova, Ilya Zhukov, Powder Technology, Structural and mechanical properties of aluminium-based composites processed by explosive compaction, 313, 5 (2017)
[8] Shuai Ma, Enze Xu, Zhifeng Zhu, Powder Technology 326, 2 (2018).
[9] A.Z. Zhuk, M.S. Vlaskin, A.V. Grigorenko, S.A. Kislenko, E.I. Shkolnikov, Journal of Ceramic Processing Research, 17, 9 (2016)
[10] A.V. Lisitsyn, L.A. Dombrovsky, V.Y. Mendeleyev, A.V. Grigorenko, M.S. Vlaskin, A.Z. Zhuk, Infrared Physics & Technology, 77 (2016)
[11] S.A. Kislenko, M.S. Vlaskin, A.Z. Zhuk, Solid State Ionics, 293 (2016)
[12] A.O. Dudoladov, O.A. Buryakovskaya, M.S. Vlaskin, A.Z. Zhuk, E.I. Shkolnikov, International Journal of Hydrogen Energy 41, 4 (2016)
[13] M.S. Vlaskin, A.O. Dudoladov, O.A. Buryakovskaya, G.N. Ambaryan, International Journal of Hydrogen Energy, 4-2 (2018)
[14] G.N. Ambaryan, M.S. Vlaskin, A.O. Dudoladov, E.A. Meshkov, A.Z. Zhuk, E.I. Shkolnikov, International Journal of Hydrogen Energy 41, 39 (2016)

Influence Analysis of Different Capacity Reduction Policies in Annual Power Generation Planning under Electricity System Reform in China

Chen Zhang[1]**, Bo Hu**[1]**, Cuifen Bai**[1]**, Huanhuan Li** [2]

[1]State Grid Energy Research Institute, Beijing, China
[2]Institute of Energy Economics and Environment, North China Electric Power University, Beijing, China

e-mail: zhangchen@sgeri.sgcc.com.cn, hubo@sgeri.sgcc.com.cn, baicuifen@sgeri.sgcc.com.cn, 942893074@qq.com

Abstract. In China, the new round of electricity market reform requires power generation to change from traditional planned power generation management into market-driven contract power generation. In the transition period, the remaining power generation plan management bears with tasks of guaranteeing power grid safety and stable operation and balancing supply and demand under the impact of electricity market uncertainty. Capacity reduction is an important policy means in plan management to inhibit power plants' vicious competition and help the electricity market steadily development. This paper researches different capacity reduction mode of the electricity market reform pilots, compare and analysis the influence of the capacity reduction modes by modeling and simulating, and proposes some suggestion for capacity reduction management in annual planning during the electricity reform transition period.

1. Introduction

Before the power system reform, the State Grid and the China Southern Power Grid as well as some local wholesale power grid provide electricity for multiple customers. From 2002 Power System Reform Program (Document No.5), power plants and grid separated into independent units. But the electricity market had not been built after that reform. Annual power generation plan is made by the corresponding grid company planning department and the scheduling department adjusts it according to the actual operation power balance constraints. Therefore, before the new round of power system reform, which started in 2015, power generation volume of each power plant is still decided by the power grid mainly in the form of power generation plan.

Power generation plan is an important link in grid dispatch and operation management, reasonable power generation plan can maximize the use of grid resources and ensure the safety and economical operation of the power grid [1]. Different cycles of power generation plans have different characteristics. Short-term power generation, namely day-ahead plan and real-time plan, depends on the short-term system forecast and scheduling and with high accuracy [2]-[4]. Long-term (annual) power generation plan is the basic of electricity supply and demand balance and electricity trading between the power grid and power plants. Medium-term (monthly) power generation plan is the connection link between the long-term and short-term plan. Under the management mode that the power grid unified purchase and sale electricity, the full cycle unit combination state and the unit load rate is the basis of coordination and optimization coordination and optimization. By coordinating different time series power generation plans, the convergence and association among different cycle power generation plan is achieved [5]-[8].

The release of the Document NO.9 (Some Opinions on Further Deepening the Reform of Electric Power System) put forward the reform objectives of carrying out effective competition in both power generation side and the sale side, cultivating independent market players, and build a power trading pattern with diversified body and orderly competition. Then the release of a series of supporting documents not only influenced the interest pattern in the sale side, but also breaking the original power generation plan management cycle, as well as proposed new requirements for the power grid to adjust their plan management mode to fit the new development situation. The trading objects of power plant changes from the power grid and local government into terminal customers and independent sales company. The former trading object pursuits power grid safety and stable operation, local power plants economic balance and social sustainable development. The latter trading object mainly pursuits its own access to high quality safe and reliable power supply. The change of trading object's demand and the oversupply relationship make the power plants show the trend of polarization. The utilization hours of efficient units would increase rapidly, along with a substantial increase in profits. Small, old or inefficient units, lacking of market competitiveness, would face with great economic losses or directly out of the market. In the transition period of the traditional planning mode into the market mode, maintaining the basic stability of the power industry is very important. Especially in the early period of building electricity market, vicious bidding, interest conflict among power plant, customer and the power grid world hinder the orderly electricity market development and bring uncertainty for power supply and demand balance.

Currently, the reform pilots are avoiding the above problems by controlling the released power generation volume, namely market trading cap. For example, in the Beijing - Tianjin - Tangshan regional power grid the market trading cap is 20% of the local electricity consumption in 2016 and the actually implementation ratio is less than 10% at the end of the year. The market process requires the market trading cap increase, except release the power generation volume step by step, plan capacity reduction is another effective measure to adjust different power plants' interest distribution and maintain the basic stability in the power generation side.

The concept of capacity is proposed in the management of large-customer and power plant directly trading. In November 2009, the National Electricity Regulatory Commission published the document of "Power customer and power plan directly trading pilot basic rules (Trial)" [9], which stipulates the power plants which obtained direct trade qualification should reduce their capacity in annual plan according to their contract power volume, and the corresponding dispatch organization would not allocate power plan to this part of capacity. However, this rule has not been implemented nationwide. Some provinces cancelled the capacity reduction or replaced it with different generation power reward rules [10]. Reduce or not reduce the corresponding capacity makes big difference on the allocated power generation volume in annual plan of each power plant directly and influence market supply and demand relationship indirectly. Unreasonable capacity reduction mode may bring arbitrariness and chaos into local power system reform and hinder power industry orderly development.

Currently, the attitudes and measures of the electric reform pilots on this point are not the same. This paper combings different capacity reduction mode in the power reform transition period and compare the influences of capacity reduction on power plant, customer and the power grid by building annual planning model and benefit analysis mode.

2. Market first and plan First Management Mode

As of the end of August 2016, there are 15 electric power system reform comprehensive pilots, 18 transmission and distribution price pilots and 5 sales-side reform pilots all around the country.

Electric generation plan release is the foundation of pilot operation. Therefore, actually every pilot is facing the problem of plan reduction problem. But they took different measures on it, which can be divided into market first and plan first management modes.

2.1. Market First Management Mode

In market first management mode, the power generation plan steps are drawn in Fig. 1:

- Partial power is preempted for market transactions;
- Plan reduction according to the market trading power;
- Calculate renewable and heating power as the guaranteed power generation, then deduct the guaranteed power generation from the remaining planned electricity.
- Assign the remaining planned electricity to each unit according to the principles of power generation planning, and in accordance with the reward and punishment mechanism, the unit maintenance arrangements to adjust the unit hours of power generation.

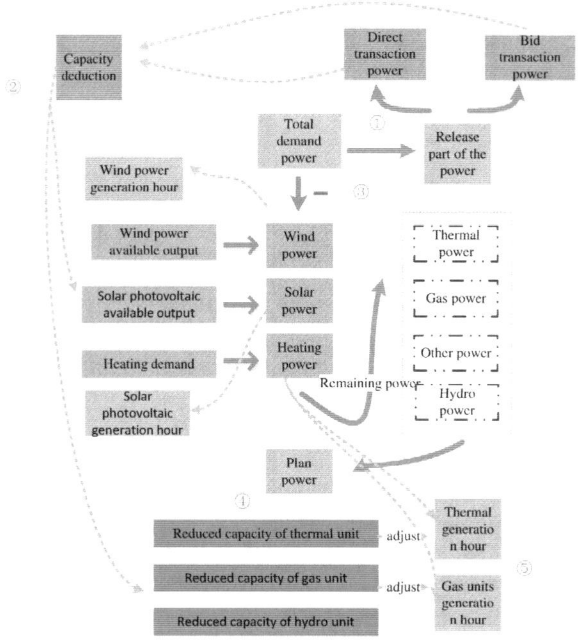

Figure 1. Market first management mode.

2.2. Plan First Management Mode

In plan first management mode, the power generation plan steps are drawn in Fig. 2:

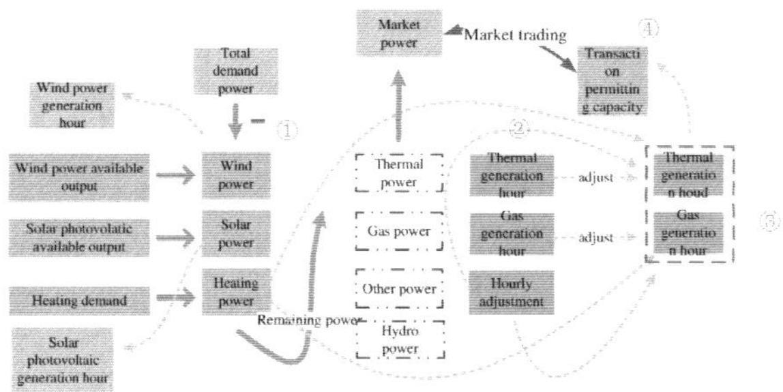

Figure 2. Plan first management mode.

- Calculate the necessary power generation hours of each unit according to the government policies, inter - provincial tie line transaction, wind power, solar photovoltaic power generation as well as heating power.
- Limit the market transaction permitting capacity of the units that participated in annual planning according to their allocated power generation hours.

- The remaining power generation, namely the total demand minus the planned power, is placed in the market for competition.

For the current Chinese electricity market, the market first management is more suitable, from the aspect of power grid security, power balance and the acceptance of the parties. Therefore, this paper discusses the different reduction policies under the market first management mode.

3. Different Plan Reduction Policies

Under the market first management mode, the transaction results would influence annual plan, which reflects as plan reduction. Because thermal units with big installed capacity are generally more efficient than that with small installed capacity, the plan allocation is made according to the effective capacity level of each unit, which is called differential generation hour (1).

$$T_i^{PLAN} = f_1\left(C_i^{PLAN}\right) \tag{1}$$

where in T_i^{PLAN} is the allocated power generation hour of unit *i*. C_i^{PLAN} is the effective capacity of unit *i* that participate into power generation plan. $f_1()$ is the calculation function to get the allocated power generation hour of unit *i* according to its effective capacity.

According to the actual planning principle units are differentiated into different level, and the allocated hour is determined the capacity level of each unit. A level would be chose as the basic level and other levels cloud calculate their power generation hour based on the allocated power generation hour of the basic level. Then (1) could be detailed as (2):

$$T_i^{PLAN} = \begin{cases} \left[INT\left(\dfrac{C_i^{PLAN} - C_0}{\Delta C}\right) - 1\right] \times \Delta T + T_0 & , C_i^{PLAN} \leq C_0 \\ INT\left(\dfrac{C_i^{PLAN} - C_0}{\Delta C}\right) \times \Delta T + T_0 & , C_i^{PLAN} > C_0 \end{cases} \tag{2}$$

where in T_0 is the allocated power generation hour of the basic level units, which is calculated by supply and demand balance. ΔT is the hour difference between two adjacent levels. ΔC is the level step of power generation capacity.

Different plan reduction method makes different influence on. As introduced in chapter 1, market transaction result would influence the effective capacity in plan. And may also be influence by other units in the same area, total power generation volume and total market transaction power volume and so on. The relationship between the effective capacity in plan and the factors could be expressed as (3).

$$C_i^{PLAN} = f_i\left(C_i, Q_i^{MARKET}, Q^{MARKET}, C^{MARKET}\right) \tag{3}$$

where in $f_i()$ is the calculation function that determined by local policy. is the installed capacity of unit *i*. Q_i^{MARKET} is the market transaction volume of unit *i*. Q^{MARKET} is the total market transaction volume C^{MARKET} is the total installed capacity of the units that participated into market.

According to different plan reduction policies, (3) would be different in different areas. Therefore, (3) is one of the core point in this model, this paper would detail equation in chapter IV. And the following of this chapter introduces the calculation of T_0 in (2) to complete the annual planning method. The value of T_0 is determined by power supply and demand balance.

Recent years, to response to the energy-saving policies, encourage renewable energy development, and encourage power generation units stable output, many of the plan management departments have already given priority to renewable power units like wind power and nuclear power, peak-valley shifting units like pumped storage power station. And reserve some power plan for reward and punish. Therefore, the annual plan could be divide as (4):

$$Q^{TOTAL} = Q^{RENEWABEL} + Q^{RESERVE} + Q^{PLAN} + Q^{MARKET} \tag{4}$$

where in Q^{TOTAL} is the total power generation volume that the plan management department used for allocates. $Q^{RENEWABEL}$ is the total protected acquisition of renewable energy, which is determined to the protected power generation hour of the corresponding resource area and the actual power generation capacity. $Q^{RESERVE}$ is the reserved power plan determined by the plan management department according to the historical data Q^{PLAN} is the total power plan for other units that dispatched by the State Gird and not have the priority.

In (4) and would be forecasted before annual planning and market transaction would finish at the end of last year. Then Q^{TOTAL}, $Q^{RENEWABEL}$, $Q^{RESERVE}$ and Q^{MARKET} could be regard as constant, and Q^{PLAN} could be calculated. The relationship between Q^{PLAN} and T_i^{PLAN} is:

$$Q^{PLAN} = \sum_{i=1}^{I} T_i^{PLAN} \times C_i^{PLAN} \tag{5}$$

where in I is the number of the units to allocate plan.

There are three main kinds of measures on plan reduction, namely, the same proportion reduction (SPR) method, hours by user reduction (HUR) method and coefficient reduction (CR) method. The SPR method means the capacity of all the power generation units that participated into the market would be reduced with the same proportion. The reduction production is determined by total trading electric volume, market equivalent utilization and total capacity of the power generation units that participated into the market. The HUR method means the plan reduction degree varies with the trading electric volume and the customer equivalent utilization hour last year of each power generation enterprise. The CR method means the capacity would be reduced according to the trading electric volume and the given coefficient, with is determined by the local government.

In SPR method a hypothetical parameter is defined, which is called market equivalent, as (6):

$$C^{MARKET} = \frac{Q^{MARKET}}{T^{MARKET}} \tag{6}$$

where in T^{MARKET} is the market equation hour, which is determined by local government according to the historical data, for example in 2016 the Shandong province set it to be 6000h.

All the units that participated into market would reduce their effective capacity with the same proportion. The proportion is calculated by (7):

$$p^{SPR} = \frac{C^{MARKET}}{\sum_{i=1}^{I} C_i} \tag{7}$$

Then in this method (3) in chapter changes into:

$$C_i^{PLAN} = C_i \times (1 - p^{SPR}) = C_i \times (1 - \frac{\frac{Q^{MARKET}}{T^{MARKET}}}{\sum_{i=1}^{I} C_i}) \quad (8)$$

In HUR method customers' utilization hour is the main factor that influences C_i^{PLAN}. Assume unit *i* reached deal with *J* customers. The definition of customer *j*'s utilization hour is:

$$T_j = \frac{Q_j}{C_j^{\max}} \quad (9)$$

where in Q_j is the total power consumption volume last year of customer *j*. $C_j^{\max}$ is the max demand capacity last year of customer *j*.
Unit *i* should reduce its effective capacity in plan considering all the transactions among *J* customers:

$$\Delta C_j = \sum_{j=1}^{J} \frac{Q_{ij}^{MARKET}}{T_i \times \gamma} \quad (10)$$

where in Q_{ij}^{MARKET} is the transaction power volume between unit *i* and customer *j*. γ is an adjustment coefficient that determined by government to adjust the reduction degree, for example in 2016 the Sichuan province set it to be 2.
Then in this method (3) changes into:

$$C_i^{PLAN} = C_i - \Delta C_i = C_i - \sum_{j=1}^{J} \frac{Q_{ij}^{MARKET}}{\frac{Q_j}{C_j^{\max}} \times \gamma} \quad (11)$$

In CR method plan reduction only influenced by policy and themselves market transaction volume and last year power generation hour.

$$\Delta C_j = \frac{Q_i^{MARKET}}{T_i^{last}} \times \mu \quad (12)$$

where in T_i^{last} is the power generation hour of unit *i* in the last year. μ is the capacity reduction coefficient and $\mu \in [0,1]$,which is determined by local policy, for example in 2016 the Beijing , Tianjin and North Hebei area set it to be 0.5.
Then in this method (3) changes into:

$$C_i^{PLAN} = C_i - \Delta C_i = C_i - \frac{Q_i^{MARKET}}{T_i^{last}} \times \mu \quad (13)$$

4. Multi-stakeholder benefit Analysis

The electric power system reform directly changes the benefit of the power grid, power user and power unit. According to the above model different capacity reduction policy would make differences on the annual plan. This chapter builds benefit analysis modes for power user, power unit and the power grid to analysis the influence of different capacity reduction policy on those stakeholders.

4.1. Benefit Analysis of Power User

Benefit change of power user mainly reflected in electricity purchase change, which can be described as:

$$\Delta\pi^{user} = (Q^{p0} - Q^{s0})\cdot p_b - \sum_{i=1}^{I} Q_i^m \cdot (p_i^m + p^{trans}) - (Q^{p0} - Q^m - Q^s)\cdot p_b \quad (14)$$

$$Q^s = (Q^{p0} - Q^m) * \left[1 - (1-\eta_l)\cdot(1-\eta_p)\right] \quad (15)$$

where in p^{trans} is the average transmission and distribution price. Q^{s0} and Q^s are the loss power before and after the power system reform, respectively. η_l and η_p are the average line loss and plant power consumption rate in the corresponding area, respectively. p_b is the average catalog price.

4.2. Benefit Analysis of Power Unit

Before the power system reform, the interest of power unit is entirely up to the plan, which can be calculated by:

$$\pi_i^0 = R_i^0 - O_i^0 = Q_i^{p_0} \times \left(p_i^p - o_i\right) \quad (16)$$

where in p_i^p is the on-grid tariff of unit *i*. $Q_i^{p_0}$ is the total power generation volume of unit *i*. o_i is the average power generation cost per kwh.

When part of the planned electricity released into the market the interest of power unit changes to be:

$$\pi_i' = R_i' - O_i' = Q_i^m \times p_i^m + Q_i^p \times p_i^p - (Q_i^m + Q_i^p) \times o_i \quad (17)$$

where in Q_i^m is the market trading power of unit *i*. p_i^m is the average market trading price of unit *i*.

Then benefit change of unit *i* is:

$$\begin{aligned}\Delta\pi_i &= \pi_i^0 - \pi_i' \\ &= Q_i^m \times (p_i^m - o_i) + (Q_i^p - Q_i^{p0}) \times (p_i^p - o_i)\end{aligned} \quad (18)$$

5. Case Study

5.1. Basic Data

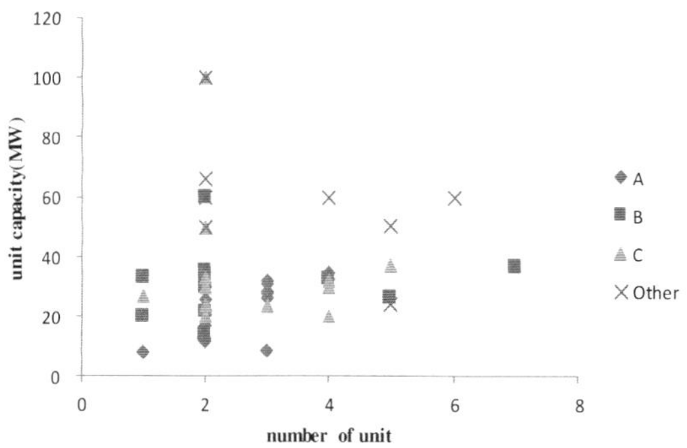

Figure 3. Power plant unit number and capacity distribution.

This paper chooses a regional power grid as the study subject, which contains three provinces and cities, namely A, B and C. The total power generation volume is allocated to 60 power plants by annual plan and market competition. According to the unit number and average capacity, power plants in this regional power grid were classified into different types, as drawn in Fig. 3. In A province, the units are mainly small capacity units with 30 million kW or less installed capacity. Power plants in B province and C are mainly large and medium-sized units, especially C province has some big capacity units with 100 kW or more installed capacity. Other units in Fig. 3 means the power plants are not built in A, B or C province, but still provide power to this area. On this basis, this paper chooses three power plants as typical subjects to analysis, with 30 million kW, 60 million kW and 100 million kW capacities. Parameters of the three plants are list in Table I. Line loss and plant electricity rate are respectively 6% and 8% in this area.

Table 1. Parameters of the Typical Power Plants

Power Plant	Average capacity (million kW)	Unit coal consumption (g/kwh)	Unit cost (Yuan/kwh)	Unit number
#1	100	362.33	0.126816	2
#2	60	454.16	0.158956	4
#3	30	602.43	0.210851	2

The annual transaction is mainly for power generation companies and power companies or electricity users signed a bilateral agreement. The transaction volume and piece are influenced by both power generation company and power purchase decisions, which are difficult to predict without long-term historical data. Until now this area does not have meaningful market transaction reference data. Therefore, this paper assumes low-cost units priority access to electricity trading because of its stronger competitiveness, and simulate market trading result.

This paper simulates the plan release degree, capacity reduction coefficient and market equivalent hour, and sets the different scenarios as Table II. Respectively under 10% and 30% plan release ration, analysis capacity reduction result and multi-stakeholder benefit with different capacity reduction coefficients and market equivalent hours. Accessibility of power plants to participate into the market and transaction limit are set according to local policy.

Table 2. Parameters in Different Scenario

Plan release ratio	**10%**	**30%**
Capacity reduction coefficients and market equivalent hours in different scenario	Scenario 1:μ=0	Scenario 4:μ=0
	Scenario 2:μ=0.5	Scenario 5:μ=0.5
	Scenario 3: μ=1	Scenario 6:μ=1
	Scenario 7: T^m=5000h	Scenario 9: T^m=5000h
	Scenario 8: T^m=6000h	Scenario 10: T^m=6000h

5.2. Result and Analysis

1) Power planet benefit analysis

Changes of the three power plants power generation volume in scenario 1~6 and 7~10 are list in Table III and Table IV, respectively. The percentage value in Table IV means proportion of power generation volume in different scenarios to the allocated power generation volume before the power system reform. According to Table III and Table IV, power generation volume change trends of the three power plants are different with different plan release ratios and different capacity reduction coefficients, while the change trends reflect certain regularity with different market equivalent hours.

Table 3. Changes of the Three Power Plant Power Generation Volume in Coefficient Reduction Mode

Power plant	Scenario 1			Scenario 2			Scenario 3		
	plan	*market*	*total*	*plan*	*market*	*total*	*plan*	*market*	*total*
#1	68.8%	94.7%	163.5%	36.6%	94.7%	131.4%	3.7%	94.7%	98.4%
#2	65.1%	36.4%	101.5%	53.9%	36.4%	90.3%	42.1%	36.4%	78.5%
#3	77.2%	0.0%	77.2%	78.1%	0.0%	78.1%	78.4%	0.0%	78.4%
Power plant	**Scenario 4**			**Scenario 5**			**Scenario 6**		
	plan	*market*	*total*	*plan*	*market*	*total*	*plan*	*market*	*total*
#1	44.0%	94.7%	138.7%	24.0%	94.7%	118.7%	2.4%	94.7%	97.1%
#2	41.6%	94.7%	136.3%	22.7%	94.7%	117.4%	2.3%	94.7%	97.0%
#3	49.4%	0.0%	49.4%	51.1%	0.0%	51.1%	51.6%	0.0%	51.6%

According to Table III, power plant #1 has the maximum average capacity and minimal power generation cost. Therefore, in the cost-oriented electricity market power plant #1 can obtain more market share than power plant 2 and 3. In scenario 1, 2, 4 and 5, if the capacity reduction coefficient is small, the total power generation volumes of power plant #1 are more than the allocated power generation volume before the power system reform. But when the capacity reduction coefficient is big (in scenario 3 and scenario 6), the total power generation volumes of power plant #1 are less than the original allocated power generation volume. The obtained power generation volume from market trading is less than the reduced planed power generation volume.
Power plant 2# has medium capacity and when the plan release proportion is small (scenario 1, scenario2 and scenario 3), power generation cost of power plant #2 is more than the power plants with large-capacity units. Therefore, whether the capacity reduction coefficient is big or small, power plant #2 is hard to get power generation volume in the market, and its total power generation volume is always less than the original allocated power generation volume. With the increase of plan release proportion, the power plants with large-capacity comes to their market trading upper limit, the remaining market share would be obtained by the power plants with medium power generation cost. When the capacity reduction coefficient is small, the total power generation volumes of power plant #2 are more than the original allocated power generation volume (scenario 4 and scenario 5). On the opposite, when the capacity reduction coefficient is big, the plan allocated power generation volume would be greatly reduced and the total power generation volumes of power plant #2 are less than the original allocated power generation volume.
Power plant #3 represents the power plants with high power generation costs, which makes them lack of market competitiveness and hard to obtain market power generation volume. In the above situations, the total power generation volumes of power plant #3 are less than the original allocated power generation volume. With the increases of the capacity reduction coefficient, this type can hardly get market trading that their plan allocated power generation volumes almost do not influence by the capacity reduction coefficient. Because of power plants with large ore medium capacity would get less plan-allocated power generation volume, the plan allocated power generation volume of power plant #3 increased (scenario 4, scenario 5 and scenario 6).

Table 4. Changes of the Three Power Plant Power Generation Volume in the Same Proportion Reduction Mode

Power plant	Scenario 7			Scenario 8		
	plan	*market*	*total*	*plan*	*market*	*total*
#1	9.5%	160.0%	169.5%	7.9%	160.0%	167.9%
#2	5.6%	38.1%	43.7%	4.7%	38.1%	42.8%
#3	45.2%	0.0%	45.2%	45.2%	0.0%	45.2%
Power plant	**Scenario 9**			**Scenario 10**		
	plan	*market*	*total*	*plan*	*market*	*total*
#1	18.9%	156.3%	175.3%	15.7%	160.0%	175.7%
#2	11.1%	99.2%	110.3%	9.2%	99.2%	108.5%
#3	29.9%	0.0%	29.9%	29.9%	0.0%	29.9%

According to Table IV, power plant #1 has the advantage of low power generation costs. Although the plan allocated power generation volumes significantly reduced in different scenarios, the total power generation volumes are more than the original allocated power generation volume. And with the increases of plan release proportion, the total power generation volume increases. The bigger the market equivalent utilization hour the more plan allocated power generation volume and the more total power generation volume.

Power plant #2, with the increases of plan release proportion, power plants with big capacity reaches their market trading upper limit. Power plants like power plant #2 can obtain the remaining market share and their total power generation volume increases. And the bigger the market equivalent utilization hour the more the deducted planned capacity and the less plan allocated power generation volume.

Power plant #3 has high power generation costs that hard to obtain market share and dose not influence by the market equivalent hour. Therefore, the total power generation volume decreases with the increasing plan release proportion.

Assuming the market-clearing price is 5% less than the benchmark price, calculate and compare the interest of the three power plants in different scenarios, as drawn in Fig. 4. Power plant 1#, with big capacity and low power generation costs, gets a large growth of interest in different scenarios. Only in scenario 3 and scenario 6 the total profit is small because of the big capacity reduction coefficient makes its plan allocated power generation volume significantly reduced. The interest of power plant #3 decreases with the plan release proportion increases because it is hard to obtain market share. And influenced by other plants' plan allocated power generation volume, when capacity reduction coefficient is big or market equivalent hour is small, its profit decrease extent would be less. Power plant #2 is influenced obviously by both plan release proportion and capacity reduction mode. When the plan release proportion is big while capacity reduction coefficient is small, its profit is more than the original profit before the power system reform. And under the same proportion reduction or big capacity reduction coefficient management mode, the profit is always less than the original profit before the power system reform.

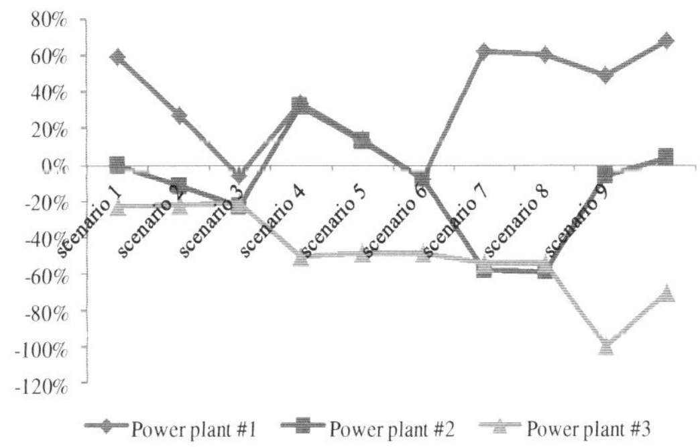

Figure 4. The profit change of the three power plants in different scenarios.

2) Customer benefit analysis

Considering the actual market transactions and related policies, this paper assumes the average transaction price is ±5% (0.38 Yuan/kWh and 0.42 Yuan/kWh), more or less than the average transmission and distribution price is 0.2 Yuan/kWh and the average catalog price is 0.6 Yuan/kWh (large industrial customers). Customer benefit change with different plan release proportion is drawn in Fig. 4. On the one hand, the trading reference base is the benchmark price, which is obviously less than catalog price on customer side. The difference between the trading price and the benchmark price is more than the transmission and distribution price. When the trading price is 5% less than the benchmark price, customer benefit increased significantly. When the trading price is 5% more than the benchmark price, customer benefit decreased slightly because of the different between the benchmark

IOP Conf. Series: Earth and Environmental Science **168** (2018) 012022 doi:10.1088/1755-1315/168/1/012022

price and the catalog price are close to the transmission and distribution price. On the other hand, according to Fig. 5, customer total benefit is not directly related with the corresponding capacity reduction management mode. The capacity reduction mode would influence power plant decision making. Therefore, the capacity reduction mode would affect the relationship between supply and demand and the trading piece to an extent. In this way the capacity reduction mode influences customer individual customer's benefit obviously. For example, power plants like power plant #1 would change their decision under different plan release proportion. On the one hand, they may reduce market trading power generation volume to increase planed capacity, plan allocated power generation volume and total profit. On the other hand, they may increase their trading prices on the basic of obtain enough market shares.

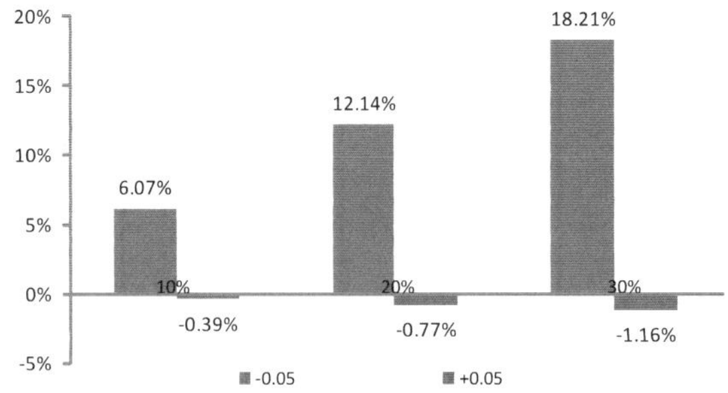

Figure 5. Customer benefit change under different plan release proportion.

6. Conclusions

This paper sorted out different capacity reduction mode in plan management during the transitional stage of the power system reform. According to the specific provisions of different provinces (cities) and pilots, put forward three capacity reduction modes and their corresponding models as well as the basic power generation hour calculation model in annual planning. The case study compared and analyzed the benefit changes of power plant, power customer and the power grid, the conclusions are as follows:

- This bigger the plan released proportion, the greater the advantage of plants with big capacity and low power generation costs, and the greater the disadvantage of the plants with small capacity and high power generation costs. But the profits of those plants do not show a positive/negative proportion or positive/negative correlation between with the plan release proportion. Their profit is mainly influenced by three functions, namely, the plan release proportion, the capacity reduction mode and their own decision making.
- Customer total benefit change rate increases with the increase of plan release proportion. It is not directly influenced by capacity reduction mode, but still indirectly influenced by the capacity reduction mode. The capacity reduction mode influences power plant decision making and supply and demand relationship, customer total benefit and individual customer benefit would be influenced by the capacity reduction mode.

7. Acknowledgment

This work was supported in part by The Soft Connection mechanism and model of Smart Grid in Global Energy Inter-connection.

8. References

[1] ZHANG Zhi-gang, XIA Qing. Architecture and Key Technologies for Generation Scheduling of Smart Grid. Power System Technology, 2009, 33 (20):1-8

[2] Ju Liwei, Tan Zhongfu, Yuan Jinyun, etal. A Bi-level Stochastic Scheduling Optimization Model for a Virtual Power Plant Connected to a Wind–Photovoltaic–Energy Storage System Considering the Uncertainty and Demand Response. Applied Energy,2016,71: 184-199

[3] Tan Zhongfu, Ju Liwei, Li Huanhuan, eta. A Two-stage Scheduling Optimization Model and Solution Algorithm for Wind Power and Energy Storage System Considering Uncertainty and Demand Response. International Journal of Electrical Power and Energy Systems
[4] TANG Wei. Research on methods of making monthly power trade schedule for generation units . Harbin: Harbin Institute of Technology, 2009.
[5] WEN lIngfeng, ZHU Ze-lei, YAN Cui-hui, etal. Short-term Generation Schedule Optimization for Electricity Market. North China Electric Power, 2015,09: 6-10.
[6] WANG Jian-fcng, GU Xin-jian, GAO Guo-ning. Rcscarch and Application of Grid Planning Scheduling Method. Electric Power,1997,09:27-29.
[7] A Multi-Objective Clustering-Based Method to Draft Medium- and Long-Term Schedules for
[8] WANG Liang-yuan , LIU Wen-bin , FANG Ri-sheng, etal. Power Generation and Purchase and Its Application. Power System Technology, 2011, 06: 167-170.
[9] LI Yu-fa, YANG Guo-xin. Generation Planned Management System of Jilin Electric Power Grids. Jilin Electric Power, 2006,34 (04):50-53.
[10] XIE Li-rong, ZHANG Yong, WANG Gang, etal. Coordination Method and Closed-loop Control of Medium-short Term Multi-cycle Generation Scheduling. Automation of Electric Power Systems,2014,10:126-13.

IOP Conf. Series: Earth and Environmental Science **168** (2018) 012023 doi:10.1088/1755-1315/168/1/012023

Research on Modeling of Participating Entities in Demand Response System

Kun Shi[1], Gaoying Cui[2], Jie Fan[2], Chao Lu[3], Bin Li[3], Bing Qi[3]

[1]China Electric Power Research Institute, Beijing, China
[2]State Grid Jiangsu Electric Power Research Institute, Nanjing, China
[3]School of Electric and Electronic Engineering, North China Electric Power University (NCEPU), Beijing, China
e-mail: direfish@163.com

Abstract. Under the opened retail electricity market, the operation rules and business process of automated demand response service become more complicated. In order to deploy automatic demand respond(ADR) business, it is necessary to define the functions and operational framework of each participant in the demand response system. This paper analyzes the main participants of the demand response system, including their responsibilities, functions, behavioral models, communication architecture and logic architecture. and focuses on the modeling of demand response aggregators and power users. Moreover, the modeling of demand response aggregator and power user is emphasized.

1. Introduction

With the rapid growth of the world economy, the consumption rate of fossil energy is accelerating dramatically, and the phenomenon of imbalance between power supply and demand appears frequently0-[3]. At the same time, the problem of environmental pollution is becoming more and more serious. To tackle these problems, people proposed the conception of demand response[4]. As a novel way to balance supply and demand of electricity, consume renewable energy and promote energy saving and emission reduction, demand response has received widespread attention from all countries[5].

Demand response in China has developed for many years, forming a pattern of government-led, power grid enterprises to implement and power users' participation, and has achieved certain economic and social benefits. Automatic demand response is playing an important strategic role in power industry, energy economy development and environmental protection[6].

The implementation mechanism and process of demand response can be summarized as follows. The grid uses the advanced measurement infrastructure (AMI) to obtain the real-time load information of the user. By analyzing the state of the grid, the DR signal, including electricity price, electricity instruction suggestion, etc., is generated and transmitted to the power terminal equipment [7]. After receiving the DR signal, the terminal devices can automatically trigger the pre-set program and respond to the DR plan from grid, i.e. performing the recommended or mandatory operation[8]. After response, the users can obtain the subsidy from grid. If the user does not accept the response strategy, they can opt out freely.

The rest of this paper is organized as follows. Section 2 analyzes the communication structure and logical architecture of demand response system. The models of demand response aggregator and electricity consumer are constructed in Section 3. Then Section 4 concludes the paper.

IOP Conf. Series: Earth and Environmental Science **168** (2018) 012023 doi:10.1088/1755-1315/168/1/012023

2. Communication and Logical Architecture of Demand Response System

The OpenADR 2.0b specification defines virtual top node (VTN) and a virtual end node (VEN) as the two roles of the communication. VTN is defined as the upstream nodes that make DR plans and transmit DR signals to other nodes such as aggregators, large industrial users and electricity markets. VEN is defined as the downstream nodes that receive and response to the DR signals. It should be noted that an entity may function as a VEN to a superordinate VTN and as a VTN to subordinate VENs[9]-[10]. The communication architecture between VTN and VEN is illustrated in Figure .

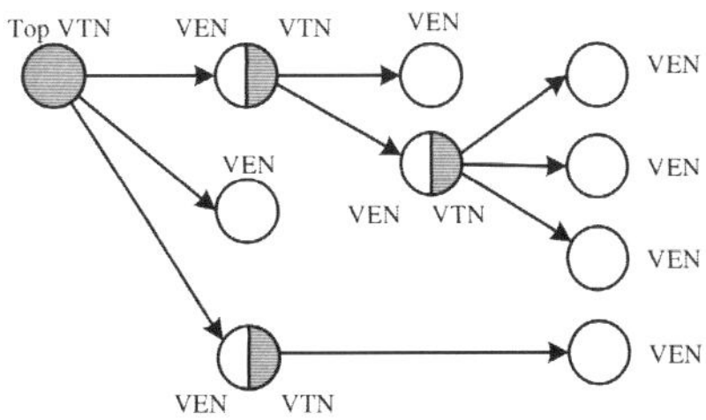

Figure 1. Communication architecture between VTN and VEN

The power demand response system has five participants, namely, power supplier, demand response service manager, demand response aggregator, demand response regulator and power user[11], the relationship between which is shown in Fig[12].

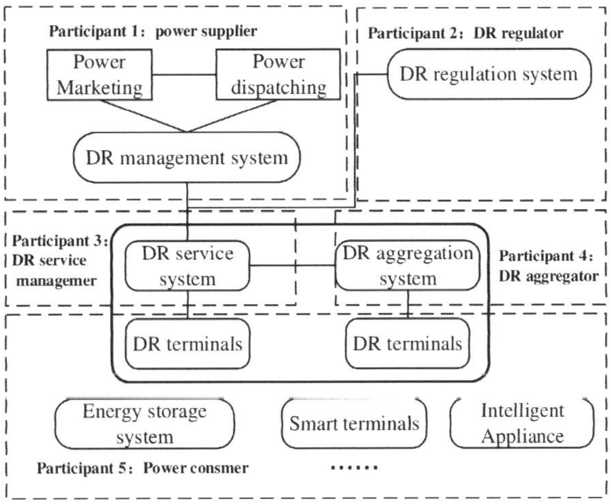

Figure 2. Relationship between five participants in DR system

The power supplier participates in the demand response through the demand response management platform, and interacts with the power users and the demand response service providers. It is responsible for releasing the dynamic price of electricity, decomposing and releasing load demand, publishing the demand response event notification, monitoring the system operation and evaluating the demand response results. Besides, it can quickly receive the feedback from the user in response to the demand and analyze the data.

The demand response aggregator participates in the demand response through the demand response aggregation system, which is connected with the demand response management platform and the demand response terminal. It is responsible for the allocation of scattered resources and the redistribution of the load reduction required by the demand response event.

The demand response service manager is responsible for managing and executing the demand response service within a region, which issues information such as electricity price, load demand and so on, and monitors the operational status of demand response system, and evaluates the effect of power users ' participation in demand response.

The demand response regulators are involved in demand response by making demand response mechanisms and rules, which include the development of requirements response plans, dynamic pricing, and demand side bidding rules.

Power users, including resident users, business users and industrial users, participate in demand response through demand response terminals, connect users' energy management systems and intelligent devices, and interact with the demand response management platform and aggregation system respectively.

3. Modeling of Demand Response Aggregator and Power Consumer

Due to the demand response aggregator and the power user are the more important entities in the demand response system, they are also the key links of DR execution, so in this section, we will analyze them in detail.

3.1. Demand Response Aggregator Modeling

Demand-response aggregators centralize power users with DR capabilities to participate as a whole in DR pilot work and act as agents for business-related matters[13]. Because the demand-response aggregator has the ability to integrate DR resources, it can be used as an active initiator of DR, a performer or only a terminal consumer with power functions. It can also act as an intermediary between the grid company and the user to promote the interaction between the supply and demand side. The main functions of the demand response aggregator can be divided into grid company setting, demand response resource management, demand response signal receiving, demand response programming, demand response signal release, power data aggregation, power data analysis and demand response evaluation, as shown in Fig.

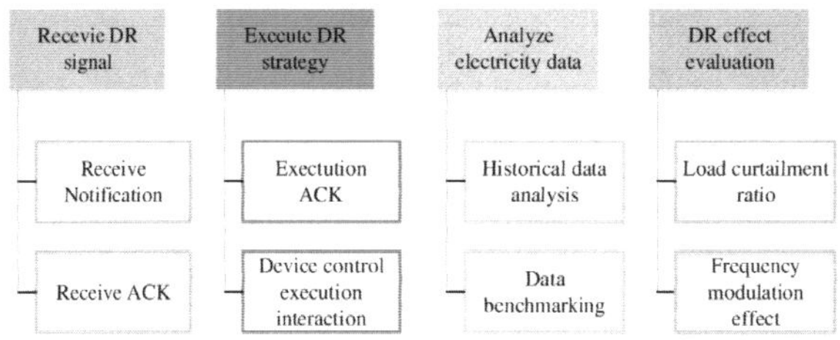

Figure 3. Functions of DR aggregator

After the DR aggregator receives the reduction metric, it needs to decompose the load and then distribute the decomposed task to the Power user. In this paper, an entropy value method is used to determine the decomposition coefficients. Entropy method is used to describe the disorder degree of a system, and when the information entropy of the index is small, the proportion of the index in the comprehensive evaluation should be greater. The calculation formula of entropy is as follows.

$$E_k(j) = -\frac{1}{\ln m}\sum_{i=1}^{m} p_{ik}(j)\ln p_{ik}(j) \tag{1}$$

$$p_{ik}(j) = s_{ik}(j) / \sum_{i=1}^{m} s_{ik}(j) \tag{2}$$

where $E_k(j)$ indicates the information entropy corresponding to power user with index k when performing the jth peak shifting task decomposition.

$$W_k(j) = [1 - E_k(j)]/[n - \sum_{k=1}^{n} E_k(j)] \tag{3}$$

where $W_k(j)$ indicates the weight of the index corresponding to the decomposition of the kth task.

In order to determine the evaluation value of each power user, it is necessary to calculate:

$$Z_i(j) = \sum_{k=1}^{n} W_k(j) \cdot y_{ik}(j) \tag{4}$$

where $Z_i(j)$ indicates that the ith user's weighted evaluation value of each index when performing the jth peak load shifting task decomposition.
Finally, we can obtain the decomposition coefficient of user *i* by equation (5).

$$R_i'(j) = Z_i(j) / \sum_{i=1}^{m} Z_i(j) \tag{5}$$

3.2. Modeling of Electricity Consumer

Power users are the carriers of power consumption and the direct participants in the implementation of DR. Since the power users are direct consumers of electricity, both and the implementation of the power demand response lunched by government, power suppliers, or power generation enterprises must rely on the power users[14]. They should update the electrical equipment of the power users and adjust the way the electricity users use electricity. Only the active participation of power consumers, could it be possible to improve the way of using electricity, improve the power efficiency of terminal, realize the reduction of installation of installed power generation and power transmission equipment, reduce power construction investment, reduce the power supply cost, saving energy resources, reduce pollutant emissions, so as to realize the overall cost of the whole society to the minimum target. Therefore, the power users are the social basis for the actual response of the power demand response, and the most important participant in the power demand response.
The power users who participate in DR can be classified into four main categories: resident user, big industry user, general industry and commerce user, the total electricity load amount of which have maintained a growth situation, while the basic structure of electricity consumption maintains stability. The biggest proportion of electricity structure is the large industrial users, the average accounted for 75%, followed by the general Industry and commerce, accounting for 13%, and finally, the resident electricity, accounting for 12%. Because of the different proportion of the power consumption of the three types of users, so is the status of participating in the implementation of DR.
As large industrial users are the biggest consumers of power loads[15], it can respond quickly to the price or incentive signals in the implementation of DR, cutting down a great deal of power demand, so it occupies a very important position in the implementation of DR. Although the electricity consumption of industrial and commercial is less than that of the industry, the general industrial and commercial load of lighting, air-conditioning load in the peak period of power system is one of the main factors affecting the power system peak load, therefore, the participation of general industrial and commercial users in DR implementation is of great significance. Resident users, including urban and rural residents, have the widest coverage and the largest number of consumers involved in electricity load. Residential users load mainly include lighting, air conditioning, washing machine, TV and other household appliances, residential electricity peak and peak power system often overlap, has an important influence on the peak load of the power system. Therefore, it is necessary to implement the residents involved in DR.
We focus on resident power users, which can be classified into five categories according to the different levels of electricity usage, whose base load lines is shown in Fig.4. Type A is vacant room users, whose electricity consumption characteristics is that the user's electricity consumption maintains a very low level, and there are no big fluctuations, the low power consumption derives from line losses. Type B users are elderly families, the electricity consumption during the day maintains a certain level, and the electricity consumption began to rise at 14:00, but the downward trend appears early. Type C is office users, whose power consumption has obvious peaks and troughs, daytime power consumption is almost equal to the night electricity consumption. Besides, evening electricity peak is later than B class, and the trend of decline is also later than Type B. Type D is office users combined with elder persons and is basically a combination of B and C. This type of user may be a mixed family of the el-

derly and the office worker. Type E is a commercial user whose power consumption has remained at a high level throughout the day.

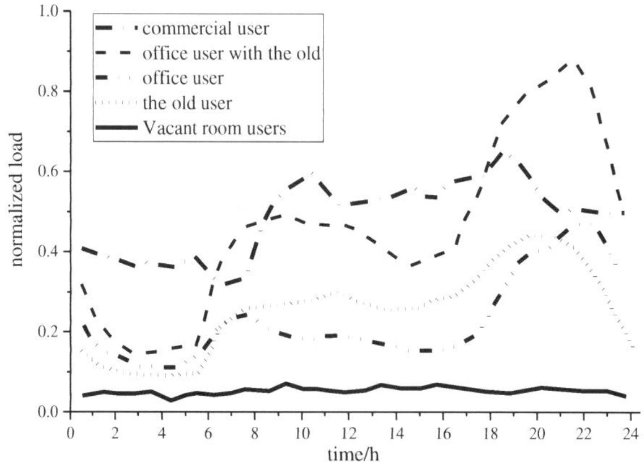

Figure 4. Different base lines of resident power users

Due to the behavior of power users is uncertain, the user's behavior is modeled according to price and incentive based demand response

1) Behavior of power users under price based DR

Based on consumer psychology model, in the death zone, price difference is small, then the user can shift electricity usage, and they may also use more electricity because the price difference is not sensitive to increased use of electricity. Therefore, the regional load transfer rate ranges in a range of positive and negative value. In the linear zone, the price difference is enough for the user to feel the pressure. At this time the transfer rate of the user is positive, but also within a certain interval distribution, and the range decreases with the increase of the price difference. In the saturated zone, the response potential has all been excavated, and the load transfer rate no longer increases with the increase of the electrical spread, but tends to be constant. This mechanism is described in Fig.5.

2) Behavior of power users under incentive based DR

Fig. 6 is the response characteristic curve of the residential users under the interruptible load(IL) scenario, and the response of the IL user is described by the rate of cut load, and the rate of cut load represents the ratio of the actual clipping load to the user's load shedding potential, i.e.

$$\lambda(x)=\sum_{i}^{n}P_{Ri}\lambda_i(x)/\sum_{i}^{n}P_{Ri} \tag{6}$$

When IL compensation level x reaches the P point, the user begins to reduce the load, but it may also do not cut its load due to low compensation. When x reaches the Q point, the user generally responds, while the cutting load rate is greater than 0, and the load reduction expected to increase, the fluctuation range is also reduced. When x reaches the M point, the user load reduction is expected to reach the maximum and the fluctuation range is approximately ignored, and all response potentials have been tapped.

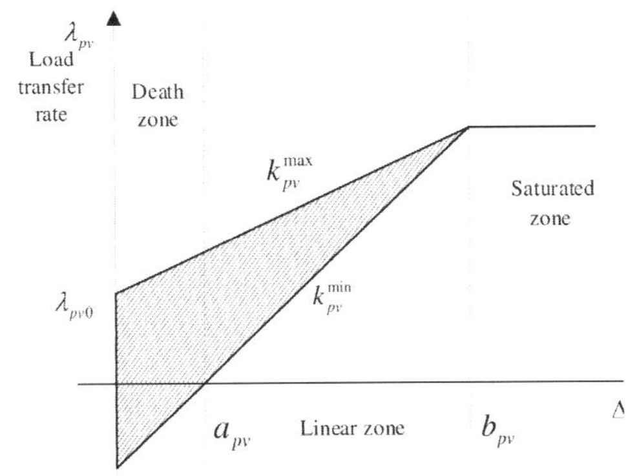

Figure 5. Behavior of power users under price based DR

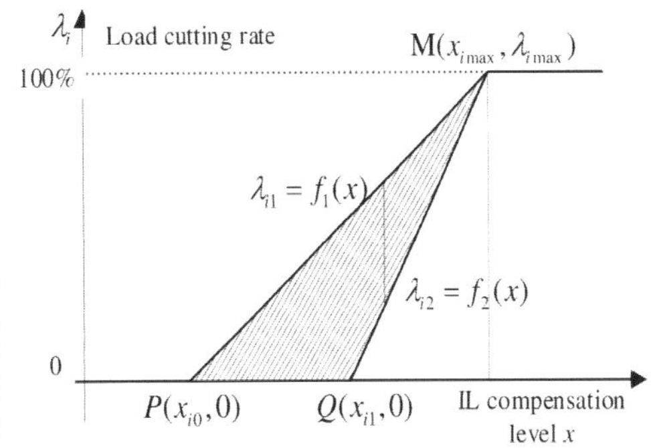

Figure 6. Behavior of power users under incentive based DR

From Fig. 6, we can conclude that under certain excitation circumstances, the user's load reduction/transfer quantity has certain randomness and thus brings certain uncertainty. And the user is driven by interest, this randomness will decrease as the level of stimulation increases. Because IL belongs to the category of non-mandatory and depend on user's voluntary response, even if the mechanism is sufficiently perfect, it will inevitably produce certain uncertainties. Thus many user agents and load-gathering operators have sprung up to provide interruptible services to reduce uncertainty through the aggregation of a large number of users.

4. Conclusion

This paper analyzes the responsibilities and functions of different participating entities in the power demand response system, as well as the cooperation mechanism and communication framework of each entity. The operation structure of demand response aggregator is analyzed emphatically, and a load decomposition strategy based on entropy method is proposed, which can be used to decompose large peak shaving task to each power user fairly and reasonably. In addition, the power consumption characteristics of several kinds of resident electricity load and the response models of power users under two scenarios, namely price based incentive based demand response, are studied. The results of this paper has certain reference meaning to the implementation of demand response.
Currently, the demand response in China is like a raging fire started, and how to use information and communication technology and mathematical modeling techniques to precisely and real-timely control the distributed load o meet the demand of the electric power market to a variety of ancillary services, is the focus of the next period of time and practical implementation challenges.

5. References

[1] OpenADR Alliance. OpenADR 2.0b Profile Specification B Profile, [2015-12-26].
[2] GAO Ciwei, LIANG Tiantian, LI Huixing, et al."Development and Application of Open Automated Demand Response," Power System Technology, vol. 37, no. 3, pp. 692-698, 2013.
[3] Siano P. "Demand response and smart grids—A survey,". Renewable and Sustainable Energy Reviews, vol. 30, no. 2, pp. 461-478, 2014.
[4] QI Bing, ZHANG Rong, LI Bin, et.al. "Design of Automated Demand Response Information Exchange Interface," Proceedings of the CSEE, vol. 34, no. 31, pp. 5590-5596, 2014.
[5] Wang L, Yang S, Yang C, et al. "A Study on Demand Response International Standard Architecture," Proceedings of the CSEE, vol. 34, no. 22, pp. 3601-3607, 2014.
[6] XU Xiaohui, XIN jianbo, FAN Ruixiang, et al. "Research and Verification of Demand Response Information Interfaction Mechanism Based on OpenADR2.0," East China Electric Power, vol. 42, no. 7, pp. 294-1298, 2014.
[7] Ghatikar G, Mathieu, Piette M A, et al. "Open Automated Demand Response Technologies for

Dynamic Pricing and Smart grid,"Grid Interop Forum. 2010G. Eason, B. Noble, and I.N. Sneddon.
[8] Ninagawa C, Iwahara T, Suzuki K. "Enhancement of OpenADR communication for flexible fast ADR aggregation using TRAP mechanism of IEEE1888 protocol," IEEE International Conference on IndustrialTechnology IEEE, 2015.
[9] Yingjie Lv, Wei Cen, Yilun Fu, et al. "Application of P-cycle in OpenADR communication network protection," 2016 7th IEEE International Conference on Software Engineering and Service Science (ICSESS), 26-28 Aug, Beijing,China.
[10] Ma O, Alkadi N, Cappers P, et al. "Demand Response for Ancillary Services," IEEE Transactions on Smart Grid, vol. 4, no. 4, pp.1988-1995, 2013.
[11] Vardakas J S, Zorba N, Verikoukis C V. "A Survey on Demand Response Programs in Smart Grids: Pricing Methods and Optimization Algorithms," IEEE Communications Surveys and Tutorials, vol. 17, no. 1, pp. 152-178, 2015.
[12] Lemay M, Nelli R, Gross G, et al. "An Integrated Architecture for Demand Response Communications and Control," Hawaii International Conference on System Sciences,"Proceedings of the. IEEE, pp.174-174,2008.
[13] Kim H, Kim Y J, Yang K, et al. "Cloud-based demand response for smart grid: Architecture and distributed algorithms," IEEE International Conference on Smart Grid Communications. IEEE, pp. 398-403,2011.
[14] Pourmousavi S A, Patrick S N, Nehrir M H. "Real-Time Demand Response Through Aggregate Electric Water Heaters for Load Shifting and Balancing Wind Generation,". IEEE Transactions on Smart Grid, vol. 5, no. 2, pp. 769-778, 2014.
[15] Giovanelli C, Kilkki O, Seilonen I, et al. "Distributed ICT architecture and an application for optimized automated demand response." IEEE Pes Innovative Smart Grid Technologies Conference Europe. IEEE, pp.1-6,2016.

Acknowledgments
This work is supported by the Science and Technology Projects from State Grid Corporation (YDB17201600115)

IOP Conf. Series: Earth and Environmental Science **168** (2018) 012024 doi:10.1088/1755-1315/168/1/012024

Economic Analysis of Power Plants Under Different Loads Based on Structural Theory

Huanhuan Bu, Pei-hong Wang, Farong Shen

School of Energy and Environment, Southeast University, Nanjing, Jiangsu210096, P.R. China

e-mail: 220150493@seu.edu.cn, phwang@seu.edu.cn

Abstract. The economy analysis of coal-fired power plants is usually based on traditional thermal balance method, which only considers the thermodynamic factors but ignores the economic factors such as cost. In this paper, to have a more comprehensive and accurate evaluation on the production performance and economical efficiency of thermal systems, the cost analysis method based on the Structural Theory of Thermoeconomic is applied to a 300MW thermal power generating unit in Yunnan province. The thermoeconomic model and the exergetic cost model for the plant based on the Fuel—Product concept have been defined to quantify the productive interaction between different devices. By calculating the thermoeconomic cost of four typical different working conditions and analyzing the distributions and change rules of generating cost, the Structural Theory of Thermoeconomic is proved scientific and useful and offers a set of feasible and reliable analysis method for the performance analyses and improvement of power plants.

1. Introduction

Thermal economics developed on the basis of exergy analytisis, focusing on economic activities based on thermodynamic analysis. Exergy analysis method takes accout of both "quantity" and "quality" of energy, which can reflect the essence of the loss of ability, and it is more accurate and objective than the thermal equilibrium method based on the first law of thermodynamics.Besides of this, thermal economics also considers the cost of energy, based on thermodynamics and economics [1], [2].

In recent years, due to the difference between peak-valley electricity load increases, thermal power plants in China often run under low load conditions [3], [4]. In order to improve the operating economy under the low load of the unit, this paper uses the structural theory of thermal economics to analyze the operating economy of a 300MW unit in Yunnan under rated working conditions of different load. In this paper, the exergy cost and thermal economics cost of each component under different working conditions are calculated, and the economic operation suggestions under the low load is put forward.

2. Structural Theory And Data

The structural theory is created by the Spanish scholar A. Valero et al. The basic idea is to divide the energy system into multiple subsystems. The sub-systems communicate the simulated element behavior through some mass flow and energy flow to obtain a series of mathematical equations of characteristic equations. Since thermal economics calculations are based on analysis, this section also introduces the concept of exergy analysis [5]-[7].

2.1. A. Fuel-Product

The concept (Fuel-Product) is used to define the functions that describe each device, where "fuel" (F) is the flow required to obtain the product,and"Product" (P) refers to the flow of production as a subsystem. In addition to the two components of the condenser and the generator, the other components in the system consume two fuels: the negative entropy (FS) generated by the resource (FB) and the condenser. The specific definitions can refer to reference 6.

2.2. B. Drawing of Physical Structure Sketch and Productive Structure Sketch

The physical structure sketch of this paper is based on the system thermal balance sketch. According to the required degree of integration, the system will be divided into 24 components, and the physical structure sketch will be drawn in accordance with the relationship between the components The productive structure sketch is based on the fuel-product definition after the completion of the physical structure drawing, using the interaction between the production elements. The physical structure sketch of this paper is shown in Fig. 1, and the productive structure sketch is shown in Fig. 2.

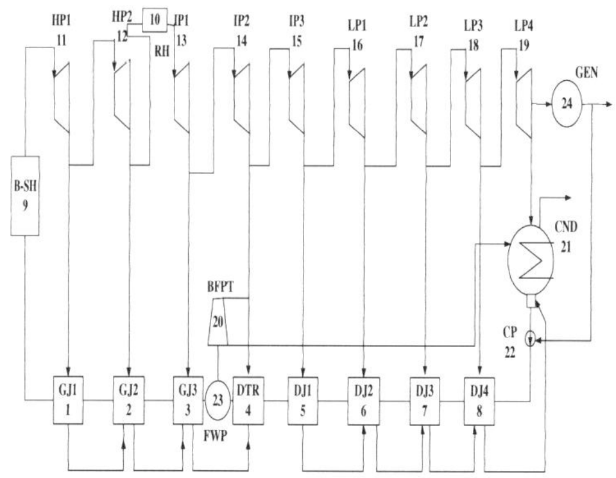

Figure 1. Physical structure sketch.

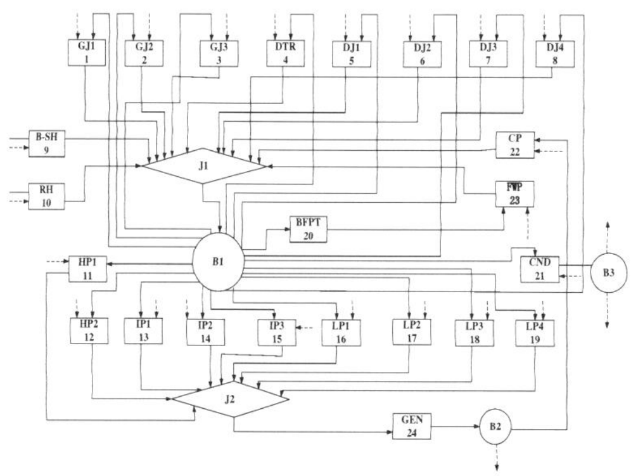

Figure 2. Productive structure sketch.

2.3. C. Indicators of Exergy Analysis

1)Loss of exergy: The actual process is irreversible, accompanied by a decline in exergy quality, and the total amount of exergy decreases, which is unavoidable. The part that is dissipated during the transfer process is called exergy loss. The formula can be expressed as: FB=P+I, where: I is the loss of the device; FB is the fuel of the input device and P is the product output of the product.

2) Exergy efficiency:it is the ratio of the resulting product to the fuel used to obtain the fuel, and is expressed by η. The expression is η = P / FB.

3) Exergy loss factor: it is the ratio of the loss of exergy (Ii) in a device to the total exergy fuel input to the system. It is expressed as: $\lambda_i = I_i / \sum F$.

2.4. D. Exergy Cost Modeling

The cost equation of the system expresses the resource B_0 consumed by the system as a function of each stream B_i, the internal parameter set of each component, and the system's final product can be gotten, that is:

$$B_0 = B_0(B_i, x, \omega) \quad i = 1, \cdots, m \tag{1}$$

The establishment of the cost equation usually uses the differential chain rule to derive the derivation of the cost equation and the characteristic equation. When the consumption of any flow resource in the system changes, the chain differential rule is used to derive the derivative of (1), and then we have the formula as below:

$$\frac{\partial B_0}{\partial B_i} = k_{0,i}^* \quad i = 1, \cdots, e \tag{2}$$

$$\frac{\partial B_0}{\partial B_i} = \sum_{\substack{j=1 \\ j \neq i}}^{m} \frac{\partial B_0}{\partial B_j} \frac{\partial g_j}{\partial B_i} \quad i = e+1, \cdots, m \tag{3}$$

The equation can be expressed as unit exergy consumption and technical product coefficient:

$$k_{p,i}^* = k_{0i} + \sum_{j=1}^{n} k_{ji} k_{p,j}^* \quad i = 1, \cdots, n \tag{4}$$

It can be seen from equation (4) that the unit cost per unit of flow is a function of the unit consumption of each component in the production structure. If the characteristic equation of each process element is known, then by solving this we can get the unit exeergy cost.

2.5. E. Thermal Economics Modeling

The thermal economic model is a mathematical expression of the system production structure diagram. Each input stream Bi is taken as the mathematical function of the output stream Bj, considering the whole product structure process and a series of internal parameters Xi, the characteristic equation can be established as follows: [8]-[11]

$$B_i = g_i(x_i, B_j) \tag{5}$$

By using the Euler and fuel-product definitions of the first-order homogeneous equation, the above equation can be converted to the following linear equation:

$$P_i = B_{i0} + \sum_{j=1}^{n} k_{ij} p_j \quad i = 0, 1, \ldots n \tag{6}$$

where: Bi0 is the product of the i-th component as the total product of the system output to the environment; kij=$\frac{\partial gi}{\partial Bj}=\frac{Bij}{Pj}$stands for the technical product coefficient [6], to express the proportion of the i-th component in order to obtain the j-th unit of the unit product.

In the end, the thermal economics costs of the product include the cost of non-energy costs and the cost cp of non-energy costs, which are economic dimensions, 10^{-6} yuan / kJ, and the cost of non-energy costs is calculated as:

$$cp = kB_n c_{FBn}^{*} + kS_n c_{FSn}^{*} + kZ_n \quad (7)$$

where kZ_n is the product investment costs for the unit (10^{-6} yuan / kJ). c_{FBn}^{*} *and* c_{FSn}^{*} is the cost of unit fuel thermal economics and unit negative entropy costs, the specific calculation method can refer to the literature [6].

3. Example And Result Analysis

Taking a 300MW thermal power unit in Yunnan as the research object, the structural theory is used to calculate the exergy cost and unit thermal economic cost of each unit at 100%, 50%, 40% and 30% THA condition. According to the changes of the cost, suggestions are proposed to improve the economic efficiency of the unit.

Table 1 below shows the values of unit exergy cost for each component in the four different operating conditions.

Table 1. Exergy Cost of each Component under Different Loads

component	100%THA	50%THA	40%THA	30%THA
	k^*_P (kW/kW)	k^*_P (kW/kW)	k^*_P (kW/kW)	k^*_P (kW/kW)
GJ1	2.26	2.41	2.48	2.58
GJ2	2.29	2.45	2.51	2.62
GJ3	2.39	2.55	2.62	2.74
DTR	2.49	2.68	2.75	2.87
DJ1	2.50	2.72	2.81	2.95
DJ2	2.43	2.64	2.73	2.87
DJ3	2.74	3.09	3.25	3.49
DJ4	3.14	3.06	2.99	2.94
BSH	2.02	2.14	2.19	2.28
RH	1.99	2.05	2.07	2.13
HP1	2.31	2.50	2.56	2.66
HP 2	2.22	2.35	2.41	2.50
IP 1	2.21	2.34	2.28	2.51
IP 2	2.17	2.31	2.37	2.45
IP 3	2.23	2.36	2.41	2.50
LP 1	2.25	2.43	2.46	2.57
LP 2	2.24	2.33	2.44	2.54
LP 3	2.43	2.53	2.60	2.68
LP 4	2.72	2.81	2.63	3.69
BFPT	2.57	2.80	2.94	3.08
CND	0.09	0.10	0.10	0.10
CP	12.95	14.69	15.74	17.35
FWP	3.56	4.86	5.75	6.65
GEN	2.31	2.45	2.48	2.62

The changes of the thermal economics cost for each component was shown in the histogram below.

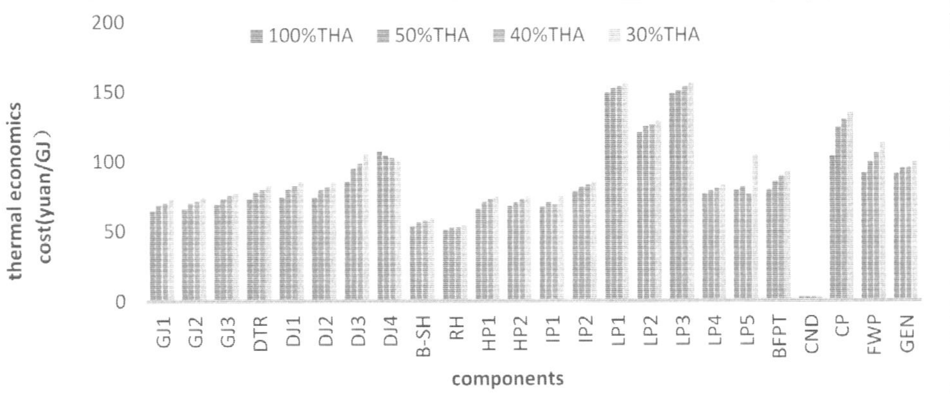

Figure 3. Thermoeconomics cost of each component under different loads.

From the table and chart above, the following conclusions can be drawn:

(1) As the load decreases, the unit's loss coefficient and cost are significantly changed, and the total exergy loss and cost are increased.

(2) As the load decreases, the thermoeconomics cost of each component is increased and the loss share of each component calculated by the thermoeconomics analysis is different from the results by previous heat balance method.

(3) After the change of the load, the change trend of the exergy cost for each unit is different. The cost of the second and third high pressure cylinder and low pressure cylinder increase obviously increased with the decrease of the load. Therefore, we should focus on its economy when the power plants are running under low load.

(4) Based on thermal economics, the place which has the larger exergy loss coefficient, have larger energy saving potential, and the boiler in the thermal system has the largest potential.

4. Conclusion

Through the calculation and analysis of the 300MW unit studied in this paper, it is found that there are still some problems in the designation of the unit, and the conclusions can be draw:

1) The load has a great impact on the economy of the unit. When the unit load is less than 100%, with the decrease of the load, the unit's exergy loss and thermoeconomics cost increase, the efficiency and economy of the unit are reduced, so the unit should be run at the load as high as possible, but less than 100%.

2) It is more scientific to evaluate the performance of the unit equipment by and thermal economics than the traditional thermal equilibrium method.

This is because the analysis of thermal economics does not only take into account the quantity of energy, but also the quality of the energy, so that it can evaluate the production performance of each component in the system
more scientificly.

3) When the 300MW unit is operated at a lower load, it should be noted that the relevant parameters and efficiency of the second and third parts of the high and low pressure cylinder , which reflects its internal irreversible resource loss is of great amount. So it has a greater energy-saving potential. In order to improve the economy, to reduce its exergy loss and its extraction steam pressure loss is necessary.

5. References

[1] Dong Baoguang, Wang Shengpeng, Liu An. "Thermal Economy Analysis of Domestic Subcritical 600MW Steam Turbine Unit". Thermal Power Generation, vol. 36 (12), 2007, pp. 1-3.

[2] Ning ding. “Feasibility Evaluation of Steam Turbine Optimization and High and Medium Pressure Cylinder for Domestic Imported 300 MW Unit”. Power Equipment, vol.5 (8), 2004, pp. 41-43.
[3] Wang Xuedong, Zhang Xuemin, Ma Baohui. “Comprehensive analysis of the scope and effect of the new steam seal of steam turbine”. Steam Turbine Technology, vol. 57 (4), 2015, pp. 297-301.
[4] Jiang Feng, Peihong Wang. “Proof and Verification of Consistency of Equivalent Enthalpy Drop and Heat Balance Algorithm”. Steam Turbine Technology, vol. 50(4),2008, pp. 261-264.
[5] Wang Chenyu. “Partial transformation of high-pressure cylinder flow in steam turbine of 215 MW unit”. Thermal Power Generation, vol. 39(10),2010, pp. 82-84.
[6] C Torres，A Valero，L Serra，J Royo. “Structural theory and thermoeconomic diagnosis: Part I. On malfunction and dysfunction analysis”. Energy Conversion & Management, vol. 43(9),2011, pp. 1503-1518.
[7] Liu Sha. “Performance analysis and optimization of large-scale steam turbine system based on thermal economics”. Southeast University, 2014.
[8] Yin Qinghua, Xin Yan, Bi Zhengyi. “Imported 300MW Turbine Drainage System Transformation and Economic Analysis”. Shanghai Electric Power, 2005 (2), pp. 208-210.
[9] Rosen M A, Dincer I. “Thermoeconomic analysis of power plants: An application to acoal fired electrical generating station”. Energy Conversion and Management. vol. 44(17), 2003, pp. 2743-2761.
[10] Valero Antonio, Correas Luis, Zaleta Alejandro, et al. “On the thermoeconomic approach to the diagnosis of energy system malfunctions: Part 1: the TADEUS problem. Energy”, vol. 29(12-15), 2004, pp. 1875-1887.
[11] Valero Antonio, Lerch Felix, Serra Luis, et al. “Structural theory and thermoeconomic diagnosis: Part II: Application to an actual power plant. Energy Conversion and Management”, vol. 43(9-12), 2002, pp. 1519-1535.

IOP Conf. Series: Earth and Environmental Science **168** (2018) 012025 doi:10.1088/1755-1315/168/1/012025

Case Analysis on Economic Efficiency of Grid-Connected Photovoltaic Power Generation Project with Carbon Emission Measurement

Peng Peng, Yubo Gao, Guoqin Yu, Xiaodong Lou,Yaping Dai, Zhonghu Chen

State Grid Shanghai Municipal Electric Power Company, Shanghai, China

e-mail: pengwyw@hotmail.com

Abstract. Taking the photovoltaic power generation project in Shanghai area as a case, economic benefit of the whole lifetime cycle of the project was analyzed by adopting the levelized cost of electricity (LCOE). Meantime in consideration of the low-carbon economic benefits, the incremental cost analysis method was applied to evaluate the unit carbon-mitigation cost of the project and sensitivity analysis was conducted, in ordering to put forward the guiding suggestions for the investment of photovoltaic power generation projects.

1. Introduction

With the intensifying exploitation of renewable energy in worldwide, solar photovoltaic industry is developing by leaps and bounds, which becomes the largest renewable energy industry with the highest investment growth over the world [1]. Compared by conventional coal-fired and oil-fired power generation, photovoltaic (PV) power generation is safe, reliable, noiseless, clean and green, which is also not bound by resource distribution areas while not consuming fuel. It's can avoid the environment damage caused by various pollution derived from conventional coal-fired or oil-fired power generation, such as CO_2、 SO_2、 NO_x and CO. So PV power generation currently is the project of green power energy supported strongly by Chinese government. The legal system about the environmental protection applied to the power industry will be improved constantly along with moving forward in China's market-oriented reform of the electric power system, so taking into consideration of the environmental cost will become the inevitable trend [2]. In this paper, economic efficiency of. PV power generation was analyzed form the carbon emission reduction point of view, meantime in consideration of the total life cycle cost.

2. Mathematical Models

The LCOE is one analytical tool that can be used to compare alternative technologies when different scales of operation, investment or operating time periods exist [3-5], which can be consider as one of the main factors to compare the cost of energy generated by a PV power plant with that of a fossil fuel generating unit or another renewable technology [6-8]. The calculation for the LCOE is the net present value of total life cycle costs of the project divided by the quantity of energy produced over the system life.

$$\text{LCOE} = \frac{\text{Total Life Cycle Cost}}{\text{Total Lifetime Energy Production}} \tag{1}$$

In consideration of the discount rate, the above LCOE equation can be disaggregated for solar generation as follows [9]:

$$\text{LCOE}=\frac{\text{Initial Investment-}\sum_{n=1}^{N}\frac{\text{Depreciation}^{n}}{(1+\text{Discount Rate})^{n}}\times(\text{Tax Rate})+\sum_{n=1}^{N}\frac{\text{Annual Cost}^{n}}{(1+\text{Discount Rate})^{n}}\times(1\text{-Tax Rate})-\frac{\text{Residual Value}}{(1+\text{Discount Rate})^{n}}}{\sum_{n=1}^{N}\frac{\text{Initial kWh/kWp}\times(1\text{-System Degradation Rate})^{n}}{(1+\text{Discount Rate})^{n}}} \tag{2}$$

The economic efficiency of PV power generation project was analyzed by taking consider into the environmental benefits in this paper, and the benefit about carbon emission reduction compared with traditional power generation's was emphasized on. Thus, this paper serves the greenhouse gas emission of traditional power generation as a basic line, and sets the benefit about carbon emission reduction as the criterion of economic efficiency of PV power generation project. So per-unit cost of carbon emission reduction can be represented as [10]:

$$\text{Per-unit } CO_2 \text{ emission reduction cost}=\frac{\text{LCOE}-\text{Per-unit cost of trational generation}}{\text{Per-unit } CO_2 \text{ emission of PV power generation}-\text{Per-unit } CO_2 \text{ emission of trational generation}} \tag{3}$$

A project of grid-connected PV power generation built in Shanghai area was taken as a case to see whether it made sound economic sense in this paper, while the sensitivity analysis of typical environment factors were made.

3. Casd Analysis

By taken grid-connected PV power generation with the installed capacity of 91.5kW built in Shanghai area as an example, per-unit cost of the carbon emission reduction was analyzed in this paper.

The basic information of the case in this paper was as follows: the installed capacity was 91.5kW, the life cycle was 25 years, the initial investment was 739800 yuan, the per-unit cost was 8.80 yuan•$W^{-1}$the annual cost was 2300 yuan, the average generating capacity was 82.1GWh， the annual load-time 896.9 hours, the discount rate was 8%, the tax rate was 17%, the system degradation rate was 0.8% and the system residual rate was 5%.

Basic information of this project and main parameters for computational analysis are listed in Table I and Table II [4, 11, 12]. The economic efficiency of this project by considering the carbon emission was calculated. The results were shown as follows.

a) LCOE of the grid-connected PV power generation project is 0.9991yuan•kWh^{-1}.

b) Per-unit cost of traditional power generating project is 0.2590 yuan•kWh^{-1}.

c) Per-unit incremental cost of the project is 0.7401 yuan•kWh^{-1}.

d) Per-unit CO_2 emission reduction of the project is 0.6784kg•kWh^{-1}.

e) Per-unit carbon emission cost of the project is 1.0909 yuan•t^{-1}.

Table 1. Major Parameters in Mathematic Analysis of Pv Project

Parameters	Capacity	Annual load-time	Initial Generating Capacity	Initial Investment	Annual Cost	Discount Rate	Lifetime	System Degradation Rate	Tax Rate	System Residual Rate
value	91.5 kW	896.9 h	82100 kWh	739800 yuan	2300 yuan	10%	25a	0.8%	17 %	5%

Table 2. Major Parameters in Mathematic Analysis of Trational Generation [10]

Energy	Ratio (%)	Coal (oil) consumption (g•kWh^{-1})	Standard coal coefficient (kg•kg^{-1})	Cost (yuan•kWh^{-1})	Average low calorific value (MJ•kg^{-1})	Carbon oxidation rate (%)	Carbon content (%)	Carbon emission coefficient (kg•kWh^{-1})
Coal-fired power	87.6	338	0.7143	0.225	20.93	65	94	0.303
Oil-fired power	1.5	328	1.4286	0.289	41.87	98	85	0.163
coal gas	3.9	0.1830 m^3•kWh^{-1}	1.7143	0.238	50.18	98	35~40	3.1013
Natural gas	7.0	0.2298 m^3•kWh^{-1}	1.3300	0.6894	35.88 MJ•m^{-3}	99	15~45	2.1622 kg•m^{-3}

According to the energy program [2016] No.13 published by Shanghai municipal development and reform commission, Shanghai government's implementation of the distributed PV power generation subsidy was classified by power consumptive, i.e. the price for industrial and commercial users was 0.25 yuan•kWh^{-1} and the price for individuals, schools and other users was 0.4 yuan•kWh^{-1}. As well as subsidies from the state (0.42 yuan•kWh^{-1}), the LCOE of grid-connected PV power generation project in this paper was 0.3291 yuan•kWh^{-1}, while Feed-in tariffs of coal-fired generation is 0.4359 yuan•kWh^{-1} in Shanghai. Thus, it can be concluded that the return on investment (ROI) of this case has certain advantages in consideration of the subsidies from the government.

4. Sensitivity Analysis

For PV power generation project, its carbon emission cost is affected mainly by the LCOE, which is related to several conditions, such as per-unit cost of the project, amount of solar radiation where the project is located, system efficiency, systematic degradation rate, maintenance costs and the lifetime of key equipment like inverter, etc. For the established grid-connected PV power generation project in this paper, various factors had been defined, for instance, the performance of key equipment and the investment of the project. Consequently, economic efficiency of this project is mainly affected by systematic generating capacity. So the sensitivity effect of environmental factors on the system generating capacity was analyzed based on the actual environment where the equipment was located in this paper.

The electrical efficiency of this PV power generation was analyzed firstly. The data of someday were chosen randomly in order to improve the accuracy of results and the corresponding curve was displayed in Fig. 1. Obviously the random variables follow the normal distribution. The maximum efficiency appeared around noon and the value was 84.34%. The average efficiency of this plant was 25.21%.

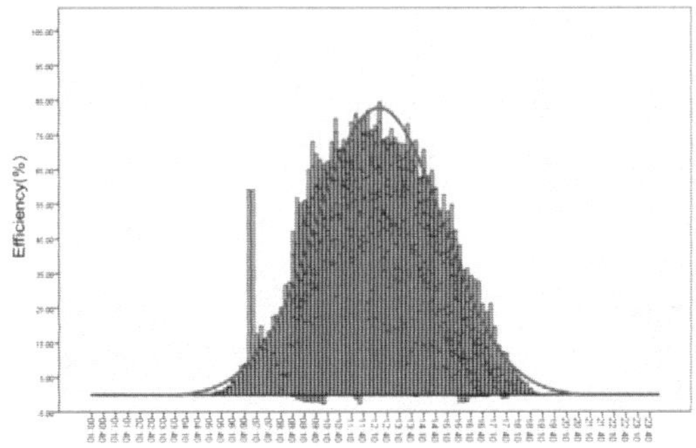

Figure 1. The curve of the electrical efficiency of the PV power generation on someday.

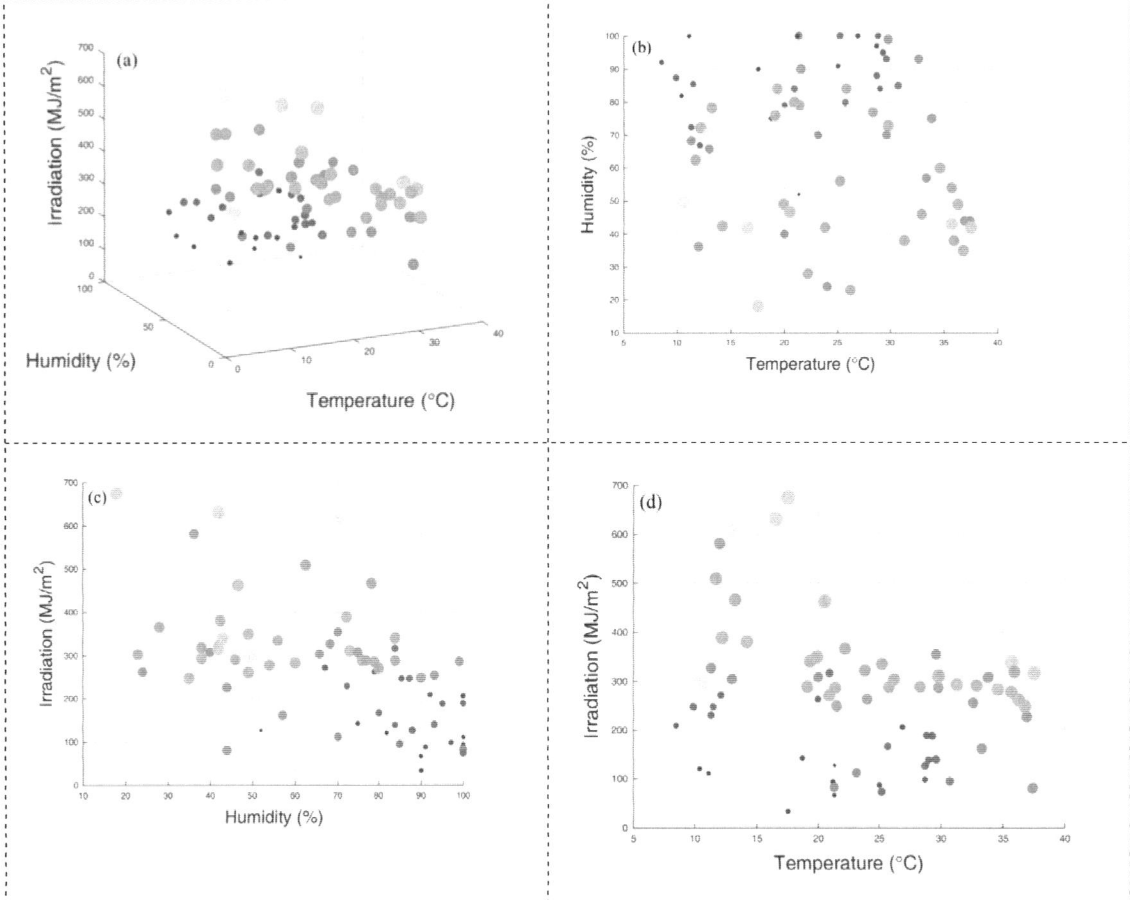

Figure 2. The relationship of the temperature, the humidity and the corresponding capacity.

According to the climatic characteristics in Shanghai, the electricity generation performance data of equipment is divided by four seasons. That is to say, the environmental data of the temperature, the humidity and the solar radiation in February, May and July, 2016 respectively were chosen to analyze the sensitivity effect on the capacity [13].

The relationship of daily environment temperature, humidity and accumulated solar irradiation, as well as daily systematic generating capacity in February, May and July, 2016 was shown in Fig. 2(a). The lighter color and larger radius of points indicated the larger systematic generating capacity. It had to be pointed out that for daily dynamic change of environmental factors, the environmental temperature and humidity at 14:00 were taken as basic data in this paper, just because those factors reach the optimal value at that moment, which was more representative. The trend of systematic generating capacities almost stayed same with the change of environment humidity in Fig. 2(b). The higher the environment humidity was, the darker and larger radius of points were, which presented the lower the systematic generating capacity. While the radius or color of the points didn't displayed any change rules with the increase of temperature in Fig. 2(b), so the environment temperature had no obvious influence on the systematic generating capacity. The energy of PV power generation is generated from the radiation of solar ray. The solar ray reaches the ground after passing through the earth's atmosphere, so that the higher atmospheric transparency is, the more solar energy transmits to the ground, accordingly which leads to the higher capacity. On the contrary, the higher environmental humidity means the more water vapor contained in the atmosphere, which would deplete the solar energy when solar ray passes through the atmosphere, so the lower systematic generating capacity is obtained. As shown in Fig. 2(c) and (d), the lighter and larger radius points were centralized in the upper of the figures. It was conclude that the higher the value of the accumulated solar irradiation was, the higher the corresponding generating capacity was. It was worth noting that the extreme value of systematic generating capacity appeared around 15°C (namely in winter) and 25°C (namely in spring). By combined with the results of the influence of environmental humidity shown in Fig. 2(b), it was deduced that the phenomenon described above was ascribed to the subtropical monsoon climate in Shanghai where abundant rain and damp air in spring, while dry air in winner. It can be concluded that environmental humidity is one of principal factors affecting the generating performance of equipment. The low environmental humidity in spring and winter displayed the high capacity. Especially, when the environmental humidity is lower than 50%, the generating performance of system is superior.

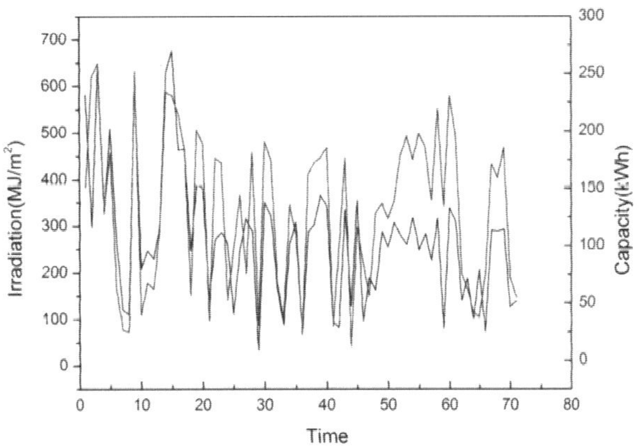

Figure 3. The relationship between accumulated solar irradiation and the corresponding capacity.

According to the results of analysis on the influences of environmental temperature and humidity, it was concluded that solar radiation exactly played a critical role in systematic generating capacity It was deduced that systematic generating capacity was almost completely consistent with the change trend of solar radiation in Fig. 3(d). The stronger the solar radiation was, the higher the systematic generating capacity was. There was a positive correlation between both of them. So the generating performance of equipment can be improved obviously by perfecting the received conditions for solar radiation and reducing the environmental humidity where the system was built.

5. Conclusions

The established grid-connected PV power generation project in Shanghai was taken as a case to calculate the economic efficiency of its whole life cycle, whose LCOE is 0.9991 yuan•kWh^{-1}, in the meantime, the increment cost analysis was adopted to evaluate the carbon emission reduction cost of the project, as well as the analysis on the sensitive factors. It's concluded that if the equipment is installed in dry areas where solar irradiation and airiness are great, it is helpful to exert the performance of systematic power generation. With the increasing popularity of distributed PV power generation, it is bound to impact on the traditional grid, so how to get a relationship of a harmony and balance will become the focus of attention.

6. References

[1] W. Wang and S. Wang, "Status and Prospect of Chinese Distributed Photovoltaic Power Generation System", Bulletin of Chinese Academy of Sciences, vol.2, pp.165-172, Feb. 2016.

[2] Y. Sun, R. Wang, L. Xiao, Jian Liu, Y. Yu and X. Zhuang, "Economical and Environmental Analysis of Grid-connected Photovoltaic Systems in China", China PoPulation, Resources and Enviroment, vol.21, No.4, pp.88-94, Apr. 2011.

[3] C. Ma, D. Shi and X. Cong, "The Research on the Genration Cost of Solar Photovoltaic Electricity and a Pathway to Grid Parity", Modern Economic Science, vol.36, No.2, pp.85-94, Mar. 2014.

[4] W. Short, D. Packey and T. Holt, "A Manual for the Economic Evaluation of Energy Efficiency and Renewable Energy Technologies", Office of Scientific & Technical Information Technical Reports, Mar. 1995.

[5] R. Chen, Y. Sun, S. Chen, H. Shen, "LCOE analysis of grid-connected photovoltaic power generation project", Renewable Energy Resources, vol.33, No.5, pp.731-735, May 2015.

[6] C. Yang, "Reconsidering solar grid parity", Energy Policy, vol.38, No.7, pp.3270-3273, 2010

[7] G. N. Doty,D. L. McCree, J. M. Doty and F. D. Doty, "Deployment prospects for proposed sustainable energy alternatives in 2020", ASME 2010 4th International Conference on Energy Sustainability, pp. 171-182, 2010.

[8] S.B. Darling, F. You, T. Veselka and A. Velosa, "Assumptions and the levelized cost of energy for photovoltaics", Energy & Environmental Science, vol.4, No.4, pp.3133-3139, 2011.

[9] M. Campbell, J. Blunden, E. Smeloff and P. Aschenbrenner, "Minimizing utility-scale PV power plant LCOE through the use of high capacity factor configurations", Photovoltaic Specialists Conference, pp. 000421-000426, 2009.

[10] J. He, X. Ma, Z. Zhao, H. Li and Y. Chen, "Study on CDM case for wind power generation", Electric Power, vol.39, No.9, pp.28-31, Sep. 2006.

[11] National Development and Reform Commission, Ministry of Housing and Urban-Rural Development of the People's Republic of China. Construction Project Pconomic Appraisal, Beijing: China Planning Press, 2006.

[12] X. Hu, K. Jiang and C. Cui, "Urban domestic waste incineration for electricity generation: a CDM case study", Energy of China, No.7, pp.21-27, 2006.

[13] L. Zhao, L. He, Q. Hao and B. Jia, "Study on the Test of Distributed Residential Grid-Connected PV Power Generation System", Energy and Energy Conservation, No.1, pp.63-67, Jan. 2017.

Research on Axial Forces Balance of Stamping-welding Canned Motor Pump

Tan Qianyan[1], Kong Fanyu[2], Bai Yuxing[3], Yuan Xin[4]

National Research Center of Pumps, Jiangsu University, Zhenjiang, Jiangsu 212013, China
e-mail: [1)]qian.tan@st.ovgu.de [2)]kongfy2918@sohu.com [3)] yxbai1012@163.com
[4)] yuan@starmatrix.cn

Abstract. In order to balance axial forces on stamping-welding canned motor pump for safe and reliable operation, the type PBCY80-65-160 pump was selected as the computational model. Combined with various reasons why axial forces of the pump were generated, the theoretical calculation and methods of numerical simulation were chosen to analyze the size, directions and rules of axial force. At the same time, the distribution and trend of the axial force components of the impeller under different working conditions are studied whether there is the balance hole or not. The results show that, in the theoretical calculation and numerical simulation calculation, the gap and internal impact loss of the internal flow field of axial forces were neglected when the internal flow of the motor was seen as an ideal flow through the resistant elements. Hence, it could not fully represent the internal pressure distribution of the pump, which caused the calculated data a little larger than simulated data. Also, the method of combining the balance hole and impeller rear seal ring could effectively balance the axial force and extend the service life of stamping-welding canned motor pump.

1. Introduction

Stamping-welding canned motor pump [1-3], named as plastic forming light pump, is a new type of pump. Stamping process manufacturing is applied for making the impellers, volutes and other parts of stamping-welding canned motor pump. Its simple structure could not only keep material and energy saving but also has high quality of flow surface and high precision hydraulic positioning, which could greatly improve the efficiency of the pump [4]. Balanced design technique of the axial force of the stamping pump is an important part of its safe and reliable operation because large axial force would lead to sliding bearing wear pump damaged. At present, the research on axial force of stamping-welding canned motor pump mainly has been learnt from the research result of Stepanov, Schultz in aspects of traditional pump. Axial force testing device of vertical stamping-welding canned motor pump which could measure the elastic modulus of normal and infiltrating resin Graphite, was designed by Luo Wen. Wang Jingquan introduced the application of axial force self-balancing in shielded pumps; Li Wei, designed a balanced axial force device generated by the auxiliary impeller in the stamping-welding canned motor pump. Zhang Liyuan researched on the measurement of axial thrust in axial shield pump [5-7]. The axial force of the shield pump has become one of the technical problems that constrain the large-scale. Therefore, it is of great significance to study the balance design of axial force for improving the reliability and economy of the shield pump.

In this paper, through the method of theoretical calculation and numerical simulation, dynamic reaction force and change rule of axial force in stamping-welding canned motor pump have been analyzed. Besides, a new device is designed with simple structure and easy installation. Furthermore, it could accurately and dynamically test the value and direction of the axial force during the operation of the stamping pump and its measured data would provide reliable basis for reducing the axial force. Meanwhile, the method of double-ring design could balance the axial force. By calculating the size of the stamping pump which is worked in safe and reliable operation double ring.

2. II Structure

Design parameters of Stamping-welding canned motor pump selected in this paper were as follows: Flow $\boldsymbol{Q}$ = 50m3 / h, head $\boldsymbol{H}$ = 32m, speed $\boldsymbol{n}$ = 2900r / min, motor power $\boldsymbol{P}$ =7.5kw.

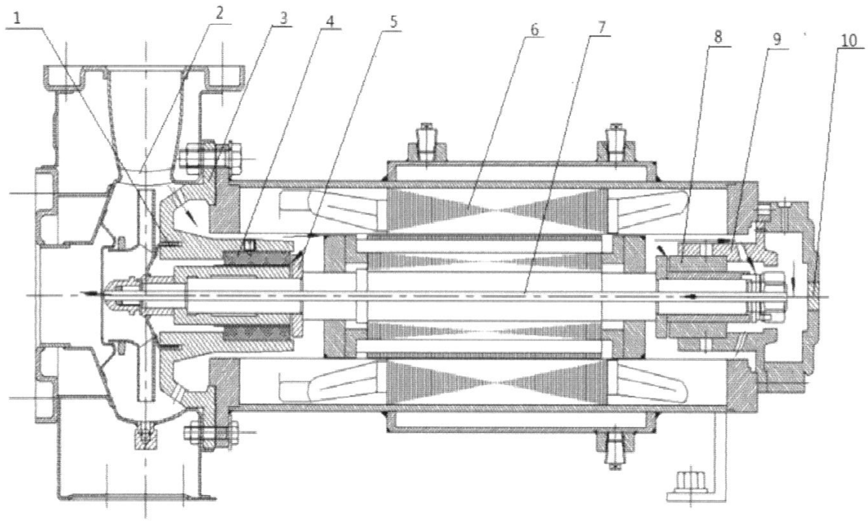

1.Wear ring 2. Volute 3. Front bearing seat 4. Front bearing seat 5. Thrust plate 6. Motor stator 7.Motor rotor 8. Rear guide bearing 9. Rear guide housing 10. Circulation tube

Figure 1. Stamping molding shield pump structure.

Combined with structural characteristics of punching Impeller, balanced hole and double ring design is applied for balancing the axial force of stamping-welding canned motor pump. In the overall design of the shielded motor, the problem of the axial force balance, cooling circuit design [8], sliding bearing lubrication and wear [9] should be considered, Fig.1 shows the structure of a typical stamping-welding canned motor pump.

3. Theoretical Calculation of Axial Force

Traditional calculation method [10] is used for the axial force of stamping-welding canned motor pump in covering force T_1, dynamic reaction T_2, the pressure difference of motor rotor at the both ends T_3, wheel hub end plate pressure T_4.

The axial force T as the whole pump rotor suffered:

$$T=T_1-T_2-T_3+T_4 \tag{1}$$

By calculation, the axial force of PBCY80-65-160 stamping shield pump under design condition is shown in Table 1:

Table 1. Axial Force of Pump 80-65-160

T_1/N	T_2/N	T_3/N	T_4/N	T/N
628.19	64.15	34.76	111.87	641.15

4. Numerical Analysis

The force in each fluid domain is extracted directly from CFD. Firstly, the numerical simulation [11-14] of the whole flow field of the stamping pump was carried out. Secondly, the cover force T_1, the motor rotor pressure difference T_3 and the wheel head end pressure difference T_4 are simulated respectively whether there is a balance hole or not.

4.1. Model

Taking the stamping molding shield pump as the research object, three-dimensional model of the entity for the main flow passage is shown in Fig.2:

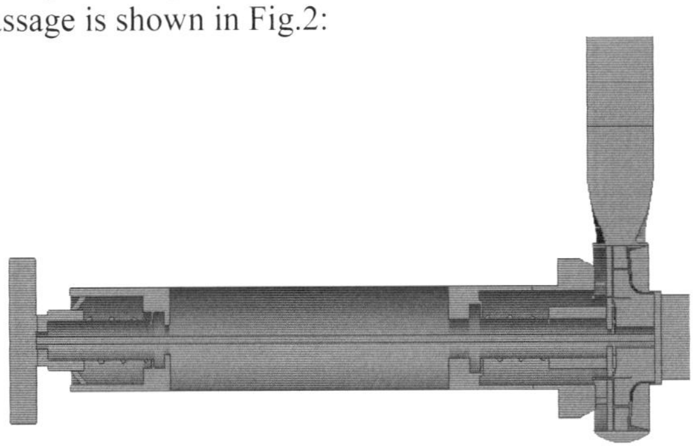

Figure 2. Three-dimensional model of the entity for the main flow passage.

4.2. Meshing

When a tetrahedron unstructured grid is used to mesh the small-sized gap flow in the fluid domain of the cooling cycle, it is difficult to ensure the grid quality and quantity at the gap. Therefore, the hexahedral structured grid is used to divide the fluid regions to ensure that there are more grid layers in the direction of the smaller gap thickness. After several numerical calculations, when the number of grid is 2.5 million, the calculation result is less affected by the number of grids. So it could be considered to satisfy the grid independence hypothesis. The number of grid cells in the pump-side calculation area is shown in Fig. 3 and Table 2:

Table 2. Number of Grid Cells in Calculated Area of Pump End

Model	Import extension	Import ring	Rear pump chamber	Impeller	Export extension	Front pump chamber	Volute	Sum
Number of Grids	160388	11000	372200	264659	177488	173844	335300	1494879

Figure 3. Grid calculation model.

5. Result Analysis

5.1. Cover force T_1 simulation

Numerical simulation could extract the internal force of the impeller cover and the surface force of the blade which could not be calculated accurately in the theoretical calculation. Impeller force mainly includes: Outside cover force F1 of impeller front and rear cover; Impeller front and rear cover plate inner surface axial force F2, blade surface axial force F3. Fig.4 shows the axial force distribution of the impeller.

5.2. Outside Cover Force F_1

The outside cover force includes: Lateral force of rear cover N_1; Lateral force of front cover N_2; Impeller seal end force N_3; Rear cover impeller of import part of the force N_4; Outside cover force $F_1 = N_1+N_2+N_3+N_4$; Fig.5 shows the change of the external force of the impeller front and rear cover plates without the equilibrium hole and the counterbalance hole in each working condition.

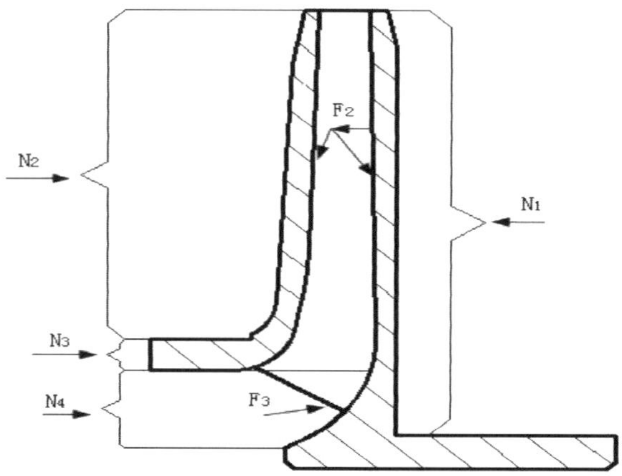

Figure 4. Impeller axial force distribution.

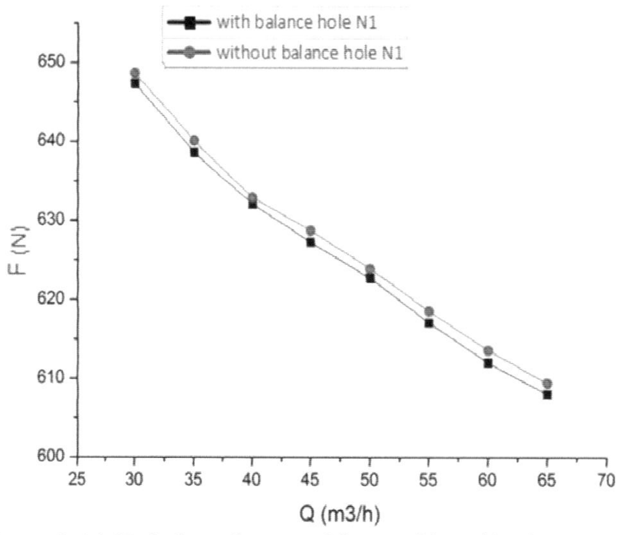

Figure 5. (a) Variation of external force of impeller front cover.

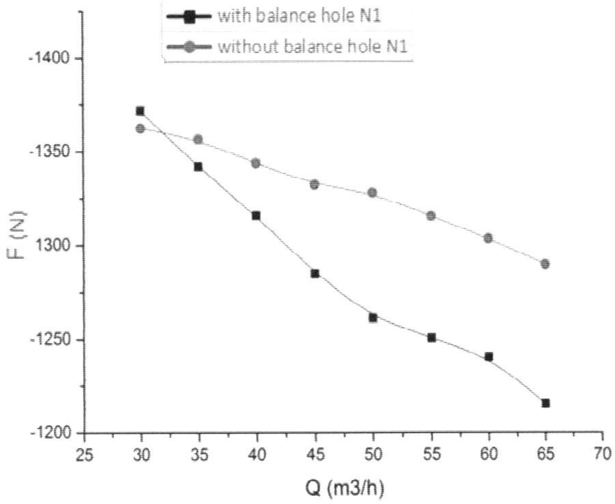

Figure 5. (b) Variation of Outside Force of Impeller Back Cover.

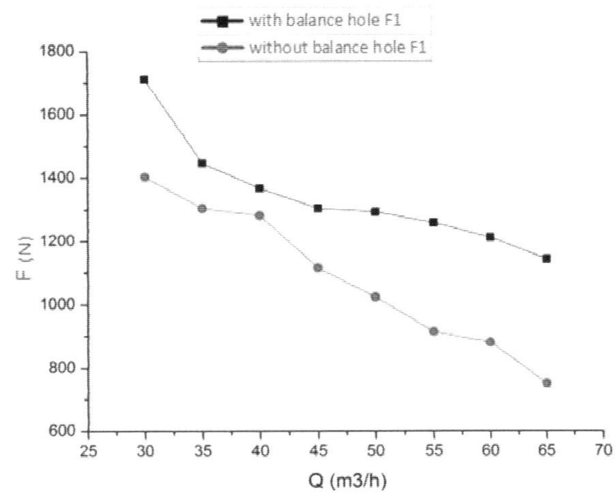

Figure 5. (c) Impeller outer cover plate force F_1.

Figure 5. Variation of the force acting on the outer side of the front and rear covers of the impeller under different working conditions

From Fig. 5(a), it could be concluded that lateral force of front cover is not changed whether there is a balance hole or not, and it tends to decrease with the rise of the flow rate. As well, the force is positive. It means that the direction of force is to the shielded motor side.

From Fig.5(b), it is showed that the change of rear panel lateral force whether there is a balance hole. In another word, the lateral force of the rear cover is significantly reduced with balance hole, so that the axial force of pump inlet direction is reduced to balance axial force. But the change of the force also makes the lateral force F_1 of the cover larger. The F_1 difference decreases first and then climbs up with the increase of q, and reaches the minimum value near the working point. The reason for the phenomenon may be that near the balance hole, The leakage of balance hole reach the maximum value and force the pressure on the lower liquid of seal ring to drop. Then, pressure difference of front and rear cover would be improved and to the most extent, the axial force on the cover plate is reduced. As a result, the force F_1 reaches the minimum value near the working point. As it is seen in Fig.5(c).

5.3. The Inner Surface Axial Force of Impeller Front and Rear Cover Plate F2

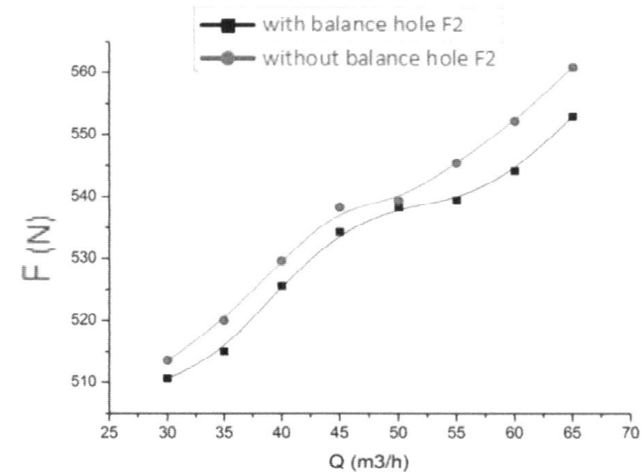

Figure 6. Axial force F_2 of inner surface of impeller front and rear cover. plate.

Fig.6 is the F_2-Q axial force curve of front and rear cover plate under different conditions in the balance hole and unbalanced hole and the force points to shield the motor side. Its value increases with the rise of flow. From the figure, the axial force of the inner and surfaces of the impeller front and rear cover plates is greater than that of the balance hole. It means the opening of the balance hole could reduce the pressure difference between the front and rear covers and achieve the purpose of reducing the axial force. However, the axial load is consistent with the flow rate in the state of the balanced and unbalanced hole.

Fig.7 shows axial force of blade surface. The blade is a cylindrical double circular blade, as shown in Fig.8, the blade is suffered by axial force in numerical simulation. Because the inlet edge of the blade has a certain angle with axial direction, so that the pressure on the inlet side produces a separated force on axial direction. Owing to low pressure and average area on the inlet edge of the blade, the axial force of the blade is smaller. A negative value in the fig.7 indicates that the force points to the pump inlet. It could be seen from the fig.8 that the axial force of the blade surface decreases with the rise of the flow rate when the equilibrium hole is added. However, when the balance hole is not added, the axial force of the blade increases with the rise of the flow rate.

5.4. Blade Surface Axial Force F3

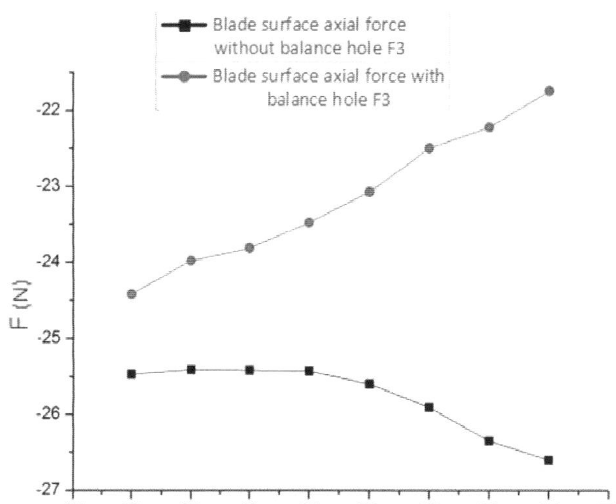

Figure 7. Axial force of blade surface.

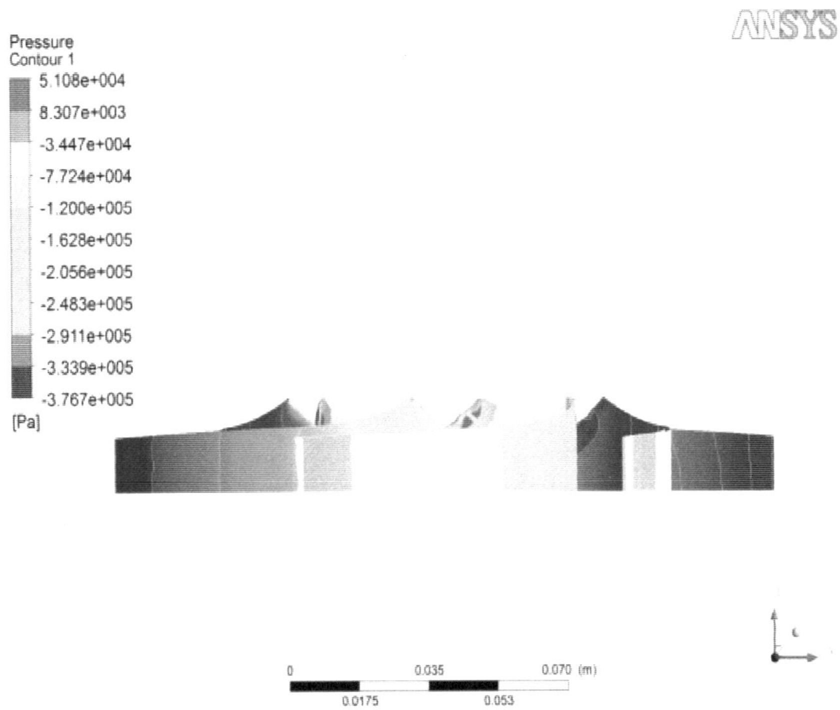

Figure 8. The Surface Pressure Diagram of Blade under Design Conditions.

6. Conclusion

1.This paper analyzes the causes of the axial force of the shielded pump, including the asymmetry of the front and rear cover of the impeller, the reaction force and the rotor structure of the shield pump motor. It is reasonable to use the method of combining the balance hole and the impeller seal ring.
2.Compared with the theoretical calculation, numerical analysis obtains the value of the axial force in the ideal situation, and the overall value is larger.

7. Acknowledgment

The author(s) disclosed receipt of the following financial support for the research, authorship, and/or publication of this article: this work was partly supported by the grants of National Natural Science Foundation of China(11602097) and the Postgraduate Innovation Foundation of Jiangsu Province (KYLX_1044).

8. Reference

[1] Qian Jiaxiang,Tong Zhicheng,Chen Wenzheng.The Characteristics of Stamping Pump and Its Application,Fluid Machinery,1999,27(8):30~33.
[2] Wang Jiaqiong, Zhao Ruijie, Kong Fanyu, Liu Yinyin, Duan Xiaohui.Research Status of Stainless Steel Stamping Welding Centrifugal Pump[J].Pump Technology,2015,04:11-15.
[3] Long Haiyong, Zhen Zechun,Luo Qiaojun.Research and Development of Stamping Welding Centrifugal Pump[J]. Chemical Equipment Technology,2011,05:35-37.
[4] Xie Rong, QiHeng,Shan Yujiao,Jin Weinan. Hydraulic design and numerical optimization of AP1000 nuclear pump guide vane. Journal of Drainage and Irrigation Machinery Engineering, 2014,12:1029-1034. m/.,
[5] Kong Fanyu, Chen Shiliang,Yuan shouqi.Expermental Research on Design of Circulating Loop of Canned Motor Pump[A].The 7th Asian International conference on Fluid Machinery[C].Japan:2003:2-8.

[6] Calum Scott.Considerations for a canned motor pump[J].World Pumps,2004,(12):22-25.
[7] Chen Ling, Cai Qi,Zhang Sheng.Grey fuzzy evaluation on technical condition of canned motor pump[J].Nuclear Power Engineering,2008,29(1):39-41.
[8] Kong Fanyu, Jiang Wanming,Wang Wenyan,Shen Xiaokai.Cavitation performance analysis for cryogenic high-speed canned pump based on coupled flow-heat[J].Journal of Huazhong University of Science and Technology(Natural Science Edition),012,05:24-28.
[9] Wu Renrong, Ling Zhiqiang,Wu Shulang.Development and Application of Water Lubricated Bearings for Pumps[J]. Pump Technology,2012,01:1-7+30.
[10] Kong Fanyu, Gao Cuilan,Ji Jiangang,et al.Axial force computation of canned motor pump based on visual C++6.0 software[J].Journal of Drainage and Irrigation Machinery Engineering, 2008,26 （6） ：15-19.
[11] Tan Linwei, Shi Weidong, Kong Fanyu, et al. Design and numerical simulation of cooling circuit in magnetic pumps[J]. Transactions of the Chinese Society of Agricultural Engineering (Transactions of the CSAE), 2015, 31(8): 84 − 91.
[12] Shi Weidong, Li Qifeng, Lu Weigang, etal .Estimation and experiment of axial thrust in centrifugal pump based on CFD[J].Transactions of the Chinese Society for Agricultural Machinery， 2009,40(1) 60-63

IOP Conf. Series: Earth and Environmental Science **168** (2018) 012027 doi:10.1088/1755-1315/168/1/012027

Analysis of Internal Flow in Boiler Water Circulating Pump

Tong Kai[1], Feng Zizheng[2], Kong Fanyu[1], Duan Xiaohui[1], Wang Kai[1], Dai Tao[1]

[1]National Research Center of Pumps, Jiangsu University, Zhenjiang, China
[2]The Eleventh Research Institute of the Sixth Academy, CASC, Xi'an, China

e-mail: tongkai94@163.com, fzz521521@aliyun.com

Abstract. As one of the core components of supercritical power station, boiler water circulating pump is a kind of glandless pump in which motor and pump are integrated. The circulating medium in the pump is high temperature and high pressure water. Understanding the internal flow characteristics is helpful to improve the hydraulic design of the pump. The numerical simulation of pressure distribution and streamline in the pump body of boiler water circulating pump were carried out with software ANSYS-CFX under three working conditions (0.6 Q, 1.0 Q and 1.4 Q). The results show that: The pressure near the outer wall of pump body is larger, and gradually decreases to the interior. With the decreases of cross section radius, the number of vortices increases. The vortices gradually decrease with the increase of flow rate, and the flow becomes more and more uniform. The pressure increases from the inlet pipe to the outlet pipe under different flow rates. The inlet and outlet pressure difference is inversely proportional to the flow rate. Under the condition of small flow rate and large flow rate, there are many low pressure zones in the inlet side of the blade.

1. Introduction

Boiler water circulating pump is an important component which forces circulation in supercritical power stations. The medium in the boiler water circulating pump is high temperature and high pressure water. Most of the current analysis of boiler circulating pump mainly is concentrated on the study of system fault [1]-[3], while the analysis of internal flow is scarce [4]-[6]. Fan Yizhang analyzed the pressure fluctuation in boiler water circulating pump with different inner diameters, the optimum inner diameter volute was confirmed and the efficiency was raised [7]. It was pointed out in Hideo Ohashi's paper that the fluid dynamic action between stator and rotor would cause partial pressure fluctuation, which would cause vibration of rotating machinery [8]. This paper analyzed the internal flow of a boiler water circulating pump under three working conditions (0.6 Q, 1.0 Q and 1.4 Q), and provide some theoretical guidance for the design and improvement of boiler water circulating pump.

2. Numerical Simulation

2.1. Calculation Model

The basic parameters of the boiler water circulating pump are: design flow Q=965m^3 /h, head H=133m, rotate speed n=2950r/min, number of blades Z=6, impeller outlet diameter D=346 mm, impeller outlet width b=40 mm, temperature T=360°C, pressure P=36MPa.

Figure 1 shows a schematic diagram of boiler water circulating pump. Modeling software Pro/Engineering is used to build 3D model of boiler water circulating pump.

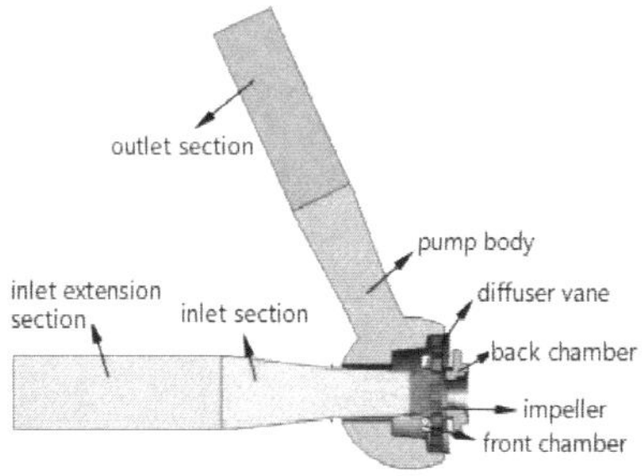

Figure 1. Schematic diagram of boiler water circulating pump.

In order to make the flow of inlet and outlet fully developed, the inlet and outlet sections were extended. The extension was three times as long as the diameter of the inlet and outlet.

2.2. Mesh Generation

The numerical simulation of the boiler water circulating pump is carried out by ANSYS. The method used is finite volume method. In order to analyze the condition of internal flow, the flow field mesh should be generated. Considering the convergence of grid and limitation of computer performance, hexahedral mesh was adopted. In addition, considering the complexity of the swirl near the wall and the effect of the blade, the mesh of each blade is encrypted. The quality of each part of the water body divided by software ANSYS-ICEM is above 0.35, and the y+ value of the boundary layer is above 200, which is in accordance with the calculation requirements of Computational Fluid Dynamics. The assembly mesh of boiler water circulating pump is shown in Figure 2.

Figure 2. Assembly mesh of boiler water circulating pump.

The mesh is checked by the grid independence which is shown in Table 1. The total number of grids is selected as 2.9×10^6 through grid independent analysis.

TABLE I. GRID INDEPENDENCE CHECKING

Mesh Number	Forecast Head/m	Forecast Efficiency /%
2.0×10^6	138.0	80.0

2.4× 10^6	139.5	81.0
2.9× 10^6	141.2	81.4
3.4× 10^6	141.2	81.4
3.9× 10^6	141.3	81.5

2.3. Numerical Calculation Method and Boundary Condition

In the actual operation of the pump, the internal flow is incompressible turbulence. In order to solve the complex flow problem under different working conditions, the k-ε model, RNG k- ε model, SST model and so on are put forward, and these models are used to calculate the corresponding practical problems. In order to obtain accurate flow in the internal flow of boiler water circulating pump, the SST model, mixed by k-ω model and k-ε model, is applied in this paper. This model has a high accuracy for the swirl near the wall calculation. The basic control equations of SST are as follows:

$$\frac{\partial(\rho k)}{\partial t}+\frac{\partial}{\partial x_i}(\rho u_i k)=\frac{\partial}{\partial x_i}\left[\left(\mu+\frac{\mu_t}{\sigma_k}\right)\frac{\partial k}{\partial x_i}\right]+P_k-\beta^{'}\rho k\omega$$

$$\frac{\partial(\rho\omega)}{\partial t}+\frac{\partial}{\partial x_i}(\rho u_i\omega)=\frac{\partial}{\partial x_i}\left[\left(\mu+\frac{\mu_t}{\sigma_k}\right)\frac{\partial\omega}{\partial x_i}\right]+D_m+\alpha\frac{\omega}{k}P_k-\beta\rho\omega^2$$

The constants in the equations above are defined as follows: β'=0.09, a=5/9, β=0.075, σ_k =2, ρ represents density, P_k represents the productivity of turbulence.

Boundary conditions should be set up according to the actual situation in the calculation. The flow field of the impeller was set up as Rotating Domain, whose speed was 2950r/min, and the other flow fields are set up as Stationary Domain, and the connection between the motion and static domain was set up as Frozen-Rotor.

The boundary condition of inlet should be set to pressure and the outlet should be set to mass flow. The residual convergence accuracy is set to 10^{-5}.

3. Analysis of Calculation Results

3.1. Analysis of Overall Pressure Distribution

The circulating medium in the boiler water circulating pump is high temperature and high pressure water, so the pump body generally adopts a hemispherical shape. However, the hemispherical shape is not conducive to the flow and diffusion of the fluid. Understanding the overall pressure distribution of the water circulating pump is helpful to improve the hydraulic design of the pump.

Figure 3. shows overall pressure distribution on cross section of shaft. The pressure increases from the inlet pipe to the outlet pipe under different flow rates. The inlet and outlet pressure difference is inversely proportional to the flow rate. Under the condition of small flow rate and large flow rate, there are many low pressure zones in the inlet side of the blade.

3.2. Analysis of Internal Flow in Pump Body

The internal flow of the pump body is analyzed by four cross sections, which is shown in Figure 4. A-A cross section is axial cross section, B-B cross section is the z=150mm plane, C-C cross section is the z=-50mm plane, and D-D cross section is z=-250mm plane near the pump inlet.

IOP Conf. Series: Earth and Environmental Science **168** (2018) 012027 doi:10.1088/1755-1315/168/1/012027

Figure 5. to Figure 8. show the pressure cloud diagram of four cross sections under three working conditions (0.6 Q, 1.0 Q and 1.4 Q). It can be seen from Figure 5, the internal pressure distribution of the pump is disordered, the pressure at the bottom of the pump close to the outer wall is greater, and there is a low pressure zone in the part of the outlet section. As can be seen from the pressure cloud diagrams of B - B, C - C and D - D cross section, the pressure near the outer wall is larger, and gradually decreases to the interior, some areas have low pressure or high pressure zone. The pressure distribution is more and more uniform with the decrease of Z value (away from the inlet of the pump). Compared with different working conditions of the same cross section, the pressure distribution under large flow rate is relatively uniform.

0.6Q 1.0Q 1.4Q

Figure 3. Overall pressure distribution on cross section of shaft.

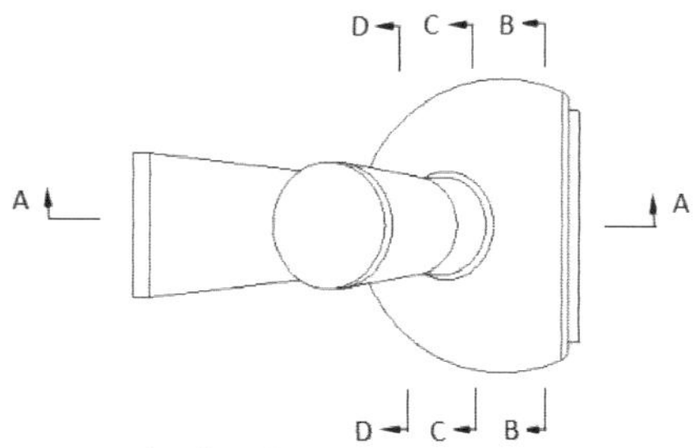

Figure 4. Distribution diagram of section position.

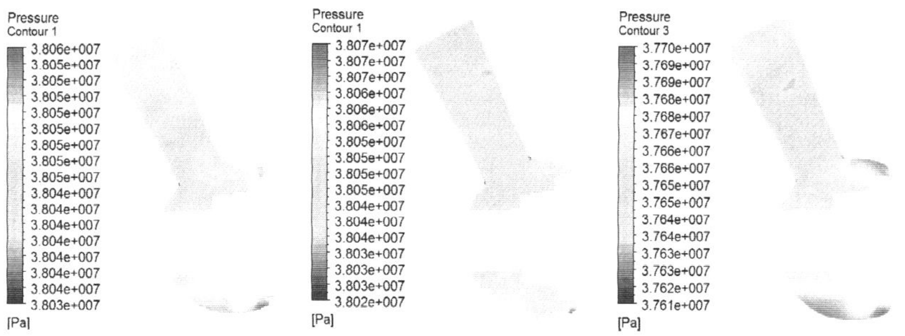

0.6Q 1.0Q 1.4Q

Figure 5. A-A section pressure distribution.

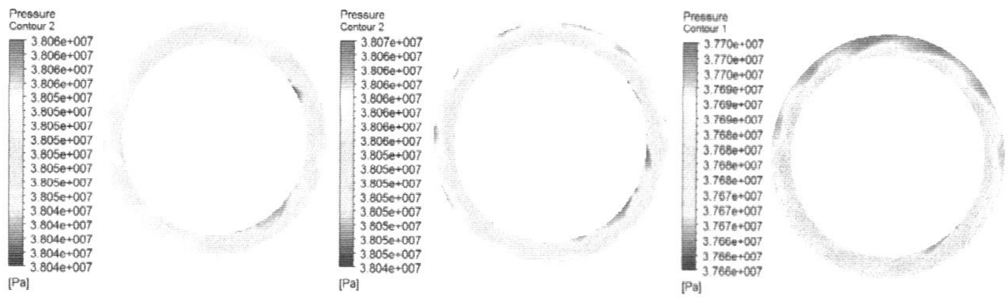

0.6Q 1.0Q 1.4Q

Figure 6. B-B section pressure distribution.

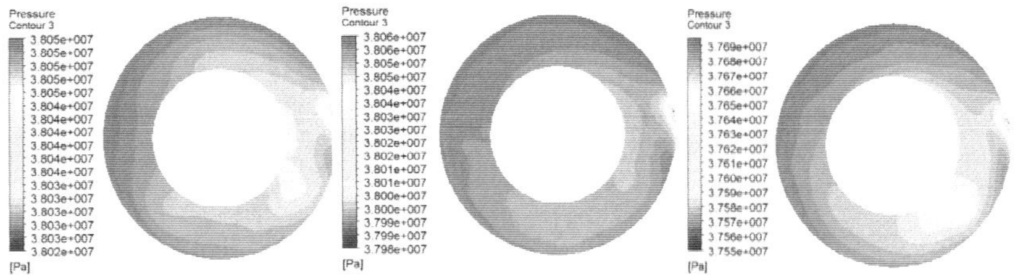

0.6Q 1.0Q 1.4Q

Figure 7. C-C section pressure distribution.

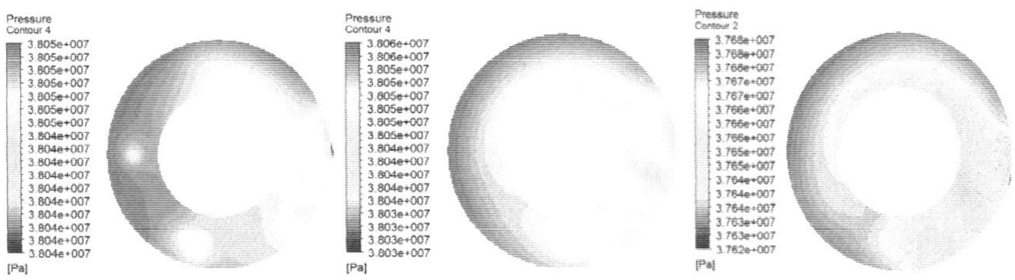

0.6Q 1.0Q 1.4Q

Figure 8. D-D section pressure distribution.

Streamlines of cross sections A-A, B-B, C-C, and D-D are shown in Figure 9. to Figure 12. According to Figure 9, on cross section A-A, there are a lot of backflow vortices at different flow rates, which are mainly concentrated in the pump body inner ring. For B-B, C-C, D-D cross sections, as the radius of cross section decreases, the number of vortices increases. Vortices on C-C cross section are mainly concentrated in inner circle, with low flow velocity and more stagnant water area. The vortices on D-D cross section are larger and more disordered. The vortices gradually decrease with the increase of flow, and the flow becomes more and more uniform.

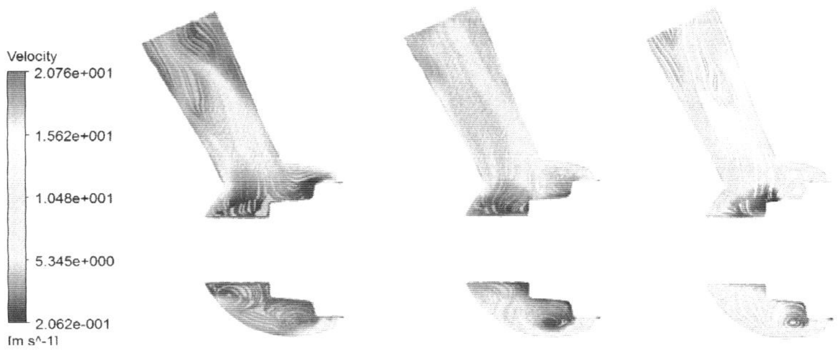

0.6Q 1.0Q 1.4Q

Figure 9. Streamline diagram of A-A cross section.

0.6*Q* 1.0*Q* 1.4*Q*

Figure 10. Streamline diagram of B-B cross section.

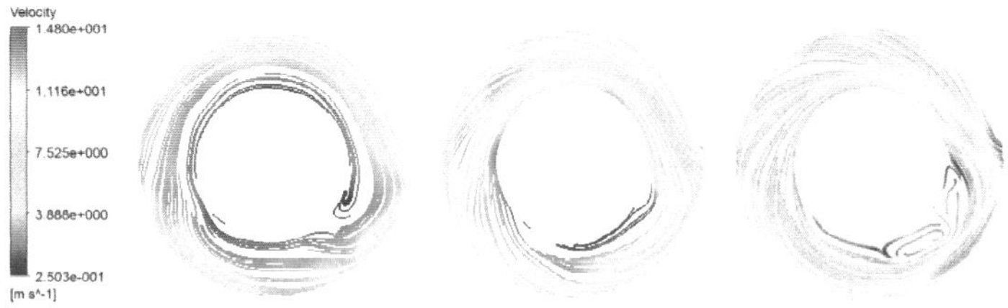

0.6*Q* 1.0*Q* 1.4*Q*

Figure 11. Streamline diagram of C-C cross section

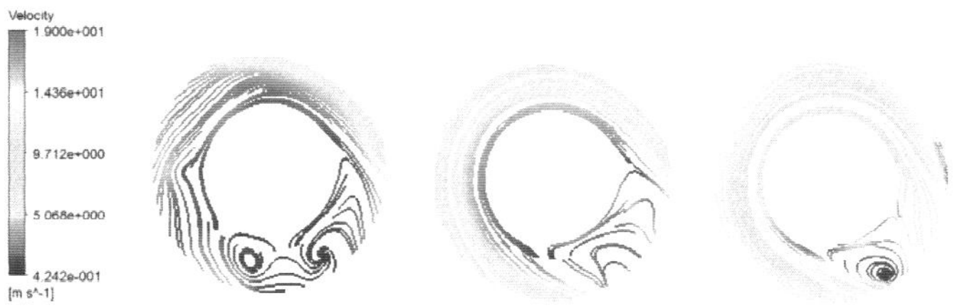

0.6*Q* 1.0*Q* 1.4*Q*

Figure 12. Streamline diagram of D-D cross section

4. Conclusion

The internal flow in the boiler water circulating pump were analyzed by numerical simulation. The results show that

1. The pressure near the outer wall of pump body is larger, and gradually decreases to the interior. Compared with different working conditions of the same cross section, the pressure distribution under large flow rate is relatively uniform.

2. As the radius of cross section decreases, the number of vortices increases. The vortices gradually decrease with the increase of flow, and the flow becomes more and more uniform.

3. The pressure increases from the inlet pipe to the outlet pipe under different flow rates. The inlet and outlet pressure difference is inversely proportional to the flow rate. Under the condition of small flow rate and large flow rate, there are many low pressure zones in the inlet side of the blade.

5. Acknowledgment

Special thanks are given to the Postgraduate Innovation Foundation of Jiangsu Province (KYLX_1044), the Project supported by the National Natural Science Foundation of China(11602097), and A Project Funded by the Priority Academic Program Development of Jiangsu Higher Education Institutions (PAPD4013000011).

6. References

[1] Xi Hui. “Boiler Water Circulating Pump Motor Failure Analysis and Treatment ,” Shanxi Electric Power, vol. 196, Feb. 2016, pp.69-72, doi:1671-0320(2016)01-0069-04

[2] Li Peihua, Dai Yunhe. “Analysis and Countermeasure on Abnormal Temperature Rising of Boiler Circulating Pump,” Electric Power Science and Engineering, vol. 26, Jan. 2010, pp.55-58.

[3] Yang Hong-min,Duan Jing-wei, Li Xin-guo. “Common Failures Analysis and Countermeasures of LUV Boiler Water Circulation Pump ,” Power System Engineering, vol. 27, Jul. 2011, p.23-25, doi: 1005-006X(2011)04-0023-03.

[4] Guo Rongxiang,Zhang Luhao,He Xiaoyue,. “The research of fuzzy control algorithm of circulating pump of heating boiler,” 2nd International Conference on Artificial Intelligence, Aug. 2011, pp.5963-5965, doi: 10.1109/AIMSEC.2011.6010113

[5] G Laschet, R Krewinkel, P Hul, et al. “Conjugate analysis and effective thermal conductivity of effusion-cooled multi-layer vane sections,” International journal of heat and mass transfer, vol. 57, Feb. 2013, pp.812-821.

[6] Baumgarten S, Friitz J,Knierim C,et al. “Numerical simulation of the flow in a boiler circulating pump,” Proceedings of the Institution of Mechanical Engineers, Part A:Journal of Power and Energy, vol. 215, Jun. 2001, pp.793-800.

[7] Fan Yizhang. “Study on Optimization of LUV Pump’s Casing Dimension Based on Numerical Simulation,” Tsinghua University,2013.

[8] Hideo Ohashi. “Case Study of Pump Failure Due to Rotor-Stator Interaction,”. International Journal of Rotating Machinery, vol. 1, Jan. 1994 , pp.53-60.

IOP Conf. Series: Earth and Environmental Science **168** (2018) 012028 doi:10.1088/1755-1315/168/1/012028

Study on Some Properties of Hydrodynamic Cavitation Damage

Zhaocun Liu[†1,2], Xiaoqing Wang[1,2], Weijia Fan[1], Kui Yv[1]

[1]Chongqing Jiaotong University,Chongqing, China
[2]Key Laboratory of Communication Civil Waterway Regulation,Chongqing, China

e-mail: Liuzc17@aliyun.com

Abstract. Some hydrodynamic cavitation damage properties were studied in this letter. From the viewpoint of cavitation damage, the flowing characters near the wall, such as three features of the turbulent structures, streak structure; vortices; and bursting phenomena, the energy distribution, and the fluctuation character were explored. The characters of cavitation and cavitation damage were investigated, the relations among the cavitation, coherent structure and viscous effect were explored, and the energy characters in cavitation damage flowing field was studied. The results show that, the cavitation damage will have close relationship with the value distribution of the Lamb vector and the helicity. The relation between the cavitation damage and the Reynolds number essentially results from the eddy effect. The conceptive formula of cavitation damage was given. The paper clearly demonstrated the affecting factors to dominate the cavitation damage to make it show stochastic characters. These results may be useful to civil engineers.

1. Introduction

Cavitation can appear in both water and any other kind of fluid. Here, we mainly concern the hydrodynamic cavitation in present letter. The cavitation will be caused by the voids or bubbles containing vapor and gas in fluid if the pressure falls locally to that of the vapor pressure corresponding to the ambient temperature. The low pressure may be caused by a high speed or vibrations or others. The essential mechanism of acoustic cavitation is identical to that of hydrodynamic ones, so does the cavitation damage.

The governing equations of the hydrodynamic bubble and that of acoustic ones are same. If a bubble collapses in the vicinity of a solid surface, then it may cause the cavitation damage, and none material will be free of cavitation erosion. The problem has challenged the mankind for more than one century. The cavitation damage is accepted at present as such, it is due to the concentration of mechanical energy on very small areas of the walls exposed to cavitation to exceed the resistance of the material resulting from the collapse of vapor structures. Usually, the cavitation contains the stages of formation, growth, oscillation, and collapse of bubbles in a liquid. As for the acoustic cavitation, besides above characters, it has effect of emission of light, dubbed as sonoluminescence (SL), which is often divided into single-bubble sonoluminescence (SBSL) and multibubble sonoluminescence (MBSL).

Both the turbulence and the cavitation damage remain open problems now, though there are many research results [1-15]. In present letter, from the viewpoint of cavitation damage, we combine the turbulence and the cavitation damage together, to give a conceptive picture to study the some hydrodynamic cavitation damage properties.

2. Flowing characters near the Wall

Here, we take the wide open channel flow as an example. In early time, i.e. from the 1910s to the 1930s, turbulence was regarded as a completely stochastic phenomenon in which a randomly fluctuating portion of velocity field is superimposed. In the 1960s, people knew that large-scale motions govern the transport properties of a fluid flow, and small-scale motions are responsible mainly for dissipative processes. Traditionally, the profile of the mean velocity is divided as: a) a linear sub-layer, $0 < y^+ < 5$, where $u^+ = y^+$; b) a buffer layer, $5 \leq y^+ < 30$, the region of maximum average production of turbulent kinetic energy; c) an overlap layer, $30 \leq y^+$, the approximate energy equilibrium; and d) a far outer layer, where the law-of-the-wake holds. Here $y^+ = (u_* y)/\upsilon$, $u^+ = (u/u_*)$, and υ is the kinematic viscosity coefficient, and u_* is the shear velocity, y is the distance from wall, and u is velocity. As for the open channel flow case, the fluctuating energy mainly come from the region near the bottom. The energy flux of mean flow transfers from the channel surface to bottom, while the energy flux of fluctuating flow transfers from bottom to surface, and this process is dynamic and reversible. From the viewpoint of phenomenology, the infinite hierarchy structure of flow is only a media in which the energy transfers between the mean flow and the fluctuating flow.

There are three features of the turbulent structures near wall, a) the forming and developing of streak structure; b) the forming and developing of longitudinal vortices, transverse vortices and horseshoe vortices; c) the bursting phenomena. The mutual influences among the three forms affect the coherent structure, such as low-speed streak, high-speed streaks, burst, ejections, sweeps, and various vortex structures. The vortex structure plays an important role, and the hairpin vortex is regarded as a basic vortex structure. The regeneration of various vortex structures is the structure source for turbulent boundary self-sustaining, and the burst is the energy source for turbulent boundary to sustain itself. The fluid flow near wall is closely related to that of the outer region to unify as a whole system.

In the region where the cavitation damage occurs, though the surface of the solid is not always to parallel to a plane, showing various appearance with different curvatures, however, the structures of the flowing field are mainly same as above.

3. Cavitation and Cavitation Damage

In case of pressure =1 atm and at temperature=25 °C, so called in normal conditions, the radius of nuclei is often scale with (~) $R = 5\times10^{-6}$ m. In cavitation study, there is a useful parameter, σ_i, which is called cavitation number, it reads:

$$\sigma_i = \frac{p_\infty - p_V(T_\infty)}{0.5\rho_L u_\infty^2} \tag{1}$$

Here, V is the total gas content, and p_∞ is the ambient pressure. T_∞ and u_∞ are ambient or referenced temperature and velocity, respectively, ρ_L is the liquid density.

Usually, the Rayleigh-Plesset equation is the governing equation of a spherical bubble in fluid:

$$\frac{p_B(t) - p_\infty(t)}{\rho_L} = R\frac{d^2R}{dt^2} + \frac{3}{2}(\frac{dR}{dt})^2 + \frac{4\upsilon}{R}\frac{dR}{dt} + \frac{2\sigma}{\rho_L R} \tag{2}$$

Where, R is bubble radius, t is time, p_B(t) connects with equation of state. Without appreciable mass transferring of gas to or from the liquid, and the bubble containing some quantity of contaminant gas with partial pressure p_{G_o} at size, R_o, and temperature T_∞. Then the equation of state (EOS) of gas in the bubble will become:

$$p_B(t) - p_V(T_B) = p_{G_o}(\frac{T_B}{T_\infty})(\frac{R_o}{R})^3 \tag{3}$$

Here, $p_V(T_B) = p_V(T_B, \rho_V)$, $\rho_V(T_B)$ and $p_V(T_B)$ are the saturated vapor density and pressure at the bubble temperature T_B respectively. Though this is only approximately, it grasp the essential characters in most cases. Sometimes, the temperature difference, $(T_B(t) - T_\infty(t))$, leads to a different $p_V(T_B)$, and this change the bubble dynamics, then, considering the equations of (2) and (3) , one has:

$$\begin{aligned} &\frac{p_V(T_\infty) - p(t)}{\rho_L} + \frac{p_V(T_B) - p_V(T_\infty)}{\rho_L} + \frac{p_{G_o}}{\rho_L}(\frac{T_B}{T_\infty})(\frac{R_o}{R})^3 \\ &= R\frac{d^2R}{dt^2} + \frac{3}{2}(\frac{dR}{dt})^2 + \frac{4\upsilon_L}{R}\frac{dR}{dt} + \frac{2\sigma}{\rho_L R} \end{aligned} \tag{4}$$

The meanings of the symbols are same as above. In equation (4), the first term denotes the instantaneous tension or driving term determined by the conditions far from the bubble. The second term is called the thermal term which may cause very different bubble dynamic behaviors.

Research results show that, the theoretical predictions based on the equation (4) are roughly correspondence well with experimental results. If the dissipation can be neglected, then the rebound of the bubble will repeat for ever with definite story. To give a concrete model, we recommend a famous example, cited by many researchers, a typical single-bubble sonoluminescence bubble containing argon with $R_0 = 4.5$ μm, driven at f=26.5 kHz and $P_a = 1.2$ atm, where, P_a is the pressure amplitude of the sound waves, and the ambient pressure $P_\infty = 10^5$ Pa ≈ 1 atm. The surrounding liquid far from the bubble is degassed to some level and maintained at room temperature. Consequently, the bubble contains about 1010 argon atoms and about 2×10^8 water molecules at the outset. The story is as follows. (1) Expansion. The bubble expansion is comparatively slow and the growth is sustained for almost half a cycle (~15 μs). In this phase, the bubble is in both thermal and mass transfer equilibrium with the liquid. Because of the falling pressure inside the bubble, it gains large numbers of water-vapor molecules (evaporating from the wall) and also some gas molecules from the liquid. (2) Turnaround at maximum radius. The driving pressure begins to increase again, and the expansion comes to a halt. At maximum radius, then there is $R_{\max} \approx 7R_0$. (3) Rayleigh collapse. As the external pressure increases, the inertial collapse of the liquid layers around the bubble begins. The radius decreases quickly (over about 4μs) from $R_{\max}$ to a value comparable to R_0. During this collapse, water vapor recondenses at the wall and the argon atoms again become the dominant species inside the bubble. The bubble dynamics near minimum radius occurs on a time scale (''turnaround time'') of ~1 ns . While the typical pulse width of SBSL 100–200 ps is still much smaller than this value. The other experimental measurements also show that this upper bound for the pulse width was much smaller than the time during which the bubble remained in its most compressed state. (4) Decoupling of water vapor. About 50 ns before the minimum radius is reached, the time scale of the bubble collapse becomes smaller than the time scale for the diffusion of water vapor. The water vapor still left inside the bubble is now trapped until the reexpansion. (5) Thermal decoupling. Only ~30 ns later, the accelerating bubble wall becomes fast enough that heat can no longer escape the bubble. (6) Onset of dissociation reactions. Once the temperature exceeds roughly 4000 K, water-vapor molecules start dissociating into OH- and H+ radicals. (7) Onset of light emission. Despite the temperature-limiting influence of water vapor, about 10000 K is finally reached in the bubble about 100 ps before maximum compression. At this temperature, with mechanisms of thermal bremsstrahlung and radiative recombination, finally, the SBSL is observed. (8) Maximum compression. At this point, the gas density reaches (almost) solid-state values. The deceleration of the bubble wall down to zero speed has begun to enhance random shape perturbations (Rayleigh-Taylor instability) and leads to massive energy loss through acoustic wave emission. The temperature and light emission peak, helped by the high densities that prevent further endothermic dissociation reactions. (9) Reexpansion. The bubble loses about 90% of its energy in the collapse, mostly due to acoustic emission. The reexpansion is

IOP Conf. Series: Earth and Environmental Science **168** (2018) 012028 doi:10.1088/1755-1315/168/1/012028

much slower than the collapse. The Rayleigh-Taylor instability grows and may overwhelm a strongly driven bubble during this stage. Only a small increase in radius and decrease in temperature are sufficient to dramatically reduce the photon absorption coefficient and quench the light emission uniformly for all wavelengths, about 100–200 ps after it begun. Thermal and diffusive equilibria are reestablished. (10) After bounces. The bubble rebounds to a much smaller size than the maximum radius before the main collapse. The afterbounces provide a parametric excitation that can accumulate and render the bubble shape unstable. The radial motion is, damped rapidly until the driving pressure dips into its negative cycle once again, and the oscillation starts anew. Over the whole cycle, shape perturbations may have been enhanced, then the bubble is parametrically unstable, or a net gain or loss of gas may have caused diffusive instability. In the correct parameter range, the bubble is stable with respect to both processes and continues to oscillate and emit light in exactly the same fashion. If in addition molecular gases such as nitrogen and oxygen are dissolved, the parameters, such as temperature may change, however, the whole mechanisms are identical. In practical civil engineering case, the cavitation phenomenon can be caused by a single bubble breakdown or/and the multi bubble breakdown.

In civil engineering case, the cavitation can be classified into different types by the bubble shapes and distribution in time and space. Then, the different researchers give different results. Such as (1) i) travelling bubble cavitation, ii) vortex cavitation, iii) attached or sheet cavitation; iiii) cloud cavitation; (2) i) travelling cavitation, ii) fixed cavitation, iii) vortex cavitation, iiii) vibratory cavitation; (3) i) bubble cavitation, ii) sheet cavitation, iii) cloud cavitation, iiii) and various forms of vortex cavitation; (4) i) bubble cavitation, ii) tip-vortex, iii) sheet cavitation, iiii) cloud cavitation, etc. The common basis among them lies that, how the low-pressure regions are generated.

The intermittency shows the small scale local moving pattern. The interactions between the vortices and waves show unsteady and nonlinear manners. The vortices and waves can moderate each other, and moreover, they can generate each other as well. So the cavitation is affected by the flowing field to show random characters. As the structures of the flowing field dominate the cavitation, there is a synchronism between the cavitation and the structures of the flowing field, though it is not so accurate sometime. The inverse U-shaped cavitation cloud is a relatively stable form to contain a strong vortex cavitation at the center surrounded by many small cavitation bubbles. The cavitation has local and random character.

Within a process of a single bubble cavitation, besides the light emission, the temperatures can be up to 20,000 K, and heating and cooling rates of $>10^{12}$ K/s, to suggest to generate high densities during bubble implosion. Cavitation can generate vortical structures, thus, it is a source to generate vorticity. In the final stages of collapse of a cavitation bubble, the pressure reach a high value (>1GPa) to force the liquid near the bubble wall briefly (~1 ns) into a metastable state of subcooling. As the coupled effect between the cavitation damage and the flowing field, this problem become more complicated.

The cavitation damage show 4 stages: 1) incubation stage. There is no detectable weight loss in this period; 2) accumulation stage. There is significant increase of erosion rate with the worn surface to become rougher with a large number of small pits and deep craters; 3) steady stage. This stage possesses a constant erosion rate; 4) attenuation stage, it appears only under certain conditions. Cavitation damage has such character, both the craters and the pits do have nothing with the material nature, grain boundaries, slip lines, or any other structure feature. The cavitation damage shows random characters.

The phenomenon of cavitation erosion connects with the gas dynamics, thermodynamics, hydrodynamic and material properties. A pit damage and a circular damage are the basic two forms. Studies show that, the velocity of the liquid jet during the bubble collapse can exceed 1.2km/s, even arrive at 7 km/s, a blast wave with peak overpressure exceeding several GPas. If a cavitation event happens, it can produce a lot of tiny bubbles. If the bubble happens to collapse vicinity near the solid surface, it can cause cavitation damage.

4. The relations among the Cavitation, the Coherent Structure and the Viscous Effect

IOP Conf. Series: Earth and Environmental Science **168** (2018) 012028 doi:10.1088/1755-1315/168/1/012028

If mass force is potential (such as the gravity), Navier-Stokes (N-S) equation can be replaced by the vorticity transport equation (here, Q, U, L are the uppercase vector forms of ω, u, l): $\frac{\partial Q}{\partial t} + \nabla \times L = \upsilon \Delta Q$, Lamb vector $L = Q \times U$, $Q = \nabla \times U$, U is the velocity vector. The space with large-magnitude Lamb vector accompanies high energy dissipation, while the space with high helicity ($h = Q \cdot U$) accompanies low energy dissipation to make the spiral structure in flow field maintain a quite long time. If we define a dimensionless parameter as $\frac{\|\nabla \times L\|}{\|\upsilon \nabla^2 Q\|}$, it behaves like the Reynolds number in cases where the denominator is non-zero, while it is feasible to apply the tiny flow element. Thus the Reynolds number can be considered as the ratio of the rate of momentum change in tangent direction to the viscous force in normal direction, which connects both the distortion in tangent direction and the dilating process in normal direction, and associates with fluctuation mechanism as well. As the Reynolds stress originates from diffusion and transport, so the Reynolds number connects with the Reynolds stress. Interactions among resonant vibration, frequency locking, and different Coriolis forces make the Reynolds stress show inhomogeneous and anisotropy properties, thus, $\mathrm{Re}_{ij} = F(\tau_{ij}, v_i, \varepsilon, u_*)$, where τ is shear stress, and ε is the dissipation rate of turbulent energy per unit mass. Traditionally, the Reynolds number is only a dimensionless parameter, it being a tensor form may cause a paradox. In fact, the Reynolds number is used to act as a parameter to describe the flowing transition state. Usually, $\mathrm{Re} = \frac{\mathrm{u_c L_c}}{\upsilon}$ = Inertial force/viscous force. Here, $\mathrm{u_c}$ is characteristic velocity and $\mathrm{L_c}$ is characteristic length. As explanation of above Reynolds number, The Lamb vector connects closely with the eddy. The Reynolds stress, $\tau_{ij} = -\rho \overline{u_i' u_j'}$, connects obviously with fluctuating motion. The more the Reynolds stress is big, the more the fluctuation is strong, and the more eddies are abundant, thus, the cavitation damage will appear easily. At some value of the Reynolds number, the cavitation damage will have close relationship with the value distribution of the Lamb vector and the helicity.

As for the vortex stability, in the direction of increasing pressure, the instability will increase. Both viscous diffusion and accelerated movement have stabilizing effect. In the regime for resistance with the square of velocity, According to Kolmogorov's theory of turbulence, $\eta = \upsilon^{0.75} \varepsilon^{-0.25}$, where η is the Kolmogorov's length, and $\varepsilon = (\bar{u})^3 / R$, R is some characteristic length. It is known that the exact equation of velocity distribution for a turbulent channel flow is

$$\frac{\partial u^+}{\partial y^+} = 1 - \frac{y^+}{H^+} + \overline{u_+' v_+'} \quad (5)$$

where u_+' and v_+' are dimensionless fluctuation velocities scaled by u^*, and H^+ is dimensionless water depth. For open channel flows, in practice, the velocity distribution can be approximately described by Dou's formula:

$$\frac{u}{u_*} = \gamma_L (\frac{u_L}{u_*}) + \gamma_T (\frac{u_T}{u_*}) \quad (6)$$

where γ_L and γ_T are the occurrence rates of laminar and turbulent flows in a specific location, respectively, and u_L and u_T are the velocities of laminar and turbulent flows, respectively. Letting $\gamma = \gamma(\gamma_L, \gamma_T)$ denote intermittence, it has $\gamma_L + \gamma_T = 1.0$. The profile of velocity distribution should have the following form:

$$\frac{u^+}{y^+} = f(\frac{u_* R}{\upsilon}, \frac{\bar{u} H}{\upsilon}, \gamma) \quad (7)$$

Because of the coupling relation of the cavitation damage and the flowing field, besides the solid properties, in the perspective of the flowing field, the cavitation damage has the similar form of equation (7). Considering the formulae of (5), (6) and (7), one can be easy to know the relations between the cavitation damage and the flowing properties to make cavitation damage show stochastic character.

5. The Energy Characters in Cavitation Damage Flowing Field

The small scale eddies have random nature and the concrete mechanism of vortex breakdown dominates the adjacent local flowing structures. The stretch of vortex plays an important role in cascade process of turbulent energy. In turbulent flow, there are spiral-like vorticity distributions, which wrap up around strained vortex tubes, in fully developed turbulence. If the Reynolds number increases to infinity, the size of bubble to cause the cavitation damage will become very small. The observation shows that, the erosion often occurs when one of the legs of a vortex cavitation touches and collapses on the solid surface, the inverse U-shaped vortex has strong cavitation ability.

The one difference between the turbulent flow and the laminar flow is that, the state of a turbulent flow at a given position depends upon upstream history and cannot be uniquely specified in terms of the local strain-rate tensor. At present, in all high Reynolds number turbulent refined flow modeling, if the turbulent model of Reynolds-averaged Navier-Stokes equation (RANS) is adopted, the turbulence kinetic energy (per unit mass), k, is developed on the phenomenon of the cascade process present in all turbulent flows involves a transfer of k from larger eddies to smaller eddies. We assume that the small scale motion occurs on a short time scale, thus such motion is independent of the relatively slow dynamics of the large eddies and of the mean flow. In fact, This is one of the spirit of Kolmogorov's (1941) universal equilibrium theory.

No matter what the final cavitation damage model and essential mechanism are, however, there are extra energy to release in the process of the bubble breakdown. Obviously, it will generate the fluctuating energy. About this issue, it deserve investigating further.

6. Conclusions

We reviewed and studied some hydrodynamic cavitation damage properties, and explored the relations between the cavitation damage and coherent structure in this paper. First, from the viewpoint of cavitation damage, the characters of the flowing structures near the wall, such as the forming and developing of streak structure; the forming and developing of longitudinal vortices, transverse vortices and horseshoe vortices; and the bursting phenomena were discussed. Then the energy distribution, especially, the relations between the fluctuation, such as Reynolds stress, with the flowing field characters was explored. The results clearly demonstrated that, it is the vortex to dominate the cavitation damage, thus it show stochastic characters both in time and in space. Not only the cavitation number, but also the Reynolds number, the Lamb vector and the helicity, are also important parameters to study the cavitation damage. Besides the solid characters, we gave a conceptive formula to describe the relation between the cavitation damage and the flowing field. The fact was also pointed out that, the process of bubble breakdown will generate the fluctuating energy, which is an important topic in turbulence modeling theory.

7. Acknowledgment

We sincerely thank the financial supported of the Natural Science Foundation of China (41601565), Foundation of Key Laboratory of Communication Civil Waterway Regulation (NHHD-201511, NHHD-201515), Natural Science Foundation of Chongqing (cstc2016jcyja0544), and Project of the mechanism, technology and application of the water jet of the Chongqing Waterway Bureau of the Yangtze River

8. References

[1] J. P. Franc and J. M Michel, Fundamentals of cavitation, Kluwer Academic Publishers, U.S.A. 2005.

[2] C. Carlos and R. Mettin, “Simultaneous High-Speed Recording of Sonoluminescence and Bubble Dynamics in Multibubble Fields” Phys. Rev. Lett., Vol.118, 064301, 2017.
[3] M. Binama, and E. B. Muhirwa, “Cavitation effects in centrifugal pumps- A review” International Journal of Engineering Research and Applications, Vol.6, pp.52–63, 2016.
[4] M. P. Brenner, S. Hilgenfeldt and D. Lohse, “Single-bubble sonoluminescence” Reviews of Modern Physics, Vol. 74, pp. 425-484, 2002.
[5] A. Vitaly, “Shneidman Time-dependent cavitation in a viscous fluid” Phys. Rev. E Vol. 94, 062101, 2016.
[6] P. F. Pelz, T. Keil, and T. F. Grob, “The transition from sheet to cloud cavitation” Journal of Fluid Mechanics, Vol.817, pp. 439-454, 2017.
[7] R. E. A Arndt, “Cavitation in vortical flows” Annu. Rev. Fluid Mech., Vol.34, pp.143–75, 2002.
[8] C. F. Delale, Bubble dynamics and shock waves, Springer-Verlag Berlin Heidelberg, Germany. 2013.
[9] G. R. Dou, “The laminar and turbulent united laws for open channel and pipe flow” Journal of China Science(A), Vol.15, pp. 472-480, 1982.
[10] Ge Gao, “A theory of dissipation and dispersion cooperation of turbulence” Science in China(Series A), Vol.18, pp. 616-629, 1985.
[11] M. Ida, “Multibubble cavitation inception” Physics of Fluids, Vol. 21, pp. 113302, 2009.
[12] D. Ranjan, J. Oakley, R. Bonazza, “Shock-bubble interactions” Annu. Rev. Fluid Mech., Vol. 43, pp. 117–40, 2011.
[13] K. S. Suslick, D. J. Flannigan, “Inside a collapsing bubble: sonoluminescence and the conditions during cavitation”, Annu. Rev. Phys. Chem. Vol. 59, pp.659–683, 2008.
[14] R. P. Taleyarkhan, C. D. West, J. S Cho, “Evidence for nuclear emissions during acoustic cavitation” Science, Vol. 295, pp. 1868–1873, 2002.
[15] D. C. Wilcox, Turbulence Modeling for CFD, DCW Industries, Inc. California, USA 2006.

IOP Conf. Series: Earth and Environmental Science **168** (2018) 012029 doi:10.1088/1755-1315/168/1/012029

Supportive Learning Tool for Electrical Machines Laboratory Using Video Clips

Somboon Nuchprayoon

Department of Electrical Engineering, Chiang Mai University, Chiang Mai, Thailand

Email: sn@eng.cmu.ac.th

Abstract. Electrical machines and electromechanical energy conversion serve as required course in most undergraduate programs in power engineering around the globe. Sophomore and junior students should attend both lecture and laboratory sessions to gain basic foundation in power engineering. The Department of Electrical Engineering, Faculty of Engineering, Chiang Mai University, Thailand, has made video clips for electrical machines laboratory and uploaded to YouTube. The main objectives of the video clips are to enable the students to review the laboratory session at their leisure time and to improve their laboratory skills. Upon completion, the students were pleased and found that the video clips were helpful.

1. Introduction

The Department of Electrical Engineering, Faculty of Engineering, Chiang Mai University, Thailand, offers both undergraduate and graduate programs in electrical engineering. The bachelor, master, and doctoral degrees are 4-, 2-, and 3-year programs, respectively. The bachelor degree program has around 80 students enrolled each year, while the master degree and the doctoral degree programs have 10-20 students enrolled each year. The department has 24 faculty members.

The bachelor degree program comprises two fields of specialization: a) electric power and b) electronics, control, telecommunication, and information technology (ECTI). On average, three-fourth of the students has chosen electric power field. The students may register their major elective courses in the third and fourth year.

Electrical machines, also known as electromechanical energy conversion, is one of the required course of both fields of specialization. There are 2 lecture courses (3+3 credits) [1] and 1 laboratory course (1 credit) [2]. Electrical Machines I involves principles of energy conversion and DC machines. Electrical Machines II involves AC machines and performance analysis. Electrical Machines Laboratory can be taken in the second semester of the third year, after passing both lecture courses.

Electrical machines can be regarded as one of the classical subjects in electrical engineering. If electric drives were excluded, it can be said that the course has not changed over a decade by considering the contents of both classical [3]-[7] and advanced [8]-[10] textbooks in the subject. As a result, the teaching way is also classical and might be boring to some extent. Electric Machines Laboratory is required to ensure that the students have good laboratory skills and practical experiences: reading diagrams, connecting circuits, using measuring instruments, as well as writing a report.

It is challenging to draw attention from students, especially in the era of social media and smart devices. Students tend to lose their attention easily and frequently. Either lecture or laboratory session must be clear and concise. It is essential for an instructor to develop a learning tool to support students

to study at their leisure time. Some supplementary materials may be removed from class and put in the internet, which are available for students to retrieve them. This learning tool has been known as e-learning [11] or virtual course [12]. For instance, the Consortium of Universities for Sustainable Power (www.cusp.umn.edu), hosted by the University of Minnesota, has provided many online course materials on power systems, power electronics, electric machines & drives, and renewable energy. Recently, blog, Line, Facebook, WhatsApp, YouTube, etc. are good example of applying social media for educational purpose.
In this paper, the instructors of the Department of Electrical Engineering, Chiang Mai University have developed video clips for Electrical Machines Laboratory and uploaded to YouTube. The video clips are provided as learning tool for students to study and review circuit connections and measuring techniques as well as prepare for the practical examination.
Section I describes electrical machines course offered by the Department of Electrical Engineering, Chiang Mai University and addresses the motivation to develop video clips as supportive learning tool. Section II describes the electrical machines laboratory – a course offered to junior students of the Department. Section III explains the process of developing video clips. Student feedback and evaluation are provided in Section IV. This work is concluded in Section V.

2. Electrical Machines Laboratory

2.1. Course Structure
Table I shows the course contents of Electrical Machines Laboratory taught at the Department of Electrical Engineering, Chiang Mai University. This course has been supervised by 3 instructors and 2 technicians. The laboratory has 15 sessions and each session is 3 hours per week. There are 9 laboratories, covering both DC machines in weeks 3-7 and AC machines in weeks 9-12. Every student is required to turn in a preliminary note at the beginning of each laboratory and turn in a report a week after completing each session. Both the note and the report must be written individually. The practical examination of each student is set in weeks 14-15. A final examination has 40 multiple-choice questions and most students would take 1-2 hours to finish up.

2.2. Laboratory Equipment and Instrument
The laboratory equipment is capable to serve up to 5 groups of students concurrently so that, the course was usually divided into 4 sections each semester. Each section has 4-5 groups. A group of 4-6 students is set for each laboratory bench. The number of enrolled students each year is 80 so that each section has up to 20 students. The laboratory equipment is illustrated in Figures 1-3. Most of the equipment shown in Figure 1 is TERCO electrical machine system (www.terco.se). The main pieces of equipment are power pack, prime mover, torque meter, and resistive load. Test machines may be added to the prime mover via a coupling shaft. The motor-control panel in Figure 2 was built by domestic supplier. Figure 3 shows common measuring tools i.e., multimeter, tachometer, voltmeter, ammeter, wattmeter, and power-factor meter.

2.3. Course Evaluation
Evaluating criteria of this course are comprehensive. The preliminary note, class participation, reports, practical examination, and final examination account for 10%, 10%, 40%, 20%, and 20%, respectively. It was found that most students have done well for preliminary note, class attention, and reports. Their scores were relatively high and not much different. But, the scores from practical and final examinations are quite different and become a grading index.
Each student must take a 15-minute practical examination. A student has up to 10 minutes to set up an experiment without circuit diagram. The experiment is randomly chosen from four topics. The test topics are separately-excited DC generator, shunt DC motor, squirrel-cage induction motor, and motor control. The remaining 5 minutes is to run the experiment, as specified by the instructor, and do a

simple measurement & calculation such as efficiency. It has been considered as the most challenging part of this course. Figure 4 shows the practical examination in April 2017.

TABLE I. LIST OF COURSE CONTENTS OF ELECTRICAL MACHINES LABORATORY

Week	Content
1	Reviews of basic concepts and principles
2	Introduction to equipment and instrument
3	Generation of electromotive force (EMF)
4	Winding factors of AC machines
5	Separately excited DC generator
6	Shunt, series, and compound DC generators
7	Shunt and series DC motors
8	Review of DC machines
9	Synchronous generator
10	Synchronous motor
11	Squirrel-cage inductor motor
12	Motor control
13	Review of AC machines
14	Training for practical examination
15	Practical examination

Figure 1. Benches in eletrical machines laboratory.

Figure 2. Motor-control panel in electrical machines laboratory.

Figure 3. Measuring instruments in electrical machines laboratory.

Figure 4. Practical examination in April 2017.

3. Developing Video Clips

The students might attempt to memorize all circuit connections of the four test topics, but it would be much effective for the students to embed their understanding into their heads. Thus, the practical examination is a driving force for developing the video clips. Besides, the student can also review other topics they are interested or missed in the class.

The instructors and technicians have planned to develop the video clips in December 2014. It took three months for scripting, shooting, and uploading. The video clips become available to the public via YouTube in March 2015. One may use the word "ee320cmu" for searching in the internet.

There are 23 video clips uploaded, as shown in Table II. Most video clips are less than 5 minutes. The video clips were recorded as .mpg file by simply using a camcorder. Video clips 1-5 review the basic equipment and instrument. The remaining video clips review all laboratory topics. The motor control laboratory has 6 video clips given that circuit connections are quite complicated. The focus of all video clips is on circuit connections & tips. The procedure of running experiment was left for self-review. Nonetheless, the video clips for running experiments might be available in the future.

4. Feedback and Evaluation

In the last week of the course, a questionnaire was given to every student and the anonymous evaluation is summarized in Table III. In general, the students were pleased and found that the video clips are helpful. Numerically, only 16.7% of the students did not watch the video clips. Half of the students watched each video clip more than once. Half of the students spent 11-20 minutes watching each time. Around three-fourth of the students found that they have better understanding by watching the video clips and they could also do better in the practical examination.

TABLE II. List of Video Clips of Electrical Machines Laboratory

Clip	Topic	Duration (minute)
1	Prime mover	4:45
2	Oscilloscope	8:38
3	Resistive load	2:48
4	Watt meter	2:49
5	Power factor meter	2:43
6	Generation of electromotive forces	3:51
7	Winding factors of AC machine	4:01
8	Separately-excited DC generator	4:26
9	Shunt DC generator	4:58
10	Series DC generator	3:41
11	Shunt DC motor	5:44
12	Series DC motor	4:02
13	Synchronous generator	4:54
14	Synchronous motor	7:39
15	Squirrel-cage induction motor	4:26
16	Squirrel-cage induction motor (maximum torque)	2:39
17	Introduction of motor control	4:37
18	Motor control – direct on line (control circuit)	3:16
19	Motor control – direct on line (power circuit)	1:40
20	Motor control – forward/reverse (control circuit)	3:20
21	Motor control – forward/reverse (power circuit)	2:41
22	Motor control – wye/delta start (control circuit)	5:52
23	Motor control – wye/delta start (power circuit)	3:33

Figure 5 illustrates the frequency distribution of number of students taking the practical examinations, compared between 2014-2017 (before and after watching the video clips). The numbers of students in 2014-2017 are 84, 81, 82 and 106 respectively. The average scores in 2014-2017 are 14.5, 15.0, 16.0, and 16.4, respectively. The standard deviation of scores in 2014-2017 are 4.5, 5.0, 3.5 and 3.8, respectively. So, it can be statically observed that, with the increasing average score and narrower spread of scores, the student performance is better off after watching the video clips.

The number of students with full score (20) in 2014-2017 are 6.0%, 13.6%, 15.6% and 16.0%, respectively. The number of students with score less than 9 has been decreasing from 14.3% in 2014 to 5.6% in 2017. In contrary, the number of students with score more than 16 has been increasing from 40.5% in 2014 to 64.2% in 2017. Thus, it is evident that the students can perform better in the practical examinations with the help of video clips.

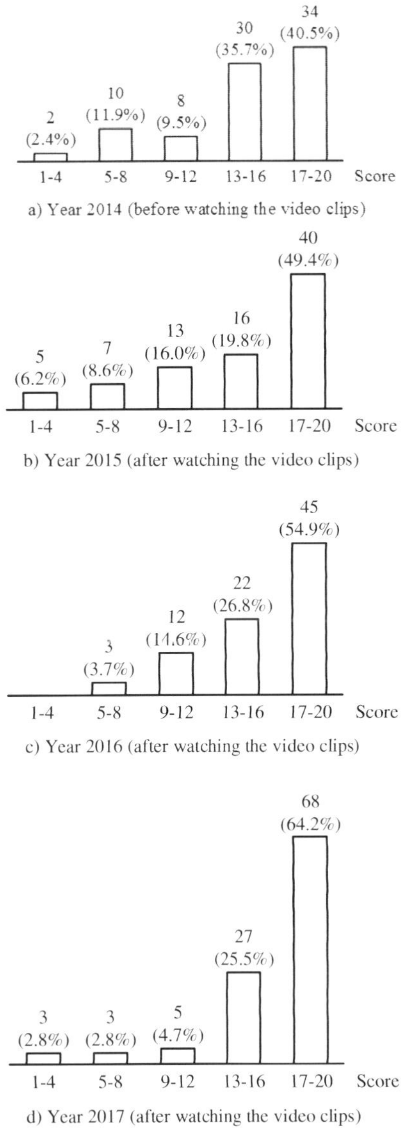

Figure 5. Frequency distributions of number of students taking the practical examinations.

Table IV shows the overall scores of the course, compared between 2014-2017. The average scores of these years are relatively constant but slightly higher, while the spread of scores tends to be narrower. Given that the practical examinations account for only 20% of the total scores for grading, it may not be able boost up the overall performance. By the way, it would be an interesting idea to develop a mechanism to increase the overall performance of the students.

TABLE III. FREQUENCY DISTRIBUTION OF NUMBER OF STUDENTS ANSWERING THE QUESTIONAIRE

Q1	**Have you ever watched the video clip?**		
	Never	All clips	Some clips
	28	10	126
Q2	**How many times did you watch each video clip?**		
	1	2-3	> 3
	62	70	4
Q3	**How long did you spend watching each time?**		
	0 – 10 min.	11 - 20 min.	> 21 min.
	29	65	42
Q4	**Did the video clips help you gain better understanding?**		
	Good	Fair	Poor
	85	48	3
Q5	**Did the video clips help you on the practical examination?**		
	Good	Fair	Poor
	104	32	-

TABLE IV. COMPARISON OF OVERALL SCORES IN 2014-2017

Year	Max.	Avg.	Min.	Stdev.
2014	82.2	69.6	50.0	6.8
2015	86.7	69.8	44.5	8.7
2016	85.5	70.2	43.8	7.6
2017	84.4	71.8	45.4	6.6

5. Conclusion

The Department of Electrical Engineering, Chiang Mai University has developed video clips, using as supportive learning tool, for electrical machines laboratory. The video clips were uploaded in YouTube. The video clips enable the students to review the laboratory sessions and help them improve their laboratory skills. The students have expressed positive feedback after watching the video clips and most of them obtained higher scores in the practical examinations. Nevertheless, the overall scores of the class were slightly better off, it is thus subject to further evaluation and requires additional mechanism to improve the students' learning performance.

6. Acknowledgment

The author would like to acknowledge the laboratory manual developed by Kumpanart Ratawessnunt for serving as a pilot lamp of the video clips and the contributions of Ramest Nuntajan and Prayoon Buarin for shooting the video clips. This work is financially supported by the Faculty of Engineering, Chiang Mai University.

7. References

[1] K. Ratawessnunt, *Electrical Machines, Class note for EE223 and EE321*, Department of Electrical Engineering, Chiang Mai University.

[2] K. Ratawessnunt, *Electrical Machines Laboratory Manual*, Department of Electrical Engineering, Chiang Mai University, 2005.

[3] A.E. Fitzgerald, C. Kingsley, Jr. and S.D. Umans, *Electrical Machinery*, 6th ed., McGraw-Hill, 2003.

[4] S.J. Chapman, *Electric Machinery Fundamentals*, 3rd ed., McGraw-Hill, 1998.

[5] P.C. Sen, *Principles of Electric Machines and Power Electronics*, 2nd ed., Wiley, 1996.

[6] P.S. Bimbhra, *Electrical Machinery*, Khanna, 2011.

[7] N. Mohan, *Electric Machines and Drives: A First Course*, Wiley, 2012.

[8] M.S. Sarma, *Electrical Machines: Steady-State Theory and Dynamic Performance*, 2nd ed., Cengage Learning, 1997.

[9] P.C. Krause, O. Wasynczuk, and D. Sudhoff, *Analysis of Electrical Machinery and Drive Systems*, 2nd ed., Wiley-IEEE Press, 2002.

[10] T. Gonen, *Electrical Machines with MATLAB*, 2nd ed., CRC Press, 2011.

[11] New web-based interactive e-learning in power electronics and electrical machines, in *Proc. of the 36th IAS Annual Meeting*, Chicago, IL, 30 Sep.-4 Oct. 2001.

[12] R.G. Beru and I.N.C. Husanu, Using a virtual platform for teaching electrical machines and power system virtual courses, in *Proc. of the 120th ASEE Annual Conference and Exposition*, Atlanta, GA, 23-26 Jun. 2013.

IOP Conf. Series: Earth and Environmental Science **168** (2018) 012030 doi:10.1088/1755-1315/168/1/012030

Online Measurement System of Pulverized-Coal Concentration in Power Plant

Pan Han, Wang Peihong

Key Laboratory of Energy Thermal Conversion and Control of Ministry of Education, School of Energy and Environment, Southeast University, Nanjing 210096, China

Email: panhan_seu@hotmail.com, phwang@seu.edu.cn

Abstract. Measurement for pulverized-coal concentration is important to the safe and economic operation in power plant. This paper presents an on-line measurement system for pulverized-coal concentration based on the method of ultrasonic attenuation in the gas-solid two phase flow. According to the non-liner error of the ultrasonic transducer, multiple reflections signal from the two ultrasonic transd ucer is used to calculate the concentration, the result of which is displayed on LabVIEW. The experiment result shows that the multiple reflections signal can reduce the non-liner error and improve the accuracy of the measurement.

1. Introduction

The importance of pulverized-coal in power plant has given considerable impetus to the development of analytical techniques to provide information about concentration, such as electrostatic sensor technology [1], light scattering [2], and light microscopy [3]. Each technique has its own advantages and limitations, and consequently has a specific application. Ultrasonic wave has been applied widely for its high penetrability and precision [4-6]. Nevertheless, ultrasonic attenuation has been applied to the measurement of liquid-solid concentration [7-8], rarely to gas-solid concentration.

The non-liner error exist in the ultrasonic transducer, and it often shows that the initial signal amplitude will change by the time. The amplitude change will seriously restricted the accuracy of measurement. With the problem in hand, multiple reflect signal is used to reduce the influence. And then, the concentration data will be displayed on computer through the LabVIEW [9].

This paper is intended to introduce the measurement system based on ultrasonic attenuation, and it is verified by the experiments.

2. Methodology

Epstein [10] found theoretical model that ultrasound propagate through two phase flow by analyzing Navier-Stokes function in hydromechanics based on law of conservation of mass. After that, Allegra and Hawley [11] developed this model and expanded its scope of application. The model is called ECAH model [12-13] in memory of those people who have made outstanding contributions to it.

The derivation process of the ECAH model, as follows. The solid must meet requirements that the solid media is of elastic isotropic heat-conducting. The wave equations are obtained from a stress-strain relation, conservation laws and thermodynamic equation of state, wave equations can be written as

IOP Conf. Series: Earth and Environmental Science **168** (2018) 012030 doi:10.1088/1755-1315/168/1/012030

$$\begin{cases} (\nabla^2 + k_c^2)\phi_c = 0 \\ (\nabla^2 + k_s^2)S = 0 \\ (\nabla^2 + k_T^2)\phi_T = 0 \end{cases} \tag{1}$$

where

$$\begin{cases} k_c = \omega / c + i\alpha_L \\ k_s = (\omega\rho / i\eta_s)^{0.5} \\ k_T = (1+i)(\omega\rho c / 2\lambda) \end{cases} \tag{2}$$

where k_c is compressional wave number, k_T is thermal wave number, k_s is shear wave number, α_L is the attenuation coefficient, ω is the angular frequency, ρ is the density, η_s is the coefficient of shear viscosity, c is the velocity, λ is the thermal conductivity and $i = \sqrt{-1}$.

The wave equations can be solved in spherical coordinates and six equations are obtained by utilizing the boundary conditions at the surface of the particle. The equations can be solved for the coefficientA_n. In condition that the effect of the individual particles can be additive, the attenuation can be written as [14]

$$\alpha = -\frac{12\varphi}{k_c^2 D^3}\sum_{n=0}^{\infty}(2\text{n}+1)\,\text{Re}\,A_n \tag{3}$$

where φ is the volume fraction of the solid phase, D is the diameters of solid phase.

McClements simplified the model on account of the complexity of the coefficient A_n. What he thinks is that the coefficient can be ignored but A_0 and A_1 [15] when the diameters of solid phase is far less than the wave length. The coefficient A_0 and A_1 can be written as

$$A_0 = \frac{ik_c^{\,3}R^3\left(\rho k_c^{'2} - \rho' k_c^{\,2}\right)}{3\rho' k_c^{\,2}} - k_c^{\,2}RcT\rho\lambda H\left(\frac{\beta}{\rho C_p} - \frac{\beta'}{\rho' C_p'}\right)^2 \tag{4}$$

$$A_1 = \frac{-ik_c^{\,3}R^3}{3} \times \frac{\rho - \rho'}{3\rho + 4R^2\left(\rho' - \rho\right) / \left[2R^2 + 3R\left(1+i\right)\delta_v + 3i\delta_v^2\right]} \tag{5}$$

where

$$H = \left[\frac{\delta_t}{\delta_t + (1-\text{i})\text{R}} - \left(\lambda / \lambda'\right)\tan z' / \left(\tan z' - z'\right)\right]^{-1} \tag{6}$$

$$z' = (1+\text{i})\,\text{R}/\,\delta_t' \tag{7}$$

$$\delta_t = \sqrt{\frac{2\lambda}{\rho\omega c_p}}\quad \delta_t' = \sqrt{\frac{2\lambda'}{\rho'\omega c_p'}}\quad \delta_v = \sqrt{\frac{2\mu}{\rho\omega}} \tag{8}$$

where the simple with superscript and none refer to the properties of the continuous phase and the emulsion droplets, respectively, T is the temperature, β is the thermal dilatation, c_p is the specific heat at constant pressure.

The coefficient α can be derived from an intensity balance for an infinitesimally thin layer of suspension:

$$dI = -I\alpha \cdot 2 \cdot \Delta s \tag{9}$$

By integration, it can be written as

$$\alpha = \frac{\ln(I^{'}/I)}{2\cdot\Delta s} \tag{10}$$

where $I^{'}$ and I are the radiation intensities transmitted in the absence and presence of particles, respectively, and Δs is the path length.
Finally, the expression of volume fraction of the solid phase can be written as

$$\varphi = -\frac{24\cdot\Delta s}{k_c^{\ 2} D^3 \ln(I^{'}/I)} \mathrm{Re}\left(A_0 + 3A_1\right) \tag{11}$$

3. Measuremental System

3.1. Hardware Design
Figure 1 illustrates the hardware structure of the system. The generation source is sinusoid burst wave which is supplied from arbitrary waveform generator (AWG). The ultrasonic measuring section consists of two piezoelectric transducers, the center frequency of which is 170 KHz, acting as emitter and receiver. The signal received is amplified by signal-amplifier and enter into the four-order active band pass filter. In order to collect and process data in real time, high-speed data acquisition is used to collect the signal to which is analyzed by computer after A/D conversion.

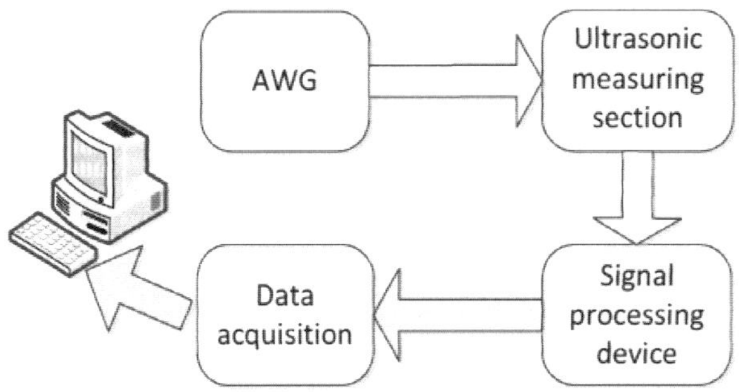

Figure 1. The hardware structure of the system.

3.2. Software Design
In order to facilitate extending and debugging, the measuring system is building with modular design. The whole software is made up of six circulation module, it were event capture module, UI processing module, data acquisition module, data processing module, logging module and display module, respectively. Each module is independent of each other and they are parallels-operating.
Upcoming articles will describe the data acquisition and data processing module.

1) Data acquisition

USB-1210 (adlinktech) is used as data acquisition device. It is convenient for quick development of applications with the help of DAQPilot.
First, as can be seen in figure 2, PLT Create Channel VI is used to create new task with local virtual channel information; The Sample Clock VI can assign samples per channel, sampling (update) rate per channel, and conversion (update) source properties; PLT Trigger VI is used to Sets whether the trigger occurs when the analog signal crosses the level; PLT Start Task VI is used to starts measurement or generation by transitioning the task to the running state;

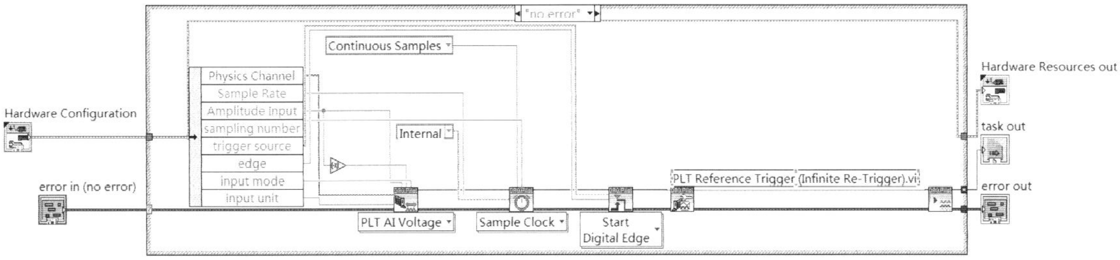

Figure 2. Device initialization.

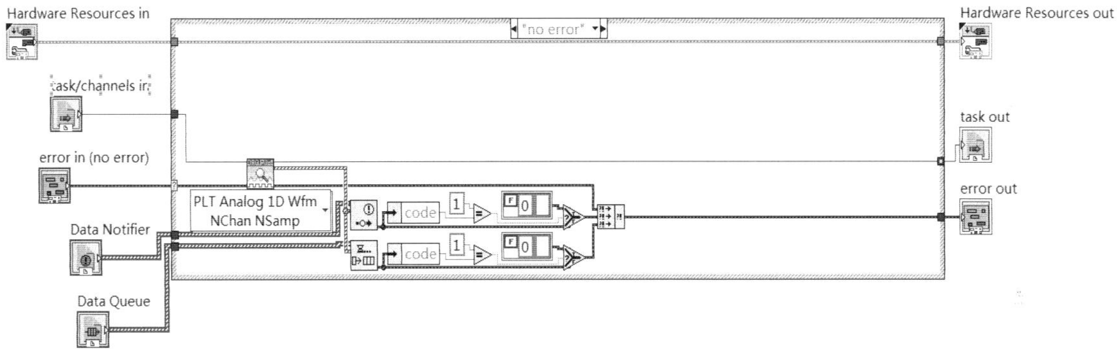

Figure 3. Data acquisition.

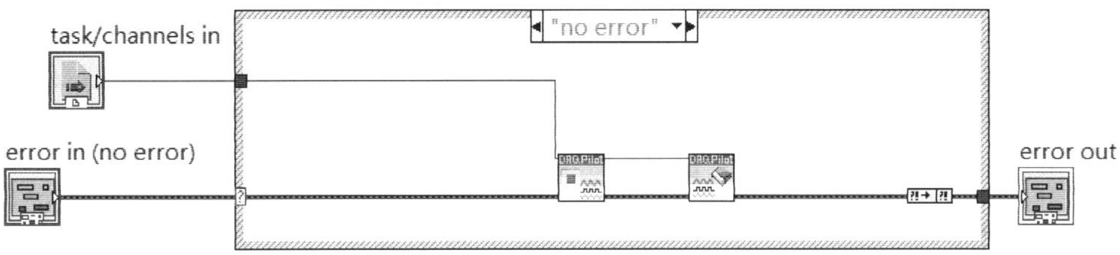

Figure 4. Device closing.

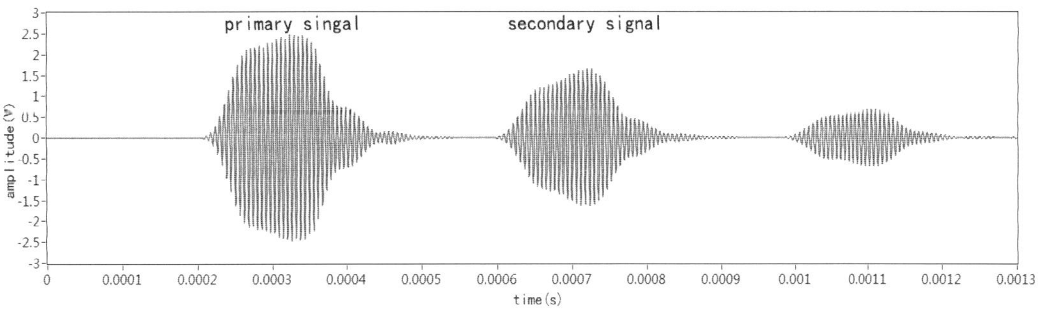

Figure 5. The signal received by the ultrasonic transducer.

Second, as can be seen in figure 3, after the device initialization, PLT Read VI is used to acquire data. In order to realize continuous measurement, PLT Read VI is called circularly. The real time data is send to data display module by notifier (real-time performance), and the real time data is send to data processing module by queue (data integrity).

Finally, as can be seen in figure 4, when the data acquisition task is finished, PLT Stop Task VI is used to stop the task, and PLT Clear Task is used to clear the task, release resources.

2) Data processing

Figure 5 illustrates the multiple signal received by the ultrasonic transducer. Generally, the primary signal amplitude is used to calculate the ultrasonic attenuation, but the result may be inaccuracy because of the non-liner error of the ultrasonic transducer.
Experiments found that the primary signal and secondary signal have the same change direction in the results of the non-liner error. So this paper uses the ratio of amplitude of primary to secondary signal to present the ultrasonic attenuation.
The flowchart of data processing is shown in figure 6.

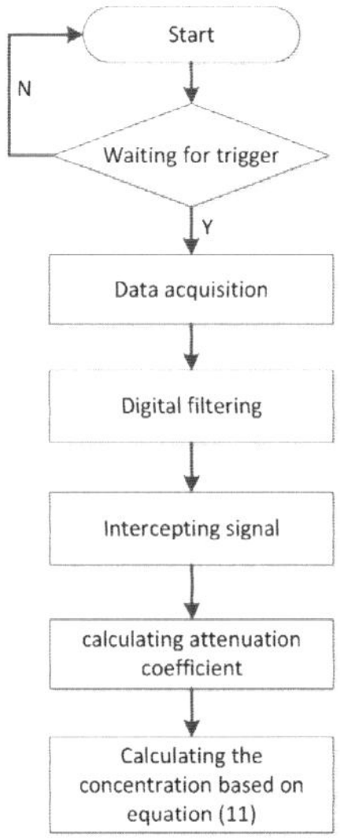

Figure 6. The software structure of the system.

4. Experimental Results and Analysis

In order to evaluate the performance of the measurement system, measurements with different concentration were conducted. For testing, glass beads were chosen, the concentration of which is controlled by the vibrating feeder. And the mass flow is calculated by the weighting sensor data.
The result of figure 7 is calculated by the primary signal, and the result of figure 8 is calculated by the ratio of the primary signal to the secondary signal. The R-Square of the fitting curve is 0.9264 and 0.9888, respectively. Obviously, the different particle volume fraction is clearly distinguishable and the correlation between particle volume fraction (measured by ultrasonic attenuation) and mass flow (determined from load cell) remains linear over an extensive range.

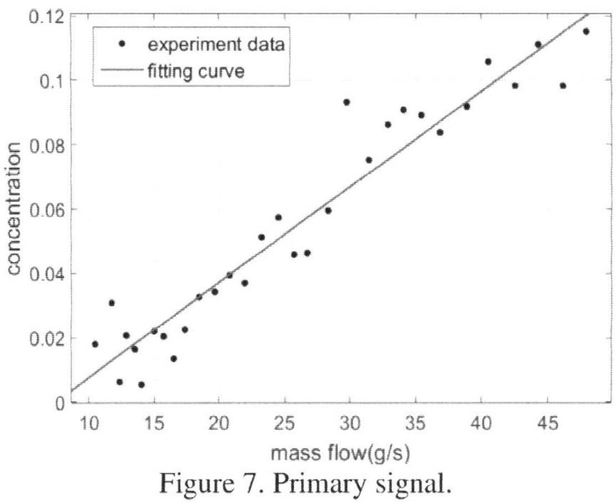

Figure 7. Primary signal.

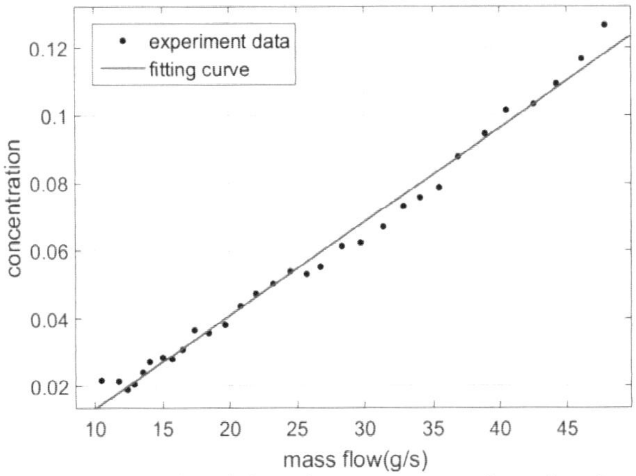

Figure 8. Ratio of the primary to secondary signal.

The results shows that the ratio of the primary signal to the secondary signal can reduce the non-liner error of the ultrasonic, and the measurement system is accuracy.

5. Conclusion
The method of ultrasonic attenuation in the gas-solid two phase flow, ECAH model, is used to calculate the concentration of pulverized-coal, and it is proved to be a powerful tool for on-line concentration. According to the non-liner error of the transducer, multiple ultrasonic wave signal is used to reduce the error, and it shows the good result.

The experimental results show that the measurement system is of accuracy, stabilization and reliability. It is certain that ultrasonic instrumentation will continue to develop and that the technique will become a powerful tool for measuring the concentration of pulverized-coal.

The particle size distribution is also an important property of the pulverized-coal, and more ultrasonic frequency is needed based on the ECAH model, it will be a good challenge in the future.

6. Acknowledgment
This work is supported partly by National Natural Science Foundation of China under Grant No. 51476028. National Key Technology R & D Program of the Ministry of Science and Technology of China No. 2015BAA03B02.

7. References
[1] Coombes J R, Yan Y. Development of an electrostatic array sensor for measuring the velocity and concentration profiles of pneumatically conveyed particles[C]// Instrumentation and Measurement Technology Conference. IEEE, 2015:138-143.

[2] Andries C, Clauwaert J. Photon correlation spectroscopy and light scattering of eye lens proteins at high concentrations[J]. Biophysical Journal, 1985, 47(5):591.

[3] Werts M H, Raimbault V, Texier-Picard R, et al. Quantitative full-colour transmitted light microscopy and dyes for concentration mapping and measurement of diffusion coefficients in microfluidic architectures[J]. Lab on A Chip, 2012, 12(4):808-820.

[4] ZHOU H, CEN K F, LI J Z, et al. Experimental investigation into the performance of pulverized-coal splitter with an elbow using fiber optic probe[J]. Chinese Journal of Scientific Instrument, 2007, 28(11): 1967-1971.I.

[5] SHI X G, ZHOU H, CEN K F. Technique for measuring concentration of gas-solid two-phase flow inside the pipe based on ultrasonic method[J]. Thermal Power Generation, 2005,34(5):37-38,64.

[6] LIU X Z, WANG Y X, et al. The solid phase concentration measurement of the gas-solid

two-phase flow by the sound velocity method[J]. Journal of Engineering Thermophysics, 2007, 28 (1):201-204.

[7] Horak G. Real-Time Ultrasonic-Spectroscopy in Suspensions[J]. Acustica, 1977, 37(1):11-20.

[8] Wei Z, Mingxu S U, Cai X. Particle size distribution measurement based on ultrasonic attenuation and phase velocity spectra[J]. Ciesc Journal, 2014, 65(3):898-904.

[9] Elliott C, Vijayakumar V, Zink W, et al. National Instruments LabVIEW: A Programming Environment for Laboratory Automation and Measurement[J]. Journal of the Association for Laboratory Automation, 2007, 12(1):17-24.

[10] Epstein P.S., Carhart R.R. the Absorption of Sound in Suspensions and Emulsion[J], TheJournal of The Acoustic Society of America,1953,25(3):553-565.

[11] Allegra J R,Hawley S,A. Attenuation of Sound in Su-spensions and Emulsions: Theoryand Experiments[J], The Journal of The Acoustic Society of America,1972,51(5):1545-1564.

[12] Su M X, Cai X S, Xue M H, et al. Particle sizing in dense two-phase droplet systems by ultrasonic attenuation and velocity spectra[J]. Sci. China Ser. E-Tech. Sci., 2009, 52(6):1502-1510.

[13] Mcclements D.J., Ultrasonic characterization of emulsions and suspensions[J], 1991,37(1-2):33-72.

[14] Yucel U, Coupland J N. Ultrasonic Attenuation Measurements of the Mixing, Agglomeration, and Sedimentation of Sucrose Crystals Suspended in Oil[J]. Journal of the American Oil Chemists Society, 2011, 88(1):33-38.

[15] Hibberd D J. Ultrasonic Techniques to Characterize Concentrated Colloidal Dispersions[J]. Acs Symposium, 1999, 737:262-275.

IOP Conf. Series: Earth and Environmental Science **168** (2018) 012031 doi:10.1088/1755-1315/168/1/012031

Modeling and Simulation of Low Carbon Supply Chain Based on the Product Attributes

Luo Ming

School of Economics and Management, Tibet University, Lhasa Tibet, China
e-mail: 2536494944@qq.com

Abstract. The influence of product attributes on low carbon supply chain is studied in this paper. In the background of global emphasis on green and low carbon economy, there are low carbon and quality attributes for the product. In order to meet the requirements of the consumer demand, the enterprise needs to coordinate the relationship between quality and low carbon products at the same time. So the enterprise should adjust of low carbon action to achieve product quality attributes and product attributes to the low carbon balance. Based on the two variable utility theory, through the construction of the supply chain demand structure model, the influence of the product attributes in the supply chain is simulated and analyzed by numerical calculation of low-carbon supply chain process.

1. Introduction

The product is the consumption target provided by the upstream and downstream enterprises of the supply chain, and is the intermediary between the consumers and the supply chain enterprises [1]. The impact of product effect and can not be simply equated with consumer awareness of low carbon, as can be observed, for most consumers, despite the catastrophic consequences of climate change to contemporary people with profound shock, but because it is not clear the date of the disaster and its possible tolerance level, so consumers generally do not at the same time strengthen low carbon consciousness weakening the basic characteristics of products such as preference, in the automobile products, carbon emissions is significantly higher than the average family car, but from the perspective of products, their sales with certain rationality [2]. It has its advantages in bad traffic conditions, and it can also meet the security requirements in the city (obviously it can be better protected in a traffic accident than a small car). In other words, for consumers, meet the quality by using the utility is the basic value. In the background of low carbon economy [3], low carbon consciousness of consumers while the increase of products to meet the requirement of low carbon low carbon utility, resulting in low carbon value of the product, the product has the double the quality and properties of low carbon [4].

Through the review of the literature, it could be found that the existing literature mainly focused on the realization of low-carbon products, including design, production, packaging, pricing and so on. And some studies pay more attention to the cost and supply chain profit pressure from the low carbon products, but there is no further discussion on product attributes bring to the low carbon supply chain problems. This paper has mainly studied the relationship between products attributes and low carbonization in supply chain.

2. Construction of Low Carbon Supply Chain and Some Assumptions

The dual attributes of the product makes it necessary for the supply chain to consider the relationship between quality and low carbon, and to coordinate the relationship between the two attributes [5].
There is contradiction between quality usually products and low carbon consumption, because more often means higher quality, such as the car will consume more than ordinary cars, steel and have more displacement; notebook computer production of Apple Corp in the United States [6], because the rice with the whole block pin cutting method to produce the result a large number of strokes is waste material, but also guarantees the sheer amount of the industry's highest level of heat dissipation efficiency and the advantages of traditional food and beverage; China soup is very exquisite, continuous cooking candle, candle will shorten the cooking time if to reduce carbon emissions, but also reduce the consumer of food quality evaluation etc..

2.1. Construction of Low Carbon Supply Chain
Firstly, choose reasonable partners. In the process of building an enterprise supply chain, one of the important factors is to choose the right partner, and it is also the main part of the whole supply chain. Therefore, enterprises must follow the premise of enterprise strategy, select the right partner. If both partners are not able to long-term development, then once the partners were changed, the supply chain will need more or less reconstruction work, which is not only a waste of time and energy of the enterprise, but also caused a serious economic loss to the enterprise.
Secondly, achieve low carbonization of the process. Under the guidance of the new society low-carbon model, in the process of building the supply chain of enterprises, it is necessary to re organize the entire supply chain work-flow. In the low carbon mode of the supply chain, many important factors such as the core enterprises, manufacturers, distributors, suppliers and so on are to be re to low-carbon vision for the effective optimization, so as to enable enterprises to maximize production efficiency. Establish a process re-engineering platform is a good method, on this platform, can the enterprise information sharing in supply chain is established in the process of various factors, thus, logistics service flow and the flow of funds of the enterprises to make good guarantee, to make it more adapt to changes in current society market. In the construction process of supply chain, to make low-carbon protection for each link in the planning stage, to make the whole planning in as soon as possible after the whole operation process to achieve synchronization and integration, and in the purchase stage, you can purchase mode widely used to make appropriate optimization to find goods sectors, thus reducing procurement of manpower and resources, thereby reducing the production cost of enterprises.
Finally, build low carbon evaluation system. The construction of low carbon evaluation system can be evaluated by the actual low carbonization degree of the enterprise. Thus, it provides a benchmark and comparability for the low carbonization process of each enterprise. And then effectively improve the low carbon progress of enterprises.

2.2. Some Assumptions
The contradiction between weak quality and low carbon properties of the products also exist, such as power service, power plant using coal, natural gas, hydro, wind, solar and nuclear energy to generate electricity for power products consumers need, there is no difference in quality, but in the low carbon level is very different. This characteristic of power services can be called "image quality" and "low carbon decoupling", but that does not mean that all services are quality decoupled from low carbon. For example, consulting services, can be considered to meet the quality and low carbon decoupling, in face-to-face consultation, transportation to the front of the customer, or by the video system with remote client conversation, and will not affect the effect of consultation, but can serve the effects of carbon emissions significantly; but for the mobile communication service, is not this is the case, taking into account the requirements including the quality of mobile communication service in the coverage, connection rate, the number of wireless base station indoor distribution system in a mobile communication network must continue to grow, and keep the hours of uninterrupted, and even prove intelligent carrier has a remarkable energy-saving effect off technology, also will be abandoned

because ensure the connection rate requirements for mobile communication services, the contradiction between quality and carbon emissions is very prominent.
Different types of product supply chain, in the process of low-carbon, is bound to be affected by the dual attributes of the product, the upstream and downstream enterprises in the supply chain must balance between the two attributes, looking for a reasonable balance point.
This is the main research contents in influencing the quality of product attributes and properties of low carbon, supply chain upstream and downstream enterprises to carry out incentive and obstacle of carbon reduction action, the quality of carbon emission intensity of products related to supply chain emissions reduction effect.
In order to simplify the analysis process, without considering the impact of carbon disclosure and the initial low carbon level of the supply chain before the emission reduction decision, it is assumed that both the low carbon level decision of the supply chain enterprises is also the carbon reduction decision. The supply chain is composed of an upstream enterprise (manufacturer) and a downstream enterprise (retailer). The upstream enterprise is established as the core enterprise, and the upstream enterprise can make a decision on the wholesale price of the intermediate variable. Low carbon parameter, expressed as $e_d = e_S - e_O$, is the difference between carbon emission and low carbon standard.
The parameters description of upstream enterprises in supply chain is shown as the p_m represents the ex-factory price/wholesale price of the upstream enterprise products, the c represents the unit cost of the product, the π_m represents upstream corporate profits.
The parameters description of downstream enterprises in supply chain is shown as the p represents the retail prices of downstream products, sets downstream enterprises only purchase cost p_m, the π_d represents downstream corporate profits.
So the profit maximization of the upstream enterprises in the supply chain is achieved after the carbon disclosure:

$$\begin{aligned} Max \quad \pi_m(p_m) = \\ Max(\mathrm{A} + t \times e_d - \mathrm{B} \times p) \times (p_m - c) \end{aligned} \tag{1}$$

The profit maximization of the enterprises in the supply chain after carbon disclosure is as follows:

$$\begin{aligned} Max \quad \pi_d(p) = \\ Max(\mathrm{A} + t \times e_d - \mathrm{B} \times p) \times (p - p_m) \end{aligned} \tag{2}$$

where $A > 0, B > 0$ are coefficient of the demand function.
In order to obtain the demand function of the supply chain product by the consumer utility function, the following assumptions are made:

(1)Different consumers have different willingness to pay for quality and low carbon, so Assume that θ and β respectively represent the random variable subject to uniform distribution in $(\theta_{\min}, \theta_{\max})$ and $(\beta_{\min}, \beta_{\max})$;

(2) θ and β are independent of each other;

(3)Assume that the total number of consumers on the market is 1, Consumers who are in demand for the product are the consumers who meet the condition $U_e(\theta, \beta) \geq 0$.

3. Demand Function of Low Carbon Supply Chain

The purpose of consumer spending is to obtain utility consumer surplus. For the products supplied by the supply chain, the following assumptions can be made on consumer behavior:

$$U = \begin{cases} \theta \times S - p \geq 0, \text{buy} \\ < 0, \qquad \text{else} \end{cases} \tag{3}$$

where U is the utility/consumer surplus; S is the quality parameter of products, on behalf of the quality of the product; p is product price; θ is consumer preference of quality parameters, is the willingness of consumers on the quality of the payment. Generally, $\theta > 0$, and the value is different for different consumers.

In the context of low carbon economy, low carbon products will increase the consumer preferences, consumer demands on the use of the content, and the optimization of the environment, so that the products provided by the utility should not only satisfy the consumer demand for the use of performance, but also to meet the consumer demand for the performance of low carbon, low carbon performance reflect the requirements of the low carbon consumer awareness. Taking into account the low carbon conscious consumers will have a certain willingness to pay for the low carbon value, the product structure of the low-carbon economy will have a two variable utility value structure.

Changes in consumer behaviors may be made in the context of a low carbon economy:

$$U_e = \begin{cases} \theta \times S + \beta \times e_d - p \geq 0, \text{buy} \\ < 0, \qquad \text{esle} \end{cases} \tag{4}$$

where: U_e is the utility/consumer surplus; e_d is the low carbon parameter of the product, which represents the low carbon level of the product; β is the low carbon preference parameter of consumption which represents the consumer's willingness to pay for low carbon, and generally there is $\beta > 0$.

For the particular S, e_d, p , the demand function of the supply chain is:

$$q = \iint_{\theta \times S + \beta \times e_d - p \geq 0} \frac{1}{(\theta_{\max} - \theta_{\min})(\beta_{\max} - \beta_{\min})} d\beta d\theta \tag{5}$$

Taking into account the relationship between product quality and low carbon levels of the product, the following assumptions are made on the model:

$$S = S_0 - g \times e_d \tag{6}$$

Which S_0 represents the quality of products and low carbon independent parts, and there is $g > 0$.

Defining $a = \frac{\theta_{\max}}{\theta_{\max} - \theta_{\min}}$, $b = \frac{1}{S_0(\theta_{\max} - \theta_{\min})}$, $t = \frac{(\beta_{\max} + \beta_{\min})}{2S_0(\theta_{\max} - \theta_{\min})}$, the demand function of the supply chain is:

$$q = a + (t - bg)e_d - bp \tag{7}$$

4. Supply Chain Demand Function with Product Attributes

Considering the upstream enterprises of supply chain as the core enterprises, the decision-making relationship between the upstream and downstream enterprises in supply chain can be analyzed according to Stackelberg game.

The reaction function of the downstream enterprise to the ex factory price is firstly obtained.

$$\frac{\partial \pi_d(p)}{\partial p} = \frac{\partial}{\partial p}[(a + (t - bg)e_d - bp)(p - p_m)] \tag{8}$$

$$p = \frac{a + (t - bg)e_d + bp_m}{2b} \tag{9}$$

Substitute the reaction function of the downstream enterprise into the maximization profit objective function of the upstream enterprise, and gets:

$$Max\pi_m = \mathrm{Max}\left[\frac{1}{2}(a+(t-bg)e_d-bp_m)(p_m-c+re_d)-he_d^2\right] \tag{10}$$

The upstream enterprises need to make a decision on the ex factory price p_m and the low carbon level e_d of the unit product at the same time, and the first order of profit maximization is:

$$\frac{\partial\pi_m(p_m,e_d)}{\partial p_m}=0, \frac{\partial\pi_m(p_m,e_d)}{\partial e_d}=0 \tag{11}$$

By simultaneous equations, the ex-factory price p_m and the low carbon level e_d of the unit product which meet conditions are substituted into the retail price, manufacturing sales volume, upstream supply chain profit, and the profit of the downstream enterprises of the supply chain could be calculated as follow.

Ex-factory price:

$$p_m^{st}=\frac{(b\cdot r+(t-bg))\cdot(a-b\cdot c)}{8b\cdot h-(b\cdot r+(t-bg))^2} \tag{12}$$

Retail price:

$$p^{st}=\frac{1}{2}\left(p_m^{st}+\frac{a+(t-bg)e_d^{st}}{b}\right) \tag{13}$$

Sales volume:

$$q^{st}=a+(t-bg)e_d^{st}-bp^{st} \tag{14}$$

Low carbon level:

$$e_d^{st}=\frac{4h(a+bc)-(br+(t-bg))(ra+(t-bg)c)}{8bh-(br+(t-bg))^2} \tag{15}$$

Profits of upstream enterprises:

$$\pi_m^{st}=\frac{1}{2}(a+(t-bg)e_d^{st}-bp^{st})\cdot(p_m^{st}-c+re_d^{st})-h(e_d^{st})^2 \tag{16}$$

Profits of downstream enterprises:

$$\pi_d^{st}=(a+(t-bg)e_d^{st}-bp^{st})(p^{st}-p_m^{st}) \tag{17}$$

Taking into account the profit and low carbon levels of the upstream and downstream enterprises in the supply chain, the form of the decision value is rather complicated, so the analysis is carried out in the following example.

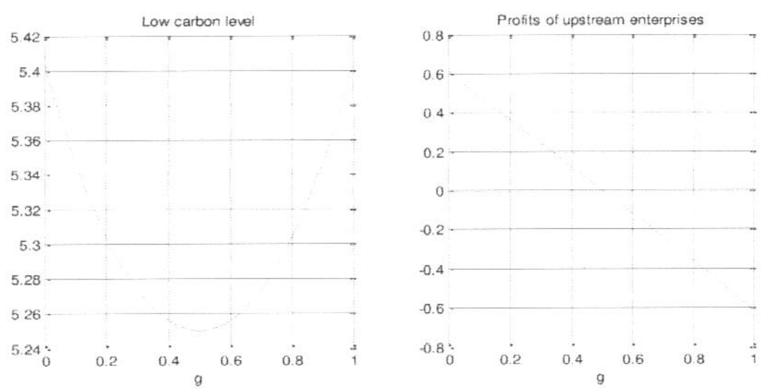

Figure 1. The influence of the correlation intensity.

5. Model Example and Analysis

For analyzing of the constants and random variables in the low carbon analysis model of the supply chain considered as: θ , β are uniformly distributed over the interval (0, 1); $S_0 = 10$; $c = 0.5$; $r = 0.001$; $h = 0.1$.

In the numerical example, the variation of the correlation intensity(g , $0 \le g \le 1$) of the product quality and carbon emission is considered only. The following calculation results can be obtained. If the upstream and downstream enterprises in the supply chain do not carry out carbon disclosure (and there is no further reduction action), then the profit of the enterprises in the supply chain will be as Figure1.

By analyzing of the calculation results, in effect the dual attributes of products, not all of the supply chain have the conditions of low carbon; and this hinders the low-carbon supply chain conditions, awareness of low carbon low carbon concept does not depend on the strength of enterprises, the level of consumer or government on carbon emissions requirements in size, but by the in the supply chain enterprises and consumers jointly formed by choice.

6. Conclusion

The final product of low carbon level is the joint decision which is decided by the duality preference between the quality and low carbon enterprise to balance the dual attributes of products. The low carbon level change products, reasonable way still from the two aspects.

From the consumer side, you can change the consumer's duality preference, especially the quality preferences, to promote the improvement of low carbon products. But consumer pursuits of improving living standards are often intense and difficult to reverse.

7. References

[1] Brandenburg M. Low carbon supply chain configuration for a new product – a goal programming approach[J]. International Journal of Production Research, 2015, 53(21):6588-6610.

[2] Kawasaki T, Yamada T, Itsubo N, et al. Multi Criteria Simulation Model for Lead Times, Costs and CO2 Emissions in a Low-carbon Supply Chain Network [J]. Procedia Cirp, 2015, 26:329-334.

[3] Shaw K, Shankar R, Yadav S S, et al. Supplier selection using fuzzy AHP and fuzzy multi-objective linear programming for developing low carbon supply chain[J]. Expert Systems with Applications An International Journal, 2012, 39(9):8182-8192.

[4] Chen C T, Lin C T, Huang S F. A fuzzy approach for supplier evaluation and selection in supply chain management[J]. International Journal of Production Economics, 2006, 102(2):289-301.

[5] Zhao H, Li Z. Dynamic game analysis of R&D investment in low carbon supply chain[C]. Chinese Control and Decision Conference, 2016:4570-4573.

[6] Jiang W, Chen X. Optimal Strategies for Low Carbon Supply Chain with Strategic Customer Behavior and Green Technology Investment[J]. Discrete Dynamics in Nature and Society,2016,(2016-2-15), 2016, 2016(5):1-13.

IOP Conf. Series: Earth and Environmental Science **168** (2018) 012032 doi:10.1088/1755-1315/168/1/012032

Wind Speed Patterns Mining Based on Multiple Views

Zongxia Xie, Jimin Jin

School of Computer Software, Tianjin University,Tianjin, China
e-mail: caddiexie@hotmail.com, jjmtju@163.com

Abstract. There are different patterns for wind speed variation in different seasons. The patterns are not only related to the statistical characteristics, but also to the trends of wind speed fluctuation and variation. However, the classical clustering can't take the trend information as an independent overall feature in the algorithm. In this paper, multi-view clustering is introduced to extract patterns of wind speed variation from both statistical information and the trends. For about 1.5-year historical data, the trends of wind speed variation and fluctuation, and other five statistical characteristics are selected as features to characterize patterns of wind speed variation. Then four patterns are obtained with multiple clustering. It is found that mean wind speed and fluctuation level are very important and the behavior of wind speed is significantly different in the past states in different clusters. Forecasting model performs worse for the cluster with large mean wind speed and fluctuation range.

1. Introduction

The energy crisis and environmental crisis caused by excessive consumption of fossil fuels become more and more serious. As renewable energy, wind power has been developed rapidly [1]. However, the inaccuracy of prediction for wind speed and wind power restricts the stability of large-scale wind power generation. Two main methods widely used for wind power prediction are numerical weather prediction (NWP) and data driven models. Based on the meteorological formula, NWP can usually obtain stable weather forecasting [2]. However, the calculation cost of NWP is expensive, and the errors in the short-term and the ultra-short term wind speed prediction are large. While data driven models are widely used in short-term and ultra-short term wind speed prediction[3]. The main algorithms are artificial neural networks, SVR [4,5], autoregressive moving average model [6], kalman filtering method[7], recurrent neural network [8], and so on.

As known, wind speed has strong randomness, and wind speed patterns are different in different seasons or days [9]. For these complicated characteristics of wind speed, research work is mainly carried out from two aspects. Firstly, hybrid models are used for wind speed and power prediction. In 2006, Sanchez proposed the short-term prediction of wind energy by assigning different weighting coefficients to several time series models for building hybrid model [10]. Since then, many hybrid models for wind speed prediction have been proposed [11-14]. DL et al decompose wind speed into different components by wavelet technique [12]. Fei et al proposed the hybrid model of empirical mode decomposition and multiple-kernel relevance vector regression algorithm (EMD-MkRVR) to decompose the data into different components by modal decomposition[13]. Ouyang et al improved the accuracy of wind speed prediction by using different kernel functions for different samples with SVR [14]. Secondly, clustering is used to extract the wind speed patterns in advance, and then different models for different patterns are constructed, respectively. The wind speed patterns were proposed in 2009 [15]. Kusiak and Li obtained different wind speed patterns by using K-means algorithm to cluster the wind speed in different scenes [16]. Hu et al pointed out that K-means which is good at finding the circular cluster is not conducive to explore the patterns of wind speed variation, and proposed generalized principal component analysis for automatically dividing wind speed into different clusters and obtaining wind speed patterns [17]. Lee and Piao proposed the projection clustering to get the patterns of wind speed variation [18]. In 2016, Wani et al used cluster based approach for mining patterns to predict wind speed [19].

However, the existing clustering methods are mainly based on the statistical characteristic of wind speed, such as the mean, variance and so on, or the simple mixture of the trend information and statistical characteristics of wind speed. It is difficult to regard trend information or statistical information as an independent overall feature in the process of clustering. Based on this, we introduce a method based on multiple views for patterns extraction. The method can take the trend information as an independent overall feature in the process of clustering with statistical characteristics at the same time, and make full use of the trend information of wind speed and statistical information to mine wind speed patterns. The method is applied to the real wind speed data, and the latent patterns of wind speed variation are extracted . What's more, decision tree is used on statistical features to get the rules of different clusters. The boxplots of trend features of different clusters are also analyzed.

2. Analysis on Variation of Wind Speed

The wind speed data is from a wind farm in Ningxia Province located in Northwest China. About 1.5-year historical data is used in this work. The observations of wind speed in the wind farm are in 10-min interval. Figure 1 shows all the wind speed. Figure 2 shows the corresponding first order difference sequence.

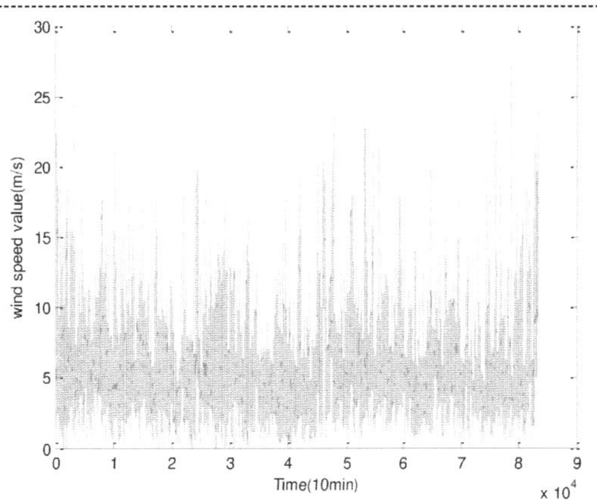

Figure 1. Wind speed data in 10-minute interval

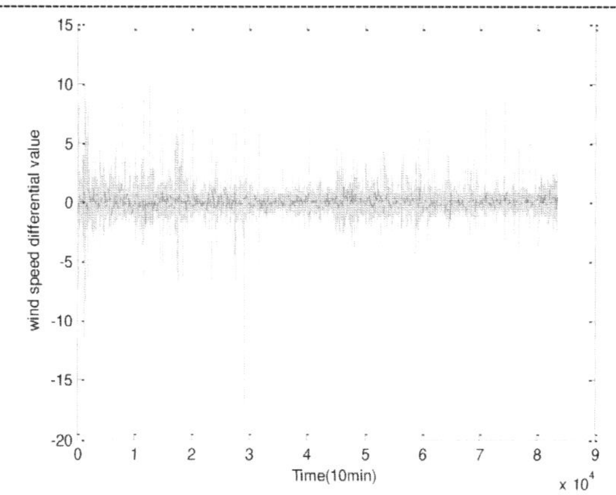

Figure 2. First order difference sequence of the original wind data

As shown in Figure 1, the wind speed is on the rise in some time period. After it climbs to a peak, it slowly or fast drops to a small value. The fluctuation of wind speed is obviously quite large. For certain period of time, the wind speed fluctuates in a certain small range with relatively steady wind speed values. And the mean wind speed is also different at different time periods.

As shown in Figure 2, the wind speed occasionally fluctuates very large, but the wind speed fluctuations are relatively small in most time. The fluctuation range of wind speed is different at different times. The variation of wind speed is very complicated.

For different time periods, the peak value of wind speed, mean wind speed and fluctuation level are also obviously different. These statistical features characterize the overall variation of wind speed. Therefore, it is necessary to consider them as the features to extract the wind speed patterns. The statistical information reflects the general information of wind speed variation. However, the trend information reflects the detailed information of wind speed variation. In order to incorporate trend information and statistical information for clustering at the same time, we fuse the information from multiple views instead of simply connecting the two types of information.

3. Methods

3.1. Defining the Features for Clustering

According to Part II, both the statistical information and trends of wind speed should be used for clustering. In this paper, two trends of wind speed are collected. One is the trend of wind speed variation, which is composed by the current wind speed v_t and the previous m-1 wind speeds $\left[v_{t-1}, v_{t-2}, \ldots, v_{t-(m-1)}\right]$. The other is the trend of wind speed fluctuation, which includes $m-1$ differences of wind speeds between the two nearest times.

At the same time, another five statistical features are computed. They are maximum, minimum and mean wind speed, absolute fluctuation level and fluctuation range of the current and previous m-1 times. Both the trends and statistical features are listed in Table I.

Table 1. Features for Clustering

Features	Definition
Trend of wind speed variation	$\left[v_t, v_{t-1}, v_{t-2}, \ldots, v_{t-(m-1)}\right]$
Trend of wind speed fluctuation	$\left[\Delta v_1, \Delta v_2, \ldots, \Delta v_{m-1}\right]$
Maximum wind speed	$\max_{i=0,1,\ldots,m-1} \{v_{t-i}\}$
Minimum wind speed	$\min_{i=0,1,\ldots,m-1} \{v_{t-i}\}$
Mean wind speed	$\operatorname{mean}_{i=0,1,\ldots,m-1} \{v_{t-i}\}$
Absolute fluctuation level	$[\sum_{t=1}^{m-1} abs(\Delta v_t)]$
Fluctuation range	$[v_{\max} - v_{\min}]$

In order to use all the features in Table I for clustering, multi-view clustering is introduced. This method can take each trend as a view and all the statistical features as a view. From Table I, we know that there are 3 views for wind speed clustering.

3.2. Multi-view Clustering Algorithm

In this part, the multi-view clustering is introduced to obtain clusters from multiple views. This method is proposed by shao et al [21], which can carry on clustering from multiple views fastly. In general, given a set of incomplete multi-view data $\{X^{(\upsilon)} \in R_{+}^{D_\upsilon \times N}, \upsilon = 1, 2, \ldots, n_\upsilon\}$, our aim is to find the latent feature matrices for each of the views and a common consensus, which represents the integrated information of all the views. The objective function can be written as :

$$\min_{\{U^{(v)}\},\{V^{(\upsilon)}\},V^*} 1 = \sum_{\upsilon=1}^{n_\upsilon} \left\| X^{(\upsilon)} - U^{(\upsilon)} Z^{(\upsilon)^T} \right\|_F^2 + \sum_{\upsilon=1}^{n_\upsilon} \alpha_\upsilon \left\| Z^{(\upsilon)} - Z^* \right\|_F^2 \quad (1)$$

$$\text{s.t.}\quad Z^* \geq 0, U^{(\upsilon)} \geq 0, Z^{(\upsilon)} \geq 0, \upsilon = 1,2,\ldots,n_\upsilon$$

where $U^{(\upsilon)} \in R_+^{D_\upsilon \times K}$ is the basis matrix, $Z^{(\upsilon)} \in R_+^{N \times K}$ is latent feature matrix for the v-th view, $Z^* \in R_+^{N \times K}$ is the consensus latent feature matrix, K is the number of clusters, and α_υ is the trade -off parameter in the objective function.

In order to eliminate the influence of the incomplete data, the instances that appear in the view are given weight 1, and the missing instances in the view are given lower weight. The objective function can be written as:

$$1 = \sum_{\upsilon=1}^{n_\upsilon} \left\| (X^{(\upsilon)} - U^{(\upsilon)} Z^{(\upsilon)^T}) W^{(\upsilon)} \right\|_F^2 + \sum_{\upsilon=1}^{n_\upsilon} \alpha_\upsilon \left\| W^{(\upsilon)} (Z^{(\upsilon)} - Z^*) \right\|_F^2 . \quad (2)$$

To enforce the sparsity of the latent feature matrix, L_1 norm is adopted. The new objective function can be written as:

$$1 = \sum_{\upsilon=1}^{n_\upsilon} \left\| (X^{(\upsilon)} - U^{(\upsilon)} Z^{(\upsilon)^T}) W^{(\upsilon)} \right\|_F^2 + \sum_{\upsilon=1}^{n_\upsilon} \alpha_\upsilon \left\| W^{(\upsilon)} (Z^{(\upsilon)} - Z^*) \right\|_F^2 + \sum_{\upsilon=1}^{n_\upsilon} \beta_\upsilon \left\| Z^{(\upsilon)} \right\|_1 \quad (3)$$

where $\|\cdot\|_1$ is L1 norm and β_υ is the trade-off parameter between the sparsity and accuracy of reconstruction for the v-th view. After optimization, we can get the best $Z^* \in R_+^{N \times K}$,which is the consensus latent feature matrix. Therefore, K-means algorithm is performed on Z^* to get the latent clusters from multiple views.

As to wind speed clustering, the number of views is 3. All the data of 3 views are used in the objective function (3) to get the optimal Z*, and then K-means is applied to Z* to obtain the final clusters.

4. Experiment

The dataset from a wind farm in Ningxia Province is used in the experiment. To predict the wind speed in the following one-hour, the current and previous $m-1$ past states of wind speed are collected as the inputs of wind speed prediction model. m is set to 23 in this work according to the mutual information [14]. Then the wind speed time series can be transformed into matrix data, each row can be regarded as a sample. The total number of samples is 83483. Among all samples, two-third of the data are used as the training set, and the left one-third as the test set.

To obtain the patterns of wind speed variation, multiple features from three views as shown in Table I are collected at first. Then multiple-view clustering of wind speed samples is conducted on the three views.

4.1. Determining the Number of Clusters

The number of clusters K is needed to determine in advance for multiple-view clustering. In order to get the suitable number of clusters, K is set to 2,3,4,5, respectively. Then, SVR is used to construct models on each cluster. The average errors on training data of all clusters with different K are presented in Table II. The mean absolute error (MAE) and the mean square error (MSE) are computed as :

$$MAE = \frac{1}{n} \sum_{i=1}^{n} \left| y_{ri} - y_{pi} \right| \quad (4)$$

$$MSE = \frac{1}{n} \sum_{i=1}^{n} \left(y_{ri} - y_{pi} \right)^2 \quad (5)$$

where y_{ri} is the real value, y_{pi} is the predicted value, and n is the number of test samples.

Table 2. Error With Different k Values

K / Error	K=2	K=3	K=4	K=5
MAE	1.0018	1.0014	1.0002	1.0019
MSE	2.0971	2.0913	2.0879	2.0936

As shown in Table II, the errors become smaller with the increase of the number of clusters K. However, the errors reach the smallest value when K is 4 and then become bigger when K is 5. Therefore, 4 is chosen as the suitable number of clusters.

4.2. Obtaining patterns based on multiple views

Multiple-view clustering is performed on the training data with the number of clusters $K = 4$. The number of samples in each cluster is listed in Table III. Cluster 2 has the most samples, cluster 3 has less samples than cluster 2, while cluster 1 and cluster 4 have the least samples. The sample number of cluster 1 and cluster 4 is only about one-ninth of cluster 2.

In order to get the rules of wind speed patterns, the decision tree is constructed based on the statistical features of these four different clusters. The tree is shown in Figure 3. In addition, some trends of wind speed variation and fluctuation of each cluster are also shown in the leaf nodes by a part of samples. $x3$ in Figure 3 represents the mean wind speed in the 24 past state of wind speed, and x4 represents the fluctuation level of wind speed in the past state. And the tree can get 82% accuracy on training data.

We can know that the mean speed and fluctuation level are partitioned into four intervals. For cluster 2, the mean speed is in [0 5.63] and fluctuation level is in [0 10.35]. The mean speeds of cluster 3 are in [0 5.63] and the fluctuation levels are larger than 10.35. The second rule for cluster 3 is that the mean speed is between 5.63 and 10.88, and the fluctuation level is smaller than 13.21. As for cluster 1, the mean speed is also between 5.63 and 10.88, but the fluctuation level is larger than 13.21. The rule for cluster 4 is just that the mean wind speed is larger than 10.88. What's more, the trends of different clusters are also different.

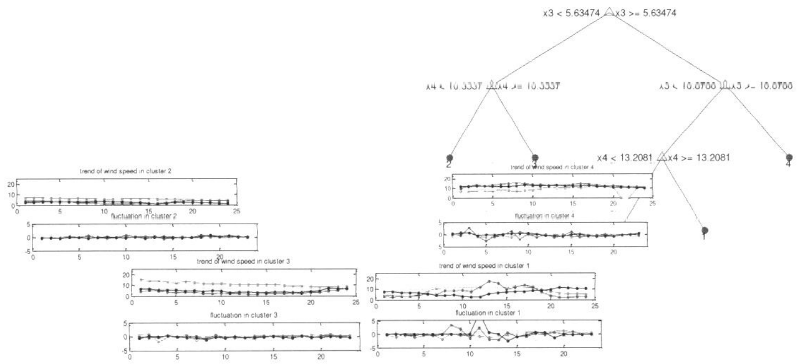

Figure 3. The rule based on Statistical characteristics for each cluster

To better demonstrate the trend rules among the different patterns, the boxplot of samples of each cluster in each 10-min intervals are plotted in Figure 4 and Figure 5. Some outliers are removed to show the main trend. Figure 4 and Figure 5 show that the trends of wind speed variation and fluctuation in different cluster are different. For cluster 1 which has median speed and large fluctuation level, the trend of wind speed becomes small at first, then it is bigger, finally it does not change a lot, the trend of fluctuation changes a lot. The trend of wind speed and fluctuation in cluster 2 stays steady. As to cluster 3, the trend of wind speed changes a lot, it becomes small at first and then becomes bigger and bigger, while the trend of fluctuation stays steady. However, both the trends of wind speed and fluctuation don't change a lot in cluster 4.

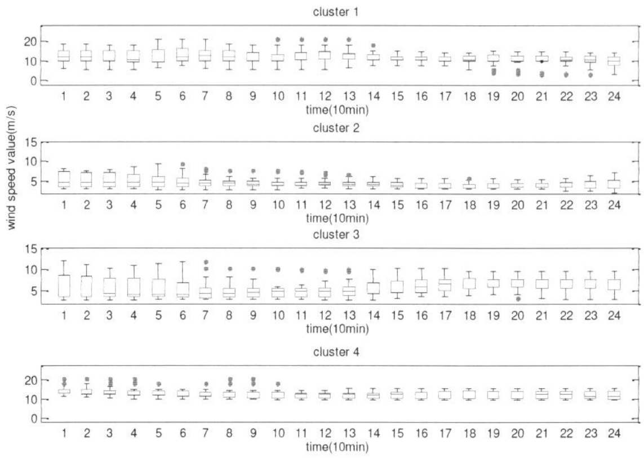

Figure 4. Wind speed variation of each cluster

Therefore, the 4 clusters obtained by multiple views are not only different on statistical information, but also different on trends information.
Finally, SVR [20] is used to build models for different clusters. While an unseen sample is assigned to a cluster according to the distances to the center of each cluster, the model on the corresponding cluster is used to predict the value of wind speed. The errors for the one-third test data are listed in TABLE III. We can know that models on cluster 2 and cluster 3 perform well, while models on cluster 1 and 4 perform worse. It also shows that it is necessary to divide the wind speed into different patterns because of the different performance of different clusters.

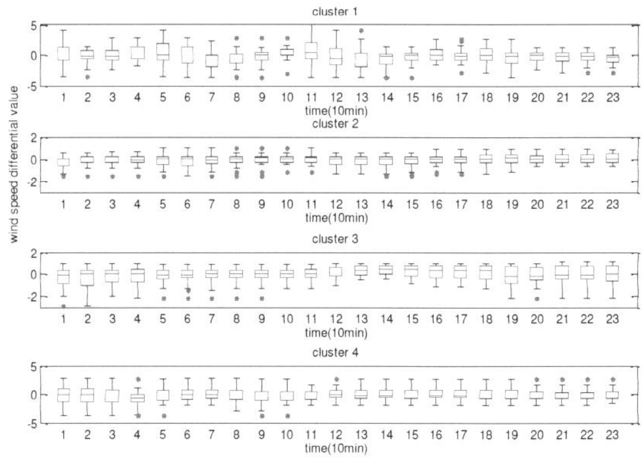

Figure 5. Wind speed fluctuation of each cluster

Table 3. Errors of Wind Speed Prediction in 4 clusters

NO. / Error	1	2	3	4
MAE	1.7957	0.7863	0.9573	1.3511
MSE	5.6466	1.265	1.7719	3.252
Number of training samples	3312	29562	18010	4771

5. Conclution

Because of the randomness, fluctuation and uncertainty of wind, great challenges are brought to the wind speed patterns mining under a single angle. In the paper, a method based on multiple views for extracting patterns of wind speed variation is proposed. In the process of clustering, it can get the latent clusters based both on statistical information and trend information. It can also take each trend as an independent overall feature. The experiment on real wind farm proves that the latent patterns of wind speed variation can be identified effectively based on multiple views. It is found that mean wind speed and fluctuation level are important in the statistical features, and wind speed variation in m past states shows different rules in different clusters. Meanwhile, wind speed can be well predicted in the cluster with low wind speed and small fluctuation. What's more, while the wind speed and its fluctuation range are larger, wind speed forecasting model performs worse. It reminds us that different models need to be constructed with different patterns, and range of error in different clusters is different. It can be seen that the multi-view clustering method can better find the wind speed patterns from a data-driven perspective, and help to improve wind speed forecasting accuracy. In future, we will collect more datasets from different wind farms to mining the patterns of wind speed variation, and extract more information of wind speed to carry out the work. In addition, how to effectively build wind speed forecasting model according to different sub-pattern is a valuable work.

6. References

[1] Council, G. W. E. (2013). Global wind statistics 2012. Global Wind Report.

[2] H G. V. Kuik, B. Ummels, and R. Hendriks(2007). Sustainable Energy Technologies. Amsterdam, The Netherlands: Springer.

[3] A Kusiak, A., Zheng, H., & Song, Z. (2009). Wind farm power prediction: a data-mining approach. Wind Energy, 12(3), 275-293.

[4] Saavedra-Moreno, B., Salcedo-Sanz, S., Carro-Calvo, L., Gascón-Moreno, J., Jiménez-Fernandez, S., Prieto, L., 2013. Very fast training neural-computation techniques for realmeasure-correlate-predict wind operations in wind farms. J. WindEng. Ind. Aerodyn.116, 49–60.

[5] Mohandes, M. A., Halawani, T. O., Rehman, S., & Hussain, A. A. (2004). Support vector machines for wind speed prediction. Renewable Energy, 29(6), 939-947.

[6] Torres, J. L., Garcia, A., De Blas, M., & De Francisco, A. (2005). Forecast of hourly average wind speed with ARMA models in Navarre (Spain). Solar Energy, 79(1), 65-77.

[7] Louka, P., Galanis, G., Siebert, N., Kariniotakis, G., Katsafados, P., & Pytharoulis, I., et al. (2008). Improvements in wind speed forecasts for wind power prediction purposes using kalman filtering. Journal of Wind Engineering & Industrial Aerodynamics, 96(12), 2348-2362.

[8] Barbounis, T. G., Theocharis, J. B., Alexiadis, M. C., & Dokopoulos, P. S. (2006). Long-term wind speed and power forecasting using local recurrent neural network models. IEEE Transactions on Energy Conversion, 21(1), 273-284.

[9] Jiang, X., Jiang, C., Peng, M., Lin, H., & Li, Z. (2010). A short-term combination wind speed forecasting method considering seasonal periodicity and time-continuity.Dianli Xitong Zidonghua(Automation of Electric Power Systems), 34(15), 75-79.

[10] Sanchez, I. (2006). Short-term prediction of wind energy production. International Journal of Forecasting, 22(1), 43-56.

[11] Tascikaraoglu, A., & Uzunoglu, M. (2014). A review of combined approaches for prediction of short-term wind speed and power. Renewable and Sustainable Energy Reviews, 34, 243-254.

[12] Faria, D. L., Castro, R., Philippart, C., & Gusmao, A. (2009, March). Wavelets pre-filtering in wind speed prediction. In Power Engineering, Energy and Electrical Drives, 2009. POWERENG'09. International Conference on (pp. 168-173). IEEE.

[13] Fei, S. W. (2016). A hybrid model of EMD and multiple-kernel RVR algorithm for wind speed prediction. International Journal of Electrical Power & Energy Systems, 78, 910-915.

[14] Ouyang, T., Zha, X., Qin, L., Xiong, Y., & Xia, T. (2016). Wind power prediction method based on regime of switching kernel functions. Journal of Wind Engineering and Industrial Aerodynamics, 153, 26-33.

[15] Wang, C., Lu, Z., & Qiao, Y. (2009, April). Modeling of wind pattern and its application in

wind speed forecasting. In Sustainable Power Generation and Supply, 2009. SUPERGEN'09. International Conference on (pp. 1-6). IEEE.

[16] Kusiak, A., & Li, W. (2010). Short-term prediction of wind power with a clustering approach. Renewable Energy, 35(10), 2362-2369.

[17] Hu, Q., Su, P., Yu, D., & Liu, J. (2014). Pattern-based wind speed prediction based on generalized principal component analysis. IEEE Transactions on Sustainable Energy, 5(3), 866-874.

[18] Lee, H. G., Piao, M., & Shin, Y. H. (2015). Wind Power Pattern Forecasting Based on Projected Clustering and Classification Methods. Etri Journal, 37(2), 283-294.

[19] Wani, M. R., Wani, M. A., & Riyaz, R. (2016, November). Cluster based approach for mining patterns to predict wind speed. In Renewable Energy Research and Applications (ICRERA), 2016 IEEE International Conference on (pp. 1046-1050). IEEE.

[20] Santamaría-Bonfil, G., Reyes-Ballesteros, A., & Gershenson, C. (2016). Wind speed forecasting for wind farms: A method based on support vector regression. Renewable Energy, 85, 790-809.

[21] Shao, W., He, L., Lu, C. T., & Philip, S. Y. (2016, December). Online multi-view clustering with incomplete views. In Big Data (Big Data), 2016 IEEE International Conference on (pp. 1012-1017). IEEE.

IOP Conf. Series: Earth and Environmental Science **168** (2018) 012033 doi:10.1088/1755-1315/168/1/012033

Influence Research of Renewable Energy Application on Power Quality Detection

Ge Beichang, Wu Xiao, Liu Yingying, Zhang Zhang, Yang Fan

State Grid Tianjin Economic Research Institute,State Grid Corporation of China, Tianjin China

e-mail: zimo_ge@163.com ,wuxiao0117@126.com, lyypower@126.com, strongzhang1988@163.com, 870415059@qq.com

Abstract. With the development of renawable energy generation systems, it not only relieves energy shortages and environmental pollution problems, but also brings harmonic pollution, which is related with power quality decetion. A modified harmonic and interharmonic analysis method is studied in this paper. By applying window interpolation, signal reconstruction, 5-point modified algorithm, and other signal handling methods, harmonic and inharmonic components are separated exactly, improving harmonic and interharmonic detection accuracy under new energy application, providing accurate signal basis and reference for further harmonic suppression and distribution network reconstruction scheme, and finally improving power quality.

1. Introduction

With the deterioration of energy shortage and environmental pollution, clean energy generation system sources, such as distributed photovoltaic power generation system, get support of national energy policy and social concern widely, for advantages of abundance, clear, and so on [1].

In recent years, photovoltaic power plants, wind power plants and other distributed energy system grow rapidly, transforming traditional distribution network with single power supply and radioactive network to be with multiple power supplies. Reneable energy power generation systems are combined with lots of non-linear electronic devices, injecting harmonics and interharmonics into distributed network, harming electrical equipments and grid safe operation [2].

Traditional FFT algorithm (fast Fourier transform, FFT) has advantages of less calculation, specific physical meaning, and good performance on integer harmonic analysis [3], [4]. However, due to interharmonics and fundamental frequency variation, it is difficult for FFT algorithm to realize synchronous sampling, causing spectrum leakage and fence effect. Although windowed interpolation FFT algorithm can release the effect, it also has limitations, which means that it is not suitable for harmonic and interharmonic analysis under energy application [5], [6].

Aimed at the above problems, a modified harmonic and harmonic analysis method, based on literature [7], is researched in this paper, effectively improving harmonics and interharmonic precision under reneawble energy application. The algorithm researched in this paper not only can provide accurate harmonic and interharmonic characteristics, but also provides signal basis and reference for radioactive, relay protection and distribution network restructuring plan under reneawble energy application.

IOP Conf. Series: Earth and Environmental Science **168** (2018) 012033 doi:10.1088/1755-1315/168/1/012033

2. Harmonic and Interharmonic Generation Mechanism Under Reneawble energy Application

When power grid runs normally, the harmonics generated by power system are mainly divided into two aspects: harmonics generated by internal power supply systems and generated by electronic components of distributed power generation system.

The harmonics generated by the grid itself are dominated by harmonics of 1, 2, 3, 5, etc. Harmonic components generated by distributed generation system are complex. Especially when it faults, lots of interharmonics are generated for nonlinear actions of power electronic devices and control system [8].

Interharmonics may cause fundamental frequency variation, making traditional FFT algorithm difficult to realize synchronous sampling. It will cause large errors in the phase calculation, hugely influencing the power system stability under reneawble energy application.

3. Harmonic and Interharmonics Detection Method Based on Modified 2-stage FFT Algorithm

3.1. Signal Expansion

For power system, signals are represented in the shape of sine [9]. For components containing both harmonics and interharmonics, the expression is as follows:

$$\begin{aligned} S(t) &= S_H(t) + S_J(t) = \sum_{h=1}^{H} S_h(t) + \sum_{j=1}^{J} S_j(t) \\ &= \sum_{h=1}^{H} A_h \sin(2\pi f_h t + \varphi_h) + \sum_{j=1}^{J} A_j \sin(2\pi f_j t + \varphi_j) \end{aligned} \quad (1)$$

Among them, A_h、φ_h and $f_h$are amplitude, phase and frequency of the *h*th harmonic, and *H* is the highest harmonic frequency; A_j、φ_j and $f_j$are value, phase, and frequency of the *j*th interharmonic, *J* is the number of interharmonics in the signal.

Spectral leakage and fence effect of fundamental and harmonics may affect the interharmonics detection effect, which can lower harmonic detection accuracy in turn [7].

In order to reduce the mutual interference between harmonics and interharmonics existed in actual signals, this paper researched a modified 2-stage FFT analysis method, which separates harmonic S_H and interharmonic S_J in signal, acquiring accurate signal characteristics under reneawble energy application.

3.2. Harmonic Analysis in modified 2-stage FFT Algorithm

Fundamental wave analysis using dual - spectral interpolation FFT method with hanning-window. Sample signal $S(t)$ (t_s duration) with the fixed sampling rate f_s, then obtain the N-point discrete sequence S(n). S(n) is analyzed by dual-spectral interpolation FFT algorithm with Hanning-Window, obtaining fundamental frequency f_0. Sample number $N=f_s \cdot t_s$, sample interval $T_s=1/f_s$.

Amplitude of fundamental wave in power signals is much larger than that of harmonics and interharmonics. And by adding window interpolation method, spectrum leakage and fence effect are further inhibited. So the obtained fundamental frequency f_0 meets the precision requirement, ensuring further synchronization accuracy.

Signal reconstruction with cubic spline interpolation. According to the calculated fundamental frequency f_0, the original sampling sequence S(n) is reconstructed by cubic spline interpolation method, obtaining a new N-point equal interval sequence, denoted as Sr(n). The sample frequency of the reconstructed sequence Sr(n) is $f_s'=n_1 \cdot f_0$, and the sample interval is $T_s'=1/f_s'$. n_1 is a positive integer, satisfying requirements for simultaneous sampling.

A suitable interpolation algorithm is a key factor in signal reconstruction. The cubic spline interpolation is a smooth curve passing through a series of shape points, with low computational complexity and high reconstruction accuracy [10]. In this paper, cubic spline interpolation is used for signal reconstruction.

Harmonic analysis of reconstructed signals. The reconstructed sequence *Sr*(n) satisfies synchronous sampling requirement of fundamental and harmonics. However, due to the uncertainty of interharmonic period, it is difficult to realize simultaneous sampling for interharmonics, and to eliminate the interference between harmonics and interharmonics. Based on literature [7], harmonics of reconstructed sequence are analyzed by odd frequency point interpolation method in this paper.
In this paper, the reconstructed sampling sequence is analyzed by 5-point modified algorithm with Rectangular-Window, acquiring amplitudeA_h, phase φ_h and frequency f_h of fundamental wave and harmonics. The reconstruction sequence *Sr*(n) is a synchronous sampling sequence, effectively inhibits spectral leakage between fundamental wave and harmonics; Rectangular-Window owns the narrowest main lobe, which can avoid the spectrum interference between harmonics and interharmonics; 5-point modified algorithm is good at suppressing spectral interference, effectively solving the spectral interference problem between harmonics and interharmonics. Through the odd frequency point interpolation method, frequency spectrum leakage, spectrum aliasing and fence effect problem are well solved, acquiring accurate fundamental wave and harmonic parameters.

3.3. Inharmonic Analysis in Modified 2-stage FFT Algorithm

Obtaining of interharmonics from the original signal. Through subtracting the fundamental wave and harmonic components $_{SH}$(n) from the original sampling signal *S*(n), interharmonic components $_{SJ}$(n) are obtained. The specific steps are as follows: According to the fundamental and harmonic components obtained in 2.2, the structure of the fundamental and harmonic components is expressed as:

$$\begin{gathered} S_H(t) = A_h \sin(2\pi f_h t + \varphi_h), \\ t = 0, Ts, 2Ts, \cdots, (N-1)Ts \end{gathered} \tag{2}$$

S_H(n) is a N-points sequence. Interharmonic component S_J(n) is abtained by subtracting S_H(n) from *S*(n).

Interharmonics analysis based on 5-point modified algorithm with Rectangular-Window. Considering spectral aliasing influence for interphonic detection, 5-point modified algorithm based with Rectangular-Window is used to analyze the separated interharmonics signal, obtaining exact interharmonic amplitude A_j, Phase φ_j and frequency f_j parameters. Compared with the traditional 2-stage FFT algorithm, the 5-point modified algorithm not only suppresses the effects of spectral aliasing, spectral leakage and fence effect, but also avoids the cumulative error and high computational complexity of peak search method.

3.4. A Modified 2-stage FFT Algorithm

In this paper, the process of modified 2-stage FFT algorithm for harmonic and interharmonic analysis is shown in Figure 1.
Compared to traditional 2-stage FFT algorithm, 5-point modified algorithm with Rectangular-Window is used in this paper, solving mutual interference problem between harmonics and interharmonics, and improving detection accuracy.

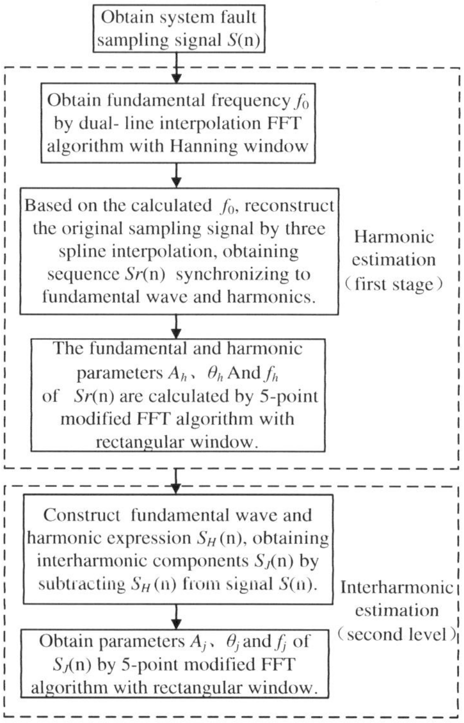

Figure 1. Flowchart of modified 2-stage FFT scheme.

4. Simulation and Experiment

Suppose that the sampling signal containing multiple harmonic and interharmonic components is:

$$\begin{aligned} S(t) = {} & 200sin(f*2\pi t - 23.1°) + 4.5sin(2f*2\pi t \\ & +115.6°) + 7sin(3f*2\pi t + 59.3°) + \\ & 5sin(5f*2\pi t + 78.7°) + 3sin(7f*2\pi t + 92.5°) \\ & +0.6sin(23*2\pi t + 73.8°) + 0.3sin(84*2\pi t + 53.7°) \\ & +0.2sin(175*2\pi + 113.4°) + 0.1sin(189*2\pi t - 23.3°) \end{aligned} \quad (3)$$

Considered fundamental wave fluctuations, set the fundamental frequency to be 49.8Hz. Harmonics are 1th, 2th, 3th, 5th and 7th. Interharmonics have frequencies of 23, 84, 175 and 189 Hz.

IEC 61000-4-7 recommends 10 basis-cycle window (50 Hz system) to analyze power system signals [11]. Therefore, this paper also uses 10 basis-cycle window. The specific steps are:

- Discreet sampling signal. Sampling frequency f_s=10kHz, sampling window length t_s=0.2s. Discrete the sampling signal to obtain the original sampling sequence S(n) with N=2000 points and sampling interval T_s=$10^{-4}s$.
- Estimate fundamental frequency. The S(n) is analyzed by dual - line interpolation FFT algorithm with Hanning-Window.
  The fundamental frequency f_0= 49.799976Hz and the error is -4.8908E - 5%.
- Synchronize sampling signal. Based on the calculated fundamental frequency f_0, the original sampling sequence S(n) is reconstructed by cubic spline interpolation method, to obtain sequence Sr(n), synchronous for fundamental and harmonics.
- The sampling points of Sr(n) are 2000, and the sampling interval T_s'=1/49.799976/200s.

- Harmonic analysis of reconstructed sampled signal. Analyze the quasi-synchronous sampling signal *Sr*(n) with 5-point modified algorithm with Rectangular-Window, which can suppress interference between harmonics and interharmonics, acquiring accurate amplitude A_h, phase φ_h and frequency f_h of fundamental wave and harmonics.
- As can be seen from Table 1, detection errors of 2th harmonic are the largest, respectively be 0.0016%、-0.00298° and 0.000256%. In general, detection of all fundamental and harmonics obtains high accuracy.
- Structure fundamental wave and harmonics expression as $s_h(t)$. Subtract $S_H(t)$ from the original signal *S*(n), obtaining interharmonic sequence $S_J(t)$.
- Interharmonic analysis. Analyze $S_J(t)$ by use of 5-point modified algorithm with Rectangular-Window, restraining spectrum leakage and aliasing, and obtaining accurate interharmonic parameters, amplitude A_j, phase φ_j and frequency f_j.
  It can be seen from Table 1 that the biggest detection error of 175 Hz is 0.5128%, - 2.49611 ° and 0.040205%, respectively.

TABLE I. ANALYSIS RESULTS AND ITS RELATIVE ERRORS OBTAINED BY THE MODIFIED 2 STAGE FFT METHOD

Analog signal parameters			**Analysis Results and Errors by Modified 2-Stage FFT Algorithm**					
Amplitude (A)	*Phase (degree)*	*Frequency (Hz)*	*Amplitude(A)*	*Error (%)*	*Phase (degree)*	*Error (degree)*	*Frequency (Hz)*	*Error (%)*
200	-23.1	49.8	200.00016	8.00E-05	23.09912207	0.000877933	49.79997635	-4.75E-05
4.5	115.6	99.6	4.50007585	0.001685549	115.5970165	0.002983468	99.59974407	0.000256961
7	59.3	149.4	7.000006806	9.72E-05	59.29950322	0.000496776	149.4000221	1.48E-05
5	78.7	199.2	4.999995719	-8.56E-05	78.69994557	-5.44E-05	249.0000014	5.77E-07
3	92.5	349	2.999990299	0.000323366	92.49999027	-9.73E-06	348.6000003	7.68E-08
0.6	73.8	23	0.600041793	0.006965558	73.82929977	0.029299772	22.99918117	0.003560111
0.3	53.7	84	0.300009971	0.003323582	53.71358186	0.013581864	83.99967662	0.000384975
0.2	113.4	175	0.201025666	0.512833066	110.9038863	2.496113707	175.0703589	0.040205067
0.1	-23.3	189	0.100000116	0.000116094	23.29887904	0.001120958	188.9999701	-1.58E-05

5. Conclusions

The modified 2-stage FFT algorithm researched in this paper, considers the interference between harmonics and interharmonics, applies window interpolation, signal reconstruction, 5-point modified algorithm, and other signal handling methods, achieving high accuracy in harmonics and inter-harmonic detection, providing accurate signal basis and reference for further researches under renewable energy application, such as harmonic suppression scheme, distribution network reconstruction plan, evaluation index of power quality, adaptability of clean energy enhancing scheme, and so on.

But the algorithm needs data window of ten fundamental wave periods, showing bad performance in quick detection. The way to compress detection data window, which can realize quick detection and analysis, will be researched in further study.

6. References

[1] Wang Chengshan, WANG Shouxiang. Study on some key problems related to distributed generation systems [J]. Automa-tion of Electric Power Systems, 2008, 32(20): 1-4,31.

[2] ISHIKAWA T. Grid-connected photovoltaic power systems: sur-vey of inverter and related

protection equipments [R]. Tokyo, Japan : Report IEA-PVPS Task , 2002.
[3] XUE Hui, YANG Rengang. Precise algorithms for harmonic analysis based on FFT algorithm[J] . Proceedings of the CSEE , 2002, 22(12) : 106 - 110
[4] ZHANG Fusheng, GENG Zhongxing, GE Yaozhong. FFT algorithm with high accuracy for harmonic analysis in power system[J] . Proceedings of the CSEE, 1999, 19(3) : 63 - 66.
[5] Pang Hao, Li Dongxia, Zu Yunxiao, et a1. An improved algorithm for harmonic analysis of power system using FFT technique[J]. Proceedings of the CSEE, 2003, 23(6): 50-54.
[6] Dariusz B, Andrzej B. Improvement of accuracy of power system spectral analysis by coherent resampling[J]. IEEE Transactions on Power Delivery, 2009, 24(1): 1004-1013.
[7] Hui Jin, Yang Honggeng. An approach for harmonic/inter-harmonic analysis based on the odd point interpolation correction [J]. Proceedings of the CSEE, 2010, 30(16): 67-72(in Chinese)
[8] HAO Jiangtao, LIU Nian, XING Jinyu, et al. Study on interharmonic of power system[J] . Electric Power Automation Equipment, 2004, 24(12) : 36 - 39.
[9] Mai R K, Fu L, Dong Z Y, et al. Dynamic Phasor and Frequency Estimators Considering Decaying. DC Components[J]. IEEE Transactions on Power Systems. 2012, 27(2): 671-681.
[10] Jin Weigang, Liu Huijin. A New Algorithm for Interharmonics Measurement Based on Time Domain Sampling and Reconstruction by Cubic Spline Interpolation[J]. Power System Technology, 2012, 36(7):74-80.
[11] Hui J, Yang H, Xu W, et al. A Method to Improve the Interharmonic Grouping Scheme Adopted by IEC Standard 61000-4-7[J]. IEEE Transactions on Power Delivery. 2012, 27(2): 971-979.

Impacts on Operation and Benefit After New Energy Generation Access to Power Grid

Yang Fan[1], Gao Pei[2], Liu Yingying[1], Zhang Zhang[1], Wu Xiao[1]

[1]Department of Distribution Network Plan,State Grid Tianjin Economic Research Institute, Tianjin, China
[2]Department of Maintenance, Tianjin Huaneng Yangliuqing Power Co., LTD Tianjin, China

e-mail: 870415059@qq.com

Abstract. As the development of social economy, the demand for energy is increasing rapidly. At the same time, energy shortage and environmental pollution problem is more and more serious, which is a big challenge for grid construction. New energy represented by wind and solar generation is flexible, safe and clean, and it provides a new idea to relieve the shortage of energy, solve the environmental pollution problem, and improve the reliability and flexibility of the grid. This paper analyzes the influences on power system operation and economic benefit by new energy generation access. The investing and construction direction for power grid enterprise in new energy industry is proposed.

1. Introduction

With the development of social economy, the traditional centralized single-mode of power supply is awkward and fragile, which can not meet the demand of power users on high quality and reliability of power supply. On the other hand, the environmental pollution problem is more seriously because of the massive consumption of fossil fuels. New energy generation represented by solar and wind generation is flexible, safe and clean, which can save the problems of energy shortage and environmental pollution.

To ensure the power supply, the combination of new energy generation and power grid is the direction in the future instead of the single-mode power supply [1]. After the access of new energy to the grid, the traditional power grid structure will be changed into a grid with power sauces and users throughout the gird. As a result, the operation of power system and the income for power grid enterprises will be influenced. This paper elaborates the effects of new energy access on the balance of electric power and energy, utilization rate of the equipment and power grid construction. And the effects of new energy access on power grid enterprises' income is analyzed by some new energy project. It can provide ideas in realizing the coordinate development between new energy generation and power grid.

2. Effects on Power Grid Operation

2.1. Balance Of Electric Power And Energy

The access of new energy generation changes the grid from radial single-terminal power source system into a system with multi-power sources [2]. The two-way load flow can break the balance of electric power and energy between load and source, therefore the grid operation will be influenced.

Take some local intensive photovoltaic project for example. This project installs photovoltaic unit on the roof of local residents. The solar generation will be used locally and the surplus will be exported to the power grid. According to the data of solar radiation and the load supply, the typical daily solar generation curve and load curve are gotten as figure 1. Area 1 represents user's power purchase; area 2 represents the power exported to the grid; area 3 represents the solar generation used locally.

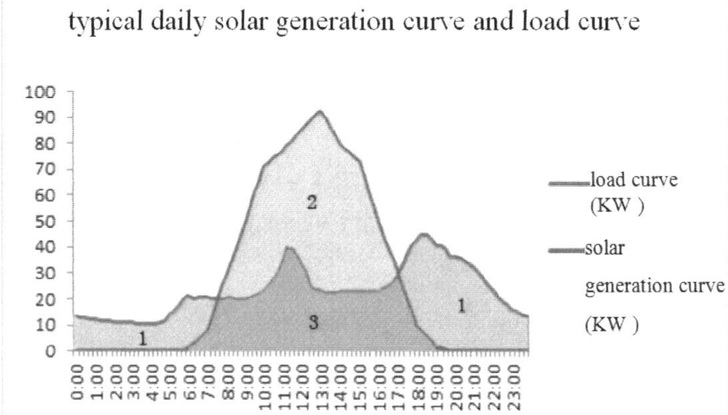

Figure 1. Typical daily solar generation curve and load curve.

The solar generation is used locally in the prior way, and if the solar generation produces more electricity than the consumption, the surplus will be exported to the power grid. The load curve will be changed as figure 2.

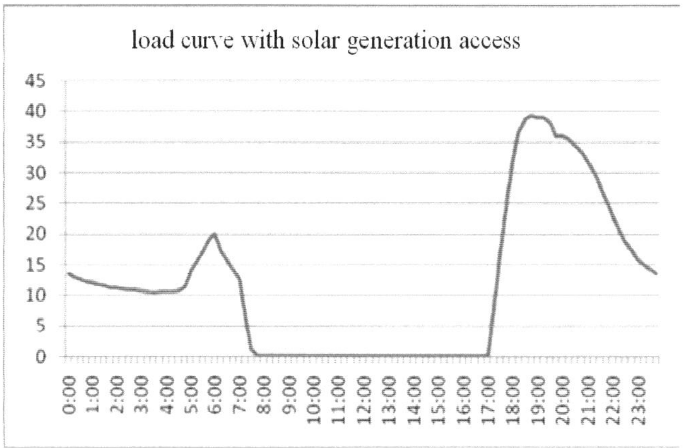

Figure 2. Load curve with solar generation access

According to the load curve, the access of new energy can take on the task of reducing peak. And the surplus exported to the grid can be used by power supply area nearby. Therefore the power shortage of the power supply area during peak energy periods can be relieved.

2.2. Equipment Utilization Rate

As the project shows, the access of new energy can be used to reduce the load peak, especially for residential customers. The load characteristic is typical and the load focuses on the noontime when is the generation peak of solar. So the solar generation can meet the demand for power supply.

The peak load before and after the access of new energy is shown as Table I. According to the table, before the access of new energy, the load peak appears at 11:15 and the peak is 40kW. After the access, the load peak at noon disappears, and there is surplus to be exported to the grid. The load of this area is very typical. The load in rural area is mainly for farming. The villagers have strong conception of resource-saving. After the access of the new energy, there is always surplus electricity to be exported to the grid during the daytime.This means that the distribution transform will be standing in no -load condition during the day. Although the capacity of the transformer in rural area is lower according to the differentiation principle, the transformer will be long-term idle because of the access of new energy. Therefore, the equipment utilization rate will be declined sharply which will increase the cost of power grid enterprises indirectly.

Table 1. THE Peak Load before and after the Access of New Energy

	Peak at noon		Peak at night		The hours of usage during peak-load (h)
	Time	Load (kW)	Time	Load (kW)	
Before	11:15	40.08	18:15	44.68	12.45
After	11:15	0	18:45	39.35	7.14

2.3. Power grid Reconstruction

The access of new energy is a challenge for the present grid's ability of power generation absorption and grid structure strength. To the area with weak network configuration, the large scale new energy connected into the power system can not be beard. When the power flow is reversed, the voltage at the terminal of power line will be increased [3], even be out of the voltage limit. Reversed power flow will also increase the line loss rate and lower the operation efficiency. But new energy access takes the responsibility of energy saving and emission reduction, therefore the power grid reconstruction caused by new energy access is of great importance for the society.

To rural area with weak grid structure, the bottlenecks and single radiation line should be solved firstly to adapt to the large-scale intensive new energy access [4]. To urban area with strong grid structure, the development of solar, wind and other new energy resources should be taken into consideration. For different area's new energy access, the weakness of the grid structure should be analyzed. Reconstructing the power grid specifically to improve the new energy access ability is necessary.

Considering the access of new energy, the no-voltage auto-recloser is necessary for overhead line. As a result, voltage transformer should be added for overhead line with auto recloser. If there is new energy access to the grid, the interconnection switch of local transformer substation should has no-voltage auto-recloser. For the access of new energy by high voltage, the grid-connected line or power line be provided with longitudinal difference protection to ensure the safe operation of power grid.

The grid-connected line, grid-connected breaker and metering device for new energy access are invested and constructed by power grid enterprise. Therefore the investment of grid reconstruction will be increased greatly. And that is the requirement to ensure the safety and reliability of new energy access to the grid.

3. Effects on Economic Benefit

The access of new energy is invaluable for social society. Although power supply enterprise takes important social responsibility that it should pay the cost of grid reconstruction, the effect of new energy access on power supply enterprise's economic income is still a concerned question.

The stakeholders in new energy access are mainly include investor and user [5]. The investor of new energy generation can be power grid enterprise or independent investor. Power grid enterprise has the characteristics of large investment amounts and good credit. As the investor, it will be convenient to coordinate the conflict of interest between new energy and power grid. The starting point for grid enterprise is not its own maximum interests, but the maximum of social benefits. Because of the large amount of reconstruction cost, the economic benefit for grid enterprise in new energy project may not be significant. Independent investor can be new energy market participants, and they want to pursue profits and expand market share. Users act as power consumer in new energy project, and they can get some compensation by providing their roof or open space for new energy generation device.

Take some rural area intensive photovoltaic project for example. The profit and cost of different operation pattern is calculated based on operating data during operation period, including purchase price of electricity, sale price of electricity, maintenance cost, etc. And the economic feasibility is analyzed.

According to the characteristics of intensive photovoltaic project, marginal utility theory should be used to calculate the economic benefit [6]. Marginal utility theory analyzes the effect when one unit is increased or reduced with other conditions unchanged. According to marginal utility theory, the incremental revenue and incremental cost should be calculated by "Availability of Comparative Law". Therefore, the incremental revenue refers to the variation of electricity purchase/sale and solar subsidies. And incremental cost refers to the cost of project construction.

It should be noted that the reconstruction of rural grid is an important task for pushing forward the urbanization, and its social benefit can not be measured by the commodity attribute and economic value of electricity. Therefore, the economic benefit can not be the only basis for evaluating if the project decision is right or not. The effects of the project on grid safety, reliability, power supply quality and social benefit should be taken into consideration. At the same time, the financial support for the project by government should be calculated.

3.1. Incremental Revenue Calculation

The incremental revenue by this photovoltaic project mainly includes the income of electricity sale (when power supply enterprise is the investor, it will be the saved power purchase cost) and the policy subsidy by state (local) government. In the calculation, benchmark price of thermal power is 0.3815 yuan/kWh, resident electricity price is 0.49 yuan/kWh, and policy subsidy by state government is 0.42 yuan/kWh. The subsidy by local government is reversely calculated as a variable.

3.2. Incremental Cost Calculation

Incremental cost refers to the cost of the photovoltaic project, including materials expense, repair expense, labor expense, depreciation expense, etc. Besides, the cost of raising funds (mainly includes loan interest) should taken into consideration.

In the calculation, the operation and maintenance cost is relatively low, and the annual cost can be 0.5% of the initial investment. The annual cost by rural grid reconstruction and new energy access project is 3% of the initial investment. Depreciation cost is 6.72% of the initial investment. The proportion of principal amount of loan is 30%, and the loan interest rate is 4.91%. The loan will be repaid in 5 years.

3.3. Statement Preparation

The photovoltaic project's financial condition during operation period can be analyzed by profit and loss statement and cash flow statement.

In the calculation, corporate income tax is paid as 25% of the tax rate. The calculation period is 15 years considering the life of photovoltaic device.

3.4. Economic Evaluation (Power Grid Enterprise as the Investor)

The economic evaluation is based on four conditions:

1) Considering the investment of new energy system construction and power grid reconstruction caused by new energy access to grid;

2) Only considering the investment of new energy system construction;

3) Considering the investment of new energy system construction and power grid reconstruction, and the finance discount policy;

4) Considering the investment of new energy system construction and power grid reconstruction, the finance discount policy and the policy subsidy by local government;
The results of economic evaluation are shown as Table II.

Table 2. Economic Evaluation results (Power Grid Enterprise as the Investor)

Calculation condition	NPV	IRR
Considering grid reconstruction	¥-40.95	0.4%
Without grid reconstruction	¥6.41	5.81%
Considering grid reconstruction and finance discount	¥-25.95	1.79%
Considering grid reconstruction and increasing policy subsidy	¥0.20	4.93%

To the investment, when the IRR(Internal Rate of Return) is greater than loan rate (which is 4.91% in this project), NPV(Net Present Value) is greater than 0, the project is feasible.
According to Table II, if the grid reconstruction cost is considered, only the increased local policy subsidy reaches more than 0.183 yuan/kWh, can IRR reach 4.93% (which is greater than loan rate 4.91%), and NPV reach ¥0.20 (which is greater than 0). And under that condition, this project will be feasible.

3.5. Economic evaluation (independent investor)
The results of economic evaluation are shown as Table III.

Table 3. Economic Evaluation Results (Independent Investor)

Calculation Object	NPV	IRR
Independent investor	¥8.94	6.21%
Power supply enterprise	¥ 58.56	

According to Table III, the IRR of independent investor is greater than loan rate and the NPV if greater than 0. Therefore, it is feasible for independent investor to invest this rural area intensive photovoltaic project. On this condition, power supply enterprise takes the investment of grid reconstruction without economic profit. Therefore, the power supply enterprise's economic benefit will suffer some loss.

4. Conclusion
New energy can meet the demand of global environmental protection and sustainable development. It is also a way to solve the problem of energy crisis and energy safety. The access of new energy to the power grid breaks the balance of electric power and energy, and the grid device's utilization rate is decreased. The grid also needs to be reconstructed to adapt to large-scale new energy access, which will bring large amount of cost. And the economic benefit of the power grid enterprise will be influenced. Therefore, new energy access to the grid is a great challenge for power grid enterprise. Power grid enterprise should take social responsibility, at the same time, the way to get economic benefit still need to be found. Power grid enterprise should not only serve for new energy industry, but also expand the new energy market actively. And the financial support from the government is of great importance. Power grid enterprise should turn the challenge into a rare opportunity, and promote the coordinated development of new energy generation and power grid.

5. Acknowledgment
The authors would like to thank the colleagues in distribution network plan department of Economic Research Institute of Tianjin Electric Power Company for information and advice.

6. References
[1] MA Cong, GAO Geng, LI Ruisheng, et al, Reactive Power Perturbation Method for Simultaneous Low Voltage Ride-Through and Islanding Detection of Grid-Tied Renewable Energy Generation Systems[J] . Power System Technology, 2016, 40(5):1406-1414.
[2] CUI J T, LI K J, SUN Y, et al. Distributed energy storage system in wind power generation[C]//2011 4th International Conference on Electric Utility Deregulation and Restructuring and Power Technologies (DRPT), 2011:1535-1540.
[3] ZHANG Yu, MENG Xiaoli, FANG Hengfu, Effect of Distributed Research on Power Loss of Distribution Network [J]. Electric Power Construction, 2011, 32(5): 67-71.
[4] Yu Jin-yi, Liu Jian, Xu Li, et al, Distribution Grids of High Reliability for Core Areas of Large Cities[J]. Automation of Electric Power System, 2014, 38(20):74-80.
[5] Li Chen, Zhou Lisha, Zeng Ming. Modeling and Simulation of the Impacts of New Energy Integration on Power Market Based on System Dynamics [J]. East China Electric Power，2012,40 (10)： 56-60.
[6] ZhangLeqin.A Study on Limit and Evolutionary Trend of Pollution Effect in Process of Urbanization Based on Marginal Model： An Empirical Study of Anhui Province [J]. Journal of Natural Resources, 2016,0(2):275-286.

IOP Conf. Series: Earth and Environmental Science **168** (2018) 012035 doi:10.1088/1755-1315/168/1/012035

Comprehensive theoretical and evaluation system of integrated energy system

Wang Shiju[1], Xu Ke[2], Xu Jing[1], Chi Fujian[2], Zhang Xuefei[1], Zhang Liang[1]

[1]State Grid Tianjin Economic Research Institute, Tianjin, 300171, CHN
[2]State Grid Tianjin Electric Power Company, Tianjin, 300010, CHN
Email: 15900361200@163.com

Abstract. In order to undertake and integrate into the global energy Internet, solve the city environmental pollution and energy security, based on the present situation and comparative analysis of city energy development, put forward the framework of city energy Internet system, and provide a theoretical basis for the construction of city energy Internet; first proposed the characteristics analysis of the target system of city energy Internet based on the energy of the Internet provides the quantitative basis for city development goals; taking Tianjin as an example, the city energy Internet development stage and target, promote the allocation of energy resources in larger scope, to promote all types of energy and electrical energy, improve the proportion of clean energy use in the power generation side and power consumption in the side, finally realize the city energy no carbon consumption.

1. Introduction

The global energy consumption pattern is divided into three stages: the first stage is the production oriented energy consumption. At the end of nineteenth Century to mid twentieth Century, the development of energy industry in the world, based on coal, the energy dispersive energy development, unreasonable structure of energy consumption, mainly rely on self-balance; the second stage is oriented by efficiency way of energy consumption. From the mid twentieth Century to the end of twentieth Century, the rapid development of world economy, the energy industry began to take the efficiency as the guidance, development and conservation of energy, a variety of common development, energy consumption in electricity and coal; the third stage is the sustainable development oriented energy consumption patterns. Since the beginning of twenty-first Century, the development and utilization of new energy is highly valued, and the proportion of energy in the final energy demand in higher power allocation of energy resources efficiency is more significant, worldwide network to interconnection direction, gradually formed to deliver clean energy as the leading global Internet, ubiquitous smart the grid - Global energy Internet [1-4]. This is an important transition period from fossil fuels to clean energy sources.

As the main body of global energy consumption, urban energy demand is huge, but urban resources are scarce. At the same time, the urban energy consumption structure is irrational, and the urban energy system has serious pollution problems. Therefore, in order to solve the city energy security, environmental pollution and energy consumption in the city cannot balance problems, the city area as the main form of global unit, construction of city energy Internet, realize the energy structure from high carbon to low-carbon, energy efficiency, to change the allocation of resources by the local balance to optimize the allocation of large range by the way, the energy service supply to the intelligent interactive by extensive model use, effectively undertake the construction and development of global energy internet.

Some scholars have been on the global energy Internet related theories have made some research on the Internet, and for the city of energy related theory research literature: less [5] to establish the basic framework of energy Internet, discusses the basic structure and composition of the energy of the Internet literature; [6] established the energy Internet system model based on distributed generation; [7] given renewable the energy of high permeability, nonlinear stochastic, multi-source data and multi-scale dynamic 4 energy characteristics of the Internet literature; [8] established the basic model of energy router; literature [9] relying on the Beijing Yanqing smart grid innovation demonstration area construction project, analyzes the basic conditions and characteristics of the new energy Yanqing area, demonstration area was discussed from the energy supply and the form of consumption, main carriers, energy transportation management decision-making and public service etc. The contents of the building, build the application framework model of the energy of the Internet literature; [10] analysis and role positioning in the smart grid energy Internet system, put forward the "generalized energy Internet source - Net - charge - storage" coordination and optimization of operation mode. Most of the above literatures put forward a theoretical framework for energy Internet, and failed to put forward the development ideas and specific targets of energy Internet application in urban areas, which cannot guide the construction and development of integrated energy system.

Therefore, according to city regional energy Internet research: first a comprehensive analysis on the current situation of city energy development and comparison; on this basis, puts forward the theoretical framework of city energy Internet, and puts forward the target system to make quantitative analysis on city energy Internet development goals; finally, taking Tianjin as an example, put forward the development city planning of the energy of the Internet, to provide theoretical support and practical basis for future energy network construction.

2. Status and Challenges of Urban Energy Development

2.1. Raging Energy Demand in Urban Area

The city is the world's major economic activity and energy consumption place. In 2014, the global total GDP reached 77 trillion US dollars, of which only 300 of the world's largest cities GDP has reached more than 38% of the global total. The same year the world's total energy consumption reached 12 billion 928 million tons of oil equivalent, equivalent to 150276 TWh of the total electricity demand, of which 80% of the energy consumption is concentrated in the city. Table 1 presents the current situation and trend data of energy demand in four typical cities.

Table 1. Energy demand in Typical Cities

Cities	Energy needs /(TW·h)		
	2014	2030	2050
New York	1062	1300	1600
Beijing	556	750	900
Shanghai	902	1250	1600
Tianjin	663	950	1200
Total major cities in the world	120 221	166 000	190 000
Global total	150 276	210 000	240 000

At this stage, the city energy demand is huge, and with the expansion of urban scale and economic development, the energy demand of some cities will steadily increase in the future. Visibly, the city is the world's major contributor to GDP, but also the main energy consumption.

2.2. Energy Shortage in Urban Areas

Renewable energy which has been developed in city is not high, the future development of the total is very limited. Relatively speaking, the city has a huge demand for energy, in addition to renewable energy, other energy sources (especially fossil energy) basically rely on delivery, and the city presents typical characteristics of energy consumption. Figure 1 shows the contrast between the amount of renewable energy, the amount of development, the amount of development and the total demand for energy in five typical cities.

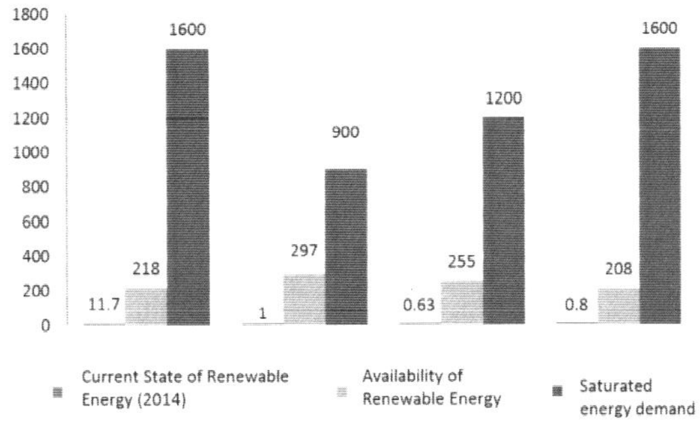

Figure 1. Renewable Energy and Energy Demand of Typical Cities (Twh/per year).

In the above figure, we can discover that there are some critical imbalances between supply and demand in these cities. There is a large shortage in local energy supply. The major part of energy needs be imported externally. Therefore, these cities are under severe energy resources deficiency.

2.3. Severe Pollution due to Urban Energy Consumption

So far, fossil fuel is still a major energy resource for urban energy consumption, and also the main source for urban pollution.

WHO global urban ambient air quality report shows that 89% of urban air quality fails to meet its health standard, where the air quality indexes of 57% of cities are 2.5 times higher beyond the standard value.

Figure 2. WHO Global Urban Air Quality Report.

The global urban pollution and CO_2 emission continue to increase year by year. A contradiction between urban development and environmental protection needs to be resolved as soon as possible.

3. Theoretical System of Integrated energy system

The integrated energy system is a connecting point for both global energy internet and Chinese energy internet. The essences of integrated energy system are electricity-oriented, cross-regional balance and low carbon.

City energy Internet is global energy Internet, Internet and China energy to undertake further extension on the load side of the subset, is the world's energy delivery to the end of the channel is an important support to protect thousands of households, with well ensure high voltage long distance transmission end, is an integral part in the city area of global energy Internet.

3.1. Core Concept

1) *Electricity-oriented.*

IOP Conf. Series: Earth and Environmental Science **168** (2018) 012035 doi:10.1088/1755-1315/168/1/012035

Considering power grid as a basic platform for energy allocation, it can support the integration of multiple energy resources (coal, oil, gas, wind, solar etc.) with electric power. Improve electric level has become to a basic measure to optimize energy structure and increase energy efficiency.

2) *Cross-regional Balance*

Gradually reducing the use of fossil energy and taking advantage of UHV long-distance transmission, it can increase the capability of receiving power externally and energy efficiency, solve the problems of large energy demand, deficient resources and environment protection, and promote optimal allocation of urban energy resources.

3) *Low Carbon*

By encouraging the large-scale integration of wind energy, solar energy and other external energy and promoting end-use alternative energy, it can boost market demand and benefit the development of low-carbon energy use.

3.2. System Architecture

Integrated energy system consists of physical layer, information layer, and service layer.

Service layer: Energy management platform of the city and service management center of integrated energy system.

Information layer: A communication and information system for all the regions and fields supplied by urban energy and the support of integrated energy system.

Physical layer: A strong, flexible, and sharing energy supply network and material basis of integrated energy system.

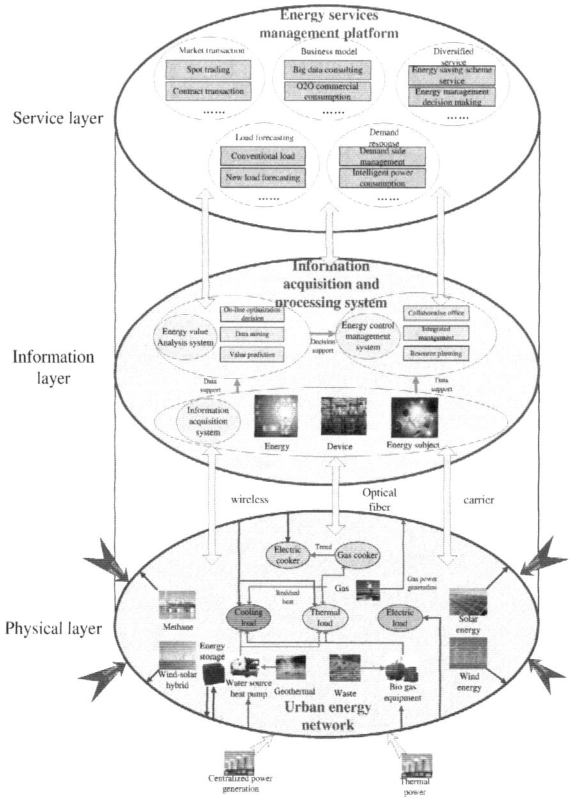

Figure 3. System Architecture of Integrated energy system.

3.3. Development Plan and Goal

Plan: By enhancing the construction of UHV external transmission and optimally coordinating different voltage level grids, large scale external electric power, wind and solar power can inject into the local network and be used properly. By encouraging smart grid construction and taking advantages of the internet properties of open, sharing, interaction, electricity, gas, heat and transportation networks can be connected and shared. By building energy service management platform, information collection and interactive business of the entities can be realized in the so called" source-grid-demand" structure. By exploring the value of each transaction of energy trade, it can stimulate the use of clean energy.
Development Goal: Constructing integrated energy system can overcome the issue of urban energy local imbalance, help to connect to global energy internet, realize large-scale energy allocation, drive the energy conversion between different types of energy resources and electric power, increase the ratio of using clean energy in the demand side, optimize urban energy pattern, increase energy efficiency, stimulate clean energy utilization, and achieve the carbon-free urban energy consumption.

4. Target System of Integrated energy system

The theoretical research on the integrated energy system has already taken shape, but there has been no research on the target system that reflects the level of integrated energy system development. Therefore, it is urgent to study a set of target system to quantitative analysis of city energy Internet development level, city guiding direction of energy development and utilization, promote clean and efficient use of energy, reduce the city carbon emissions, clean, electrification, information and network energy transformation.

4.1. Networked Target

Networking is the basic form of integrated energy system. Integrated energy system connection to external clean energy through UHV power transmission, the power is transmitted to the user through the internal implementation of strong smart grid, energy flow through the regulation of energy efficiency management platform and advanced communication technology, can realize the interconnection of various energy sources, and it is the efficient connection between energy and consumption.

Table 2. Networked Target

	Index	Unit
Networking	Energy form connectivity	%
	Sharing rate of energy service platform	%
	Energy information acquisition coverage	%

Energy form connectivity = the forms realizing energy connected with the energy management platform / total urban energy forms × 100%;
Sharing rate of energy service platform = the forms of energy sharing for service platform sharing / total urban energy forms × 100%;
Energy information acquisition coverage = energy information indicators that can be obtained / total information index × 100%.

4.2. Cleaning Target

Cleaning is the ultimate goal of integrated energy system. The construction of the Internet will be the implementation of the "energy city" development and utilization of clean energy and a high proportion of external power receiving wheel drive, to promote synergy, the eventual establishment of a clean energy lead new city energy supply system.

Table 3. Cleaning Target

	Index	Unit
Cleaning	Carbon emission per unit GDP	t/10000yuan
	Proportion of fossil energy to energy consumption	%
	Clean energy accounts for the total installed capacity	%

	External charge ratio	%

Carbon emissions per unit GDP = urban CO_2 emissions / GDP × 100%;
Fossil energy consumption ratio = fossil energy consumption / urban energy consumption × 100%;
Total clean energy to total installed capacity = total installed capacity of clean energy power generation / total installed capacity of urban power generation × 100%;
External power ratio = external battery / terminal energy consumption × 100%.

4.3. Electrification Target

Electrification is an important means of integrated energy system. With the continuous development and utilization of clean energy, the proportion of power consumption in the city's terminal consumption has been greatly improved. It has gradually realized the use of electricity instead of coal, electricity instead of gas, electricity with oil, greatly reducing fossil energy consumption.

Table 4. Electrification Target

	Index	Unit
Electrification	The proportion of electric energy to terminal energy consumption	%
	Electric vehicle ownership	%

Energy consumption in terminal energy consumption = power consumption / terminal energy consumption × 100%;
Electric vehicle ownership = electric car ownership / car ownership × 100%.

4.4. High Efficiency Target

High efficiency is the key support of integrated energy system. By improving the conversion efficiency of energy generation, realize the power supply side of multi energy complementary, improve energy efficiency; through advanced information communication, intelligent control technology, realize the city energy height of Internet communication and flexible control, realize the efficient allocation of resources among different areas.

Table 5. High efficiency Target

	Index	Unit
High efficiency	GDP energy intensity	Ton standard coal / 10000 yuan
	Primary energy input output ratio	%
	Maximum load utilization hours	%

GDP energy intensity = urban energy consumption / GDP × 100%;
Primary energy input output ratio = (coal, oil, natural gas, etc.) primary energy consumption / terminal energy consumption × 100%;
Maximum load utilization hours = annual sale / maximum load.

5. Case Study

During the "13th Five-Year" period, State Grid Tianjin Electric Power Company has become the city energy Internet construction, operation and management of the leader from the aspects of construction of all-round construction of city energy Internet.

5.1. Construction Prospective from 2030 to 2050

In 2020, Tianjin integrated energy system will be preliminarily constructed, incorporated to Chinese energy internet. The cross-national energy internet will be also realized. Green energy to primary energy rate is over 20%, electricity to end-use energy rate will reach 25%.
In 2030, Tianjin integrated energy system will be completed. Cross-continent grid connection will be preliminarily realized. It begins to integrate with global energy internet. Green energy to primary energy rate is over 40%, electricity to end-use energy rate will reach 35%.

In 2050, Tianjin integrated energy system will be well developed and extended. The close connection will be realized for cross-continent grid connection. It fully engage into the integration of global energy internet. Green energy to primary energy rate is over 80%, electricity to end-use energy rate will reach 60%.

5.2. Target system of Tianjin energy Internet
Based on the integrated energy system target system, Tianjin is used as a specific city target to quantify the phased development goals of the integrated energy system, specifically as shown in the table:

Table 6. Target System of Tianjin Energy Internet

	Index	Unit	2015 year	2020 year	2030 year	2050 year
Networking	Energy form connectivity	%	80	90	100	100
	Sharing rate of energy service platform	%	50	80	100	100
	Energy information acquisition coverage	%	70	90	100	100
Cleaning	Carbon emission per unit GDP	t/10000yuan	1.87	1.35	0.52	0
	Proportion of fossil energy to energy consumption	%	83	70	45	0
	Clean energy accounts for the total installed capacity	%	3.11	12	40	100
	External charge ratio	%	23	35	45	60
Electrification	The proportion of electric energy to terminal energy consumption	%	17	25	30	60
	Electric vehicle ownership	%	1.23	15	40	100
High efficiency	GDP energy intensity	Ton standard coal / 10000 yuan	0.5	0.42	0.32	0.18
	Primary energy input output ratio	%	38	50	70	90
	Maximum load utilization hours	%	4998	5200	5800	6500

6. Summary

To deal with the city's scarce energy resources, environmental protection pressure situation, to offer clean energy for the mission, to undertake and integrate into the global energy Internet, the definition and concept of integrated energy system is proposed, the three layer architecture, integrated energy system development ideas and objectives are put forward.
Integrated energy system theoretical framework and target system are helpful to the allocation of energy resources to achieve greater range, to promote all kinds of energy and electrical energy, to improve the proportion of clean energy use in the power generation side and power consumption in the side of the city, to achieve clean energy, electrification and information and the transformation of the Internet, to provide the theory of support and practical basis for the construction of city energy internet.

7. References

[1] C. Liu, M. Shahidehpour, Y. Fu, and Z. Li, "Security-constrained unit commitment with natural gas transmission constraints," *IEEE Trans. Power Syst.*, vol. 24, no. 3, pp. 1523–1536, Aug. 2009.
[2] K.W. Hedman *et al.*, "Co-optimization of generation unit commitment and transmission switching with N-1 reliability," *IEEE Trans. Power Syst.*, vol. 25, no. 2, pp. 1052–1063, May 2015.
[3] J. Qiu *et al.*, "A Linear programming approach to expansion coplanning in gas and electricity markets," *IEEE Trans. Power Syst.*, to be published.
[4] G. Li, Z. Bie, Y. Kou, J. Jiang, and M. Bettinelli, "Reliability evaluation of integrated energy systems based on smart agent communication," Appl. Energy, vol. 167, pp. 397–406, Apr. 2016.
[5] S. An, Q. Li, and T.W. Gedra, "Natural gas and electricity optimal power flow," in Proc.

IEEE/PES Transmiss. Distrib. Conf. Expo., 2003, vol. 1, pp. 7–12.
[6] J. M. Arroyo and A. J. Conejo, "Optimal response of a thermal unit to an electricity spot market," *IEEE Trans. Power Syst.*, vol. 15, pp. 1098–1104, Aug. 2000.
[7] R. Jabr, S. Karaki, and J. Korbane, "Robust multi-period OPF with storageand renewables," *IEEE Trans. Power Syst.*, vol. 30, no. 5, pp. 2790–2799,Sep. 2015.
[8] P. J. Hibbard, "The interdependence of electricity and natural gas: Current factors and future prospects," *Elect. J.*, vol. 25, no. 4, pp. 6–17, 2012.
[9] Y. Jia, K. Meng, and Z.Xu, "N-k induced cascading contingency screening," *IEEE Trans. Power Syst.*, vol. 30, no. 5, pp. 2824–2825, Sep. 2015.
[10] Y. Jia, Z. Xu, L. Lai, and K. Wong, "Risk based power system security analysis considering cascading outages," *IEEE Trans. Ind. Informat*, vol.12, no. 2, pp. 872–882, Apr. 2016.

Influence and Strategy Research of Cross-Regional Renewable Energy Incremental Spot Trading

Mao Hua, Li Yuanyuan, Wang Kui, Yan Dawei, Lei Zheng, Zhou Jin

Planning Review Center, State Grid Tianjin Power Economics & Technology Research Institute,Tianjin, China

e-mail: hua.mao@tj.sgcc.com.cn, yuanyuan.li3@tj.sgcc.com.cn, wangkui_hust@126.com, dawei.yan@tj.sgcc.com.cn, zheng.lei@tj.sgcc.com.cn, jin.zhou@tj.sgcc.com.cn

Abstract. Due to the incremental spot trading position for sending provinces of abandoned energy(water, wind or photovaltic power), the renewable energy increment spot transaction has obvious price advantages. At present, grid enterprises should actively participate in the spot trading of renewable energy in inter-regional provinces. With the reform of the electric power system, the power grid enterprises can participate in the spot trading of renewable energy of renewable energy in the inter-regional provinces through their own sales companies. By purchasing the peak support service from the power plant, we should deal with the peak problem caused by the spot trading of renewable energy in trans-regional provinces. The resulting peak support service costs are no longer shared by power plants, but should be compensated for by the profit generated by the incremental spot exchange.

1. Introduction

National Energy Commission issued a reply to the pilot work on the development of cross-regional renewable energy spot trading on February 14, 2017. The document pointed out cross-regional renewable energy spot trading is targeted for the incremental delivery of surplus renewable energy power generation in the province, district, or city. After the regulation resources have been used and all kinds of medium and long term decomposition plans have been implemented on the power generation side, renewable energy is still rich and may cause abandon water, wind, or solar, the renewable energy spot trading shuold carry out. The spot trading [1] includes previous and intraday forms [2], [3].

Large-scale cross-regional renewable energy spot trading will influent the grid business and power grid operation [4]-[6]. Cross-regional renewable energy has an advantage on price and will bring opportunities to the power grid, but the uncertainty of renewable energy also poses a challenge.

Based on the positioning of renewable energy spot trading and the characteristics of renewable energy power generation, this paper analyzed the influence of cross-regional spot trading and suggested power grid enterprises participate in this trading as soon as possible [7].With the advancement of the electricity market, power grid enterprises will participate actively in it with the electricity sales companies established by the grid enterprises in the future. The cost will be reduced, and the grid peaking problems will be solved through the purchase of auxiliary services [8], [9].

2. Trading Rules

Cross-regional renewable energy spot trading needs to meet the security constraints of both sides of the grid and the cross-regional channels to achieve the harmonization of the safe of the grid and the orderly operation of the market.

2.1. The Market Main Parts

The market main parts include:

1) The seller: hydropower, wind power and other renewable energy power generation enterprises in the terminal power grid. The renewable energy power generation enterprises participate in spot trading, sign and perform power transaction contracts according to the definition of electrical energy of abandoned water, abandoned wind, abandoned solar proposed by government departments.

2) The buyer: large users, electricity sales companies, power grid enterprises which agent large users and electricity sales companies initially. Power grid enterprises should ensure the safty of cross-regional transmission channels and other transmission and distribution facilities according to the rules of transmission and distribution of electricity, also the fair of transmission and distribution services in the market.

3) Transmission side: the transmission enterprises. The power dispatch department is responsible for organizing the province of large users and electricity sales companies, checking the security of power network and signing contracts.

2.2. The Trading Mode

Previous spot trading is mainly used to organize the trading of incremental delivery of surplus renewable energy after the annual, monthly cross-regional trading plans have been implemented. Recently, the spot trading is organized on a daily basis. The next day 96 points (0:15-24: 00,15 minutes as a point of time) trading will be organized on each day.

After the daily spot transaction is completed, the day spot trading is based on the change of supply and demand and the delivery plan. According to the current electricity price trading price, the current power trading result can be improved or reduced appropriately. The day trading of renewable energy shall be organized in the following day: 15-4:00, 4:15-8:00, 8:15-12:00, 12:15-16:00, 16:15-20:00, 20:15-24:00, and the six trading segments (15 minutes for a period of time).

The incremental spot transaction adopts the centralized bidding clearing mechanism, which considers the security constraints of the channel, and clears it in time.

(1) The seller's quotation is sorted from low to high, and the buyer's quotation is sorted from high to low. According to the high and low matching way, The lowest bidder and the highest bidder make the transaction first. The principle of decreasing the price difference between the buyer and seller shall be cleared in turn. When different buyers have the same price difference and the seller can not sell the quantity, then the proportion of the remaining purchase quantity of the buyer shall be distributed.

(2) Each transaction is allocated to a cross-regional channel based on the proportion of surplus space in the cross-regional channel.If the transmission line between the buyer and seller reaches the transmission capacity, it shall be regarded as the end of the transaction between the parties concerned. The seller and the buyer's quotation corresponding to the channel are deleted from the quotation sequence, but can still sell electricity to other regional market entities.

(3) The market will be cleared in order until the buyer or seller reports the full amount of electricity. But there exits the other case, when the price difference between buyers and sellers is negative, the trade ends. At the end of each delivery, the average price of the two parties shall be the system marginal price, and the total amount of the transaction shall be settled at the marginal price of the system.

3. Current Situation of Renewable Energy Generation in China

3.1. The Overall Situation

By the end of 2016, the power supply was 164.6 gigawattss. Coal installed 943 gigawatts, accounting for 57.27% of the total. Gas power generation 0.7 billion kw, accounting for 4.26 percent. Hydropower 332 million kw, accounting for 20.18%; Nuclear power was 0.34 billion kw, accounting for 2.04 percent. Wind power 1.49 billion kw, accounting for 9.03%; Photovoltaic power generation is 0.77 billion kw, accounting for 4.70%, and other installed 0.41 gw, accounting for 2.51%.

In 2016, China generated 5.99 trillion KWH of electricity. Wind power generated 24.1 billion KWH, accounting for 4.02% of total power generation. Photovoltaic power generation is 66.2 billion kilowatt-hours, accounting for 1.11% of total power generation.

At present, abandoning wind and photovoltaic situation in China is grim. In 2016, the total number of abandoned wind energy reached 49.7 billion kWh, increase of 46.6% compared to last year. The rate of abandoned wind in Gansu is 43%, Xinjiang is 38%, Jilin is 30%, and Neimenggu is 21%. In 2016, the total number of abandoned photovoltaic energy reached 7.4 billion kWh, increase of 85% compared to last year.

3.2. Relevant Conditions in North China

In 2016, the maximum load of north China power grid was 206.81 million kw, and the total social power consumption was 1227.7 billion KWH, with a total installed power generating capacity of 274.26 million kilowatts. Among them, renewable energy generation capacity is 488.2 million kilowatts, accounting for 17.8 percent of the total installed power generation, and the electricity generation reaches 7921.1 billion kilowatt-hours, accounting for 6.5 percent of the total power consumption in the whole society.

The maximum load of Tianjin power grid was 13.92 million kw, and the total power of the whole society was 76.3 billion KWH, with a total installed capacity of 14.67 million kilowatts. Among them, the renewable energy generating capacity is 989 million kilowatts, about 7 percent of the total power generating capacity, and 13.54 billion kilowatt-hours of power generation, accounting for 1.8% of the total power consumption in the whole society.

The maximum load of hebei is 2158 million kw, and the maximum load of hebei south network is 3266 million kw. Hebei power grid has 326.5 billion kilowatt-hours of electricity, with a total installed capacity of 62.75 million kilowatts. Among them, the renewable energy generating capacity is 182,000kw, accounting for 28.7% of the total installed power generating capacity and 9.9 billion kilowatt-hours of power generation, accounting for 9.2% of the total power consumption. Between 2017 and 2020, it is expected to add 22.7 gigawatts of renewable energy. In 2020, the renewable energy generation will reach 4072 mw, of which the wind power will be 208 million kw. (the hebei grid will be 19.30 million, hebei south net 1.5 million), and the solar power will generate 15.5 million kw.

The maximum load of the power grid in shanxi was 25.37 million kw, and the total power consumption of the whole society was 179.7 billion KWH. Among them, renewable energy generation capacity is 13.31 million kilowatts, accounting for 17.4 percent of total power generation and 212.3 billion kilowatt-hours of electricity generation, accounting for 11.8 percent of total social use. From 2017 to 2020, it is expected to add 17.69 million kilowatts of renewable energy. In 2020, there will be 31 million kilowatts of renewable energy, including 16 million kilowatts of wind power and 12 million kilowatts of solar power.

3.3. Relevant Conditions in Northwest CHINA

In 2016, the maximum load of the northwest power grid was 73.59 million kw, and the total social power consumption was 556.3 billion KWH. The total installed power generation was 222.95 million kw. Among them, the renewable energy generating capacity is 103.19 million kilowatts, accounting for 46.3% of the total installed power generating capacity and 173.1 billion kilowatt-hours of electricity generation, accounting for 31.1% of the total power consumption in the whole society.

Table 1. Renewable Energy in the Northwest Region of 2016(Unit: gw)

	SHAN XI	GAN SU	QIN GHAI	NINGXIA	XIN JIANG	Total

Renewable energy installations	697	2827	1943	1516	3336	10319
1.Wind power	179	1277	69	942	1776	4242
2. Solar power	246	686	682	526	893	3033
3. Biomass power	0	3	0	5	2.61	10.61
4.Water power	272	861	1192	43	665	3032

3.4. Relevant Conditions in Northeast China
In 2016, the northeast power grid had a maximum load of 524.7 million kw, and the total social power consumption was 416 billion kilowatt hours, with a total installed power generation of 131.33 million kw. Among them, the renewable energy generating capacity is 38.45 million kilowatts, accounting for 29.3% of the total installed power generating capacity, with 69.2 billion kilowatt-hours of power generation, accounting for 16.6% of the total power consumption in the whole society.

Table 2. Renewable Energy in the Northeast Region of 2016(Unit: gw)

	LIAO NING	JI LIN	HEILONG JIANG	MENG DONG	Total
Renewable energy installations	1045	979	753	1069	3845
1.Wind power	695	505	561	932	2692
2. Solar power	52	56	17	99	224
3. Biomass power	5.4	40	73	5.4	124
4.Water power	293	378	102	33	805

4. Influence of Cross-Regional Renewable Energy Incremental Spot Trading

4.1. Business Impact Analysis of Power Grid Company
In 2016, Beijing electric power trading center organized ten times "power supply XINJIANG" market transactions through the trading platform. More than 360 power generation enterprises of XINJIANG to participated, the new energy sources such as wind power and solar energy power generation enterprises is more than 90%. XINJIANG sent to JIANGSU 882 million degrees, JIANGXI 118 million degrees, TIANJIN 201 million degrees, GUANGDONG 1005 million degrees, BEIJING 301 million degrees. The total amount traded was 25.07 billion. New energy, such as wind and solar, was about 280 million. On January 14, 2017, XINJIANG launched a new energy surplus in the spot trade. The price of the power of the new power generation is 0.15 yuan per kilowatt hour. Assuming the transmission power price is 0.1 yuan/kWh, the ground side price is 0.25 yuan/kWh. Compared with the current electricity price of 0.3514 yuan/kWh in TianJin, incremental spot trading electricity price has obvious price advantage.
Due to the incremental spot trading position for sending provinces (autonomous regions and municipalities) is abandon water, abandoned wind power, or abandoned photovoltaic power, if withnot across regional renewable energy increment spot transactions between provinces, this part of the power will be wasted, so this part of the electricity price will be lower than the normal trading quantity of electricity price. At present, power grid enterprises can reduce the cost of electricity purchase and increase the profit of enterprises by purchasing the spot of renewable energy in the cross-regional province. With the reform of the electric power system, the power grid enterprises can participate in the spot trading of renewable energy of renewable energy in the inter-regional provinces through their own sales companies.

4.2. Power Grid Peak Regulation Influence

Due to uncertainty about the power of renewable energy generation, cross-regional renewable energy incremental spot trading will aggravate the problem of power grid peak regulation. Tianjin power grid has typical characteristics of urban grid, and the load peak valley is increasing year by year. The maximum peak valley gap reached 517.7 million kilowatts in 2016, accounting for 37.2 percent of the maximum load. At the same time, the main power plants in the area are basically all the heating units, the newly built thermal power units are all heat and power generation units, and the peak power capacity of the power grid is severely limited during the winter heating period.

5. Research on the Strategy of the Cross-Regional Renewable Energy Incremental Spot Trading

5.1. Business Strategy Analysis of Power Grid Company

Power grid company need to comprehensive analysis on various areas sending renewable energy increment spot trading of feed-in tariff and price of transmission channel, landing price assessments by the end of the region, so as to make the optimal electricity purchasing strategy, the specific strategies are as follows:

(1) Research on the price of electricity in each delivery area

Based on the load of present and forecast of the sending area, conventional power supply, renewable energy power generation capacity situation and planning condition, and the situation of abandoned wind or photovoltaic power, we can research on the price of electricity in each delivery area.

(2) Research on transmission power price

According to the transmission channel situation of the delivery to the affected area, such as transmission channel capacity, transmission distance, etc, the electricity price from sent to the end of the transmission can be calulated. But it needs to be pointed out, the national development and reform commission and the national energy administration will organize the cost supervision of transmission projects in trans-provincial and trans-regional transmission lines, and re-evaluate transmission prices (including line losses) according to the results of cost supervision.

(3) Calculate each delivery end to the landed price

According to the price of electricity in each delivery area and transmission power price, each delivery end to the landed price can be calculated.

(4) Determine the optimal purchasing power strategy

The optimal power purchase strategy is determined by comparing the power price and transmission channel.

5.2. Peak Regulation Strategy Analysis

At present, the cost of the compensation assistance service in north China is distributed by the power plant at the cost of the electricity [10], [11]. The specific compensation method for compensation peak service is as follows: (a) The unit will be compensated by 250 yuan/MWh for the amount of power less than the basic peak. (b)The single-machine capacity less than 100MW (including 100MW) will be compensated 1000 yuan/MW if turning off peak once. The single-machine capacity more than 100MW will be compensated 1500 yuan/MW if turning off peak once [12], [13].

According to the principle of "who benefits, who bears", it should be regarded as a paid peak, because the incremental spot transaction results in the peak of the affected area. The resulting peak support service costs are no longer shared by power plants, but should be compensated for by the profit generated by the incremental spot exchange.

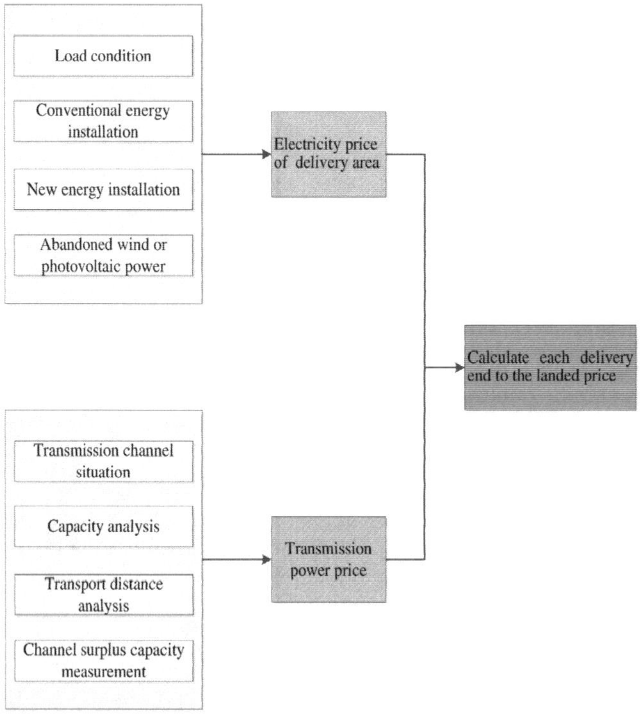

Figure 1. Business strategy of power grid company.

6. Conclusion

The renewable energy increment spot transaction has obvious price advantages, which can bring some profit growth points for the company. Due to the incremental spot trading position for sending provinces (autonomous regions and municipalities) of abandoned energy(water, wind or photovaltic power), if withnot across regional renewable energy increment spot transactions between provinces, this part of the power will be wasted, so this part of the electricity price will be lower than the normal trading quantity of electricity price. At present, grid enterprises should actively participate in the spot trading of renewable energy in inter-regional provinces. With the reform of the electric power system, the power grid enterprises can participate in the spot trading of renewable energy of renewable energy in the inter-regional provinces through their own sales companies.

In 2016, the abandonment rate of abandoned wind in Xinjiang and Gansu exceeded 30 percent, and the problem of abandoning wind and light was outstanding. It is recommended that the spot trading of renewable energy in renewable energy should be given priority at the present stage with Xinjiang and Gansu. In 2020, with the development of wind power solar power in Hebei and Shanxi, it will give priority to the spot trading of renewable energy in with Hebei and Shanxi.

By purchasing the peak support service from the power plant, we should deal with the peak problem caused by the spot trading of renewable energy in trans-regional provinces. The resulting peak support service costs are no longer shared by power plants, but should be compensated for by the profit generated by the incremental spot exchange. With the reform of electric power system and the opening of auxiliary service market, we should further study the coping strategy.

At last, in order to carry out the cross-regional renewable energy incremental spot trading, we must strengthen the operation and control of the power grid, improve the structure of the backbone network frame, and ensure the safe and stable operation of the power grid.

7. References

[1] ZHU Taihui, WU Zhongqun, "A model to prevent spot transaction risk for generation company by use of electricity futures and analysis on calculation example," Power System Technology,

vol. 32, pp. 76–80, Mar 2008.
[2] LIU Ying, MU Yinping, "Optimal ordering decision in the risk-averse supply chain in presence of option and spot market," Journal of Systems & Management, vol. 25, pp. 262–271, Mar 2016.
[3] CAI Xingguo, LIN Shiying, MA Ping, "Study on bidding price of cascaded hydro plants in spot transaction," Proceedings of the CSEE, vol. 23, pp. 56–59, Aug 2003.
[4] WANG Kui, ZHANG Buhan, YAN Dawei, LI Yuanyuan, and LUO Tao, "A multi-time scale rolling coordination scheduling method for power grid integrated with large scale wind farm," Power System Technology, vol. 38, pp. 2434–2440, Sep 2014.
[5] ZHANG Buhan, SHAO Jian, WU Xiaoshan, and WANG Kui, "Unit commitment with wind farms using scenario tree and chance-constrained programming," Power System Protection and Control, vol. 41, pp. 127–135, Jan 2013.
[6] WANG Kui, ZHANG Buhan, ZHOU Yang, and LI Junfang, "Real-time dispatch in wind power integrated system based on choas quantum-behaved particle swarm optimization," Automation of Electric Power Systems, vol. 35, pp. 141–146, Nov 2011.
[7] ZHANG Tao, LI Jiajue, ZHANG Yanfeng, WANG Yang, and XU Jianyuan, "Research of scheduling method for the wind power acceptance considering peak regulation," Power System Protection and Control, vol. 42, pp. 74–80, Nov 2014.
[8] HU Jianchen, LIU Yanhua, LI Xian, and ZHAO Dongmei, "Economic analysis on peak load regulation measure with integration of wind farms," Modern Electric Power, vol. 29, pp. 86–89, Feb 2012.
[9] CHEN Bo, XU Yiqing, HUANG Yichao, and TAO Yanfeng, "Research on peak regulation and resources sharing mechanisms of Shanghai power grid," Modern Electric Power, vol. 31, pp. 37–46, Jul 2015.
[10] WAN Shuting, CUI Yan, and YAN Lei, "Impact of wind farm connected with north China power grid on peak load regulation," Modern Electric Power, vol. 29, pp. 22–25, Aug 2015.
[11] CHEN Dezhi, DAI Ming, "Impact of grid-connected large-scale wind power on peakload regulation in Shanxi grid," East China Electric Power, vol. 39, pp. 1257–1261, Aug 2011.
[12] ZHANG Zixin, WANG Shanshan, BAI Zhe, "The study of northeast grid integrate peaking mechanism," Northeast Electric Power Technology, pp. 1–4, 2014.
[13] LV Quan, WANG Wei, QIAN Shui, HAN Shui, YUAN Shun, ZHANG Jinzhu, and LI Weidong, "A new evaluation method for wind power curtailment based on analysis of system regulation capability," Power System Technology, vol. 37, pp. 1887–1994, Jul 2013.

IOP Conf. Series: Earth and Environmental Science **168** (2018) 012037 doi:10.1088/1755-1315/168/1/012037

Carbon Footprint Analysis at the Operation Phase in a Student Residential Hall in Hong Kong

Chi-Wing Tsang[1,*], *Kar-Kit* Chu, *Gary*[2], *Yuanhao* Wang[1], *Yau Chak* Chau[1], *Xiao-Ying* Lu[1], *Bei* Wang[1], *Bin* Wang[3]

[1]Technological and Higher Education Institute of Hong Kong, Tsing Yi, Hong Kong
[2]University of Macau, Macao
[3]Hong Kong Applied Science and Technology Research Institute, Hong Kong, P.R. China

Email: ctsang@vtc.edu.hk

Abstract. According to the climate action plan published by the Environment Bureau in Hong Kong, it is expected that the absolute carbon emission to have a reduction of 36% per capita by 2030 (base year 2005). Using walkthrough combined with energy simulation, we have investigated the energy consumption pattern of a student residential hall. Due to the fluctuation nature on the number of residents in a student residential hall, we have simulated several scenario: 1) typical occupancy, 2) typical occupancy with energy optimization, 3) peak occupancy, 4) peak occupancy with energy optimization in order to estimate the range of consumption. In scenario 4, effective energy use pattern for air-conditioning system was implemented and approximately 480 tonnes (6.5%) of carbon emission can be reduced for a 50 years building life span consideration. Various energy conservation measures should be taken into consideration in order to further reduce carbon emission in the future.

1. Background

To deal with the world-wide issue of global warming or climate change, as discussed in the 2015 United Nations Climate Change Conference, there is an urgent need for the reduction of the greenhouse gas emission. In Hong Kong, the Government has proposed a target of reducing carbon emissions by 50–60% by 2020 since 2010. [1] In the *Hong Kong Climate Change Report 2015*, the Government has advised it will use China's pledge to reduce carbon intensity by 60–65% between 2005 and 2030 and formulate various measures to mitigate climate change. [2] To accede the China's climate plan and the Paris Agreement, the Government has recently outlined the long-term measures to combat climate change and attain the updated carbon intensity target for 2030 in the *Hong Kong's Climate Action Plan 2030+*. [3] The new target is to reduce Hong Kong's carbon emissions by 65%–70% by 2030 from the 2005 levels, which is equivalent to 26–36% absolute reduction and a reduction to 3.3–3.8 tonnes in per capita emissions.

To achieve the updated carbon emissions target, various strategies such as reducing coal-fired electricity generation, widen the use of renewable energy, as well as implementing energy-saving measures for buildings and transportation had been proposed. As a start to contribute to this carbon reduction mission and to devise the suitable environmental management system in our institute, we have initiated a life cycle assessment study of one of the institute halls of residence and used data simulated by eQuest for comparison. The result obtained will help to extend the life cycle assessment study to the whole institute in future.

A few literatures have mentioned the life cycle carbon emission assessment for residential buildings or similar buildings on function like hotels. However, these studies mainly focus on the material comparison or at the construction phase of the building.[4-5] Construction materials affect the energy performance of whole building. However, from the viewpoint of the whole life cycle of the building, operation stage (or use phase) constituted the largest proportion of energy consumption, i.e. carbon emission from energy generation. In terms of the characteristics of a building's operation phase, up to 85% of total carbon emissions by building type due to the use of heating and cooling energy and electrical facilities could be accounted [5]. In this study the operation stage will be focused and the electricity consumption of HVAC system and other electrical facilities such as lighting system will be evaluated.

2. Methodology

Information and parameters to predict carbon footprint through eQUEST are collected and summarized. Most of the items for simulation are referenced from the design report of the institute's residential hall including building geometry (floor plan in CAD files), building area and number of floor, floor height (floor to floor and floor to ceiling), types and dimensions of door and window, HVAC system, design temperature and air flow, etc. Figure 1 shows the typical floor plan used in this eQuest modelling study. Other parameters like number of existing residents and electricity consumptions are gathered from the hall manager. The DD (Design Development) Wizard will be used instead of SD (Schematic Development) Wizard since it can be used to create buildings with multiple shells and provide more flexibility in assigning them to building areas. Carbon emission varies with energy use and the fuel mix of energy generation. The following equation are used:

$$CO_2\ \text{Emission}(\text{kg}\ CO_2) = \text{Electricity consumption}(\text{kWh}) \times \text{Emission Factor}(\text{kg}\ CO_2\ \text{per 1 kWh}) \quad (1)$$

The net electricity consumption should be used for calculation. A net electric consumption refers to total energy consumption minus any renewable energy generation. the GHG emissions factors (EFs) is 0.540 kg CO_2 e per 1 kilowatt-hour (kWh) in 2015 in Hong Kong for electricity uses.[6] The system boundary is defined as the energy consumption of the MVAC, lighting and water heating system during the operation phase.

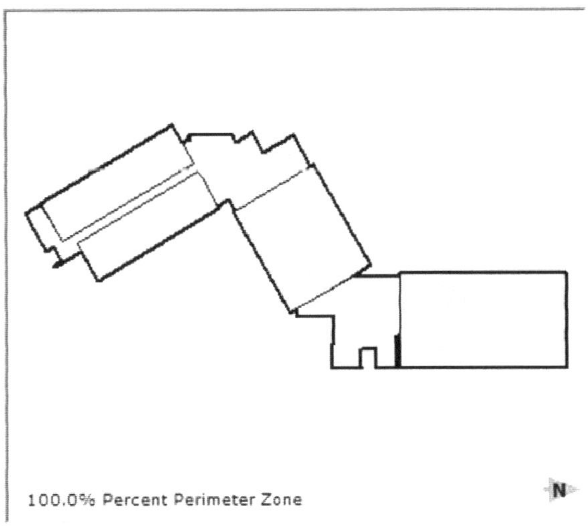

Figure 1. The floor plan (3/F) used in this eQuest modelling.

3. Results

3.1 Measurement and Validation

The residential building consists of 12 floors and the 12th floor has been unoccupied for a consecutive 12 months so it will not be considered. G/F remained in working condition in the whole year, so it can be regarded as a fixed consumption floor and will also not be considered. This scenario simulates the energy consumption of 1/F to 11/F from 1 January to 31 December 2016. 24-hour operation is set for 1 January to 31 May and 1 September to 31 December due to school day. [7] Low occupancy is set for school day due to existing low density situation. [8] In summer holiday from 1 June to 31 August, the building operation schedule is set to be closed for business. It is assumed that there is no user in the building except regular inspection which will only trigger light tube in corridor. Energy saving light tubes are installed in corridor and they will only be triggered when movement is detected, the influence should be low to this validation. Each floor has a height of 3,150 mm from floor to floor including 150 mm concrete between two floor. All floors have similar component, setting and structure. A floor consists of bedroom, mechanical/electrical room, corridor, activity room, common area. Usually more than 40% of space occupied by bedroom for student or guest, 10% for mechanical/electrical room or corridor each, others sum up to 100%. The complete common areas, each contains television, refrigerator, cooking ranges and resident controlled air conditioning system, are located on 1/F, 4/F, 7/F. There is no exterior door nor exterior window shades but only exterior window with 1500 mm width and 1550 mm height. The percentage of window against net wall area (floor to ceiling) is slightly different in each floor. Two types of air-conditioning system are installed for different areas. Variable Refrigerant Volume (VRV) System, HVAC system 1 in simulation, will be used to simulate the podium area such as lobby, corridor, common area, activity room, etc. Split-type air-conditioning unit, HVAC system 2 in simulation, will be used to simulate the bedrooms and warden's flats. Variable Refrigerant Volume (VRV) System will operate for 24 hours per day in podium area to maintain a comfortable environment. Split-type air-conditioning unit is set to operate 8 hours during students' sleeping hour. The design temperature for indoor condition is 23°C dry bulb $\pm$1°C. For the domestic water heating, an instantaneous type electric hot water heater is installed in each bedroom for bathing. Ten gallon per person per day of hot water use is assumed, the consumption is therefore approximately 50L.

As shown in Figure 2, the largest proportion of usage is space cooling. To assist space cooling, ventilation is also an important component so it occupies the second largest proportion of energy used. Typically, the trend of energy consumption in entire year should be ideally linear, a line goes up from January then reach the climax at August and becomes low again unit December. Since Hong Kong is located in subtropical regions, the weather is hot and humid during May to August. The relatively low usage in June and August is due to the low occupancy in the summer holiday. No space heating burden should be considered as no heating equipment is provided. The design parameter of outdoor temperature in winter is 7°C so no heating for warming indoor temperature is required. Water heating is mainly used to operate instantaneous type electric hot water heater for bathing. There is refrigerator in common area but consumption on refrigeration is not considered in the simulation because all of the single electrical device consumption is categorized to miscellaneous equipment. Miscellaneous equipment also include television, cooking ranges, charger, etc. either in common area or bedroom. The remaining consumption is area lighting. All the activity rooms and common areas use glass instead of concrete as an external wall to introduce day light for illumination. Automatic sensor is put in the corridor to minimize the energy consumption. The light bulb will only be triggered when the sensor detects any movement. If no movement is detected again within a minute, the light will be turned off to standby mode.

The simulated data was then used to compare with the real situation during the measured occupancy period. As seen in Figure 3, although the occupancy condition is set to zero due to summer holiday period, the simulated wage of July and August are still high and quite different from real situation. In simulation, occupancy rate affects all usage in building but only affects to a lesser extent to the majority of resident-orientated usage. the electricity usage of resident-orientated usage like lighting, water heating and miscellaneous equipment has dropped significantly in June to August however it does not make tremendous change to total consumption. The saving of air-conditioning from bedroom can cover the higher consumption of the central air-conditioning system in public area due to hot weather so the consumption level of space cooling is similar to May and September. The simulated result is higher than realistic situation during July and August, a possible reason can be attributed to the fact that the real operation schedule of air-conditioning system in summer holiday had been adjusted to a more environmental friendly mode. For example, the system was turned off at night as no resident is needed to be served. It is a massive energy saving if the system is off from 11 p.m. to 7 a.m. and it can also explain the difference between real consumption and simulated result. The coefficient of determination of the simulated and real data was calculated to be 0.9712, which demonstrated that the simulated model agrees quite well with the actual situation.

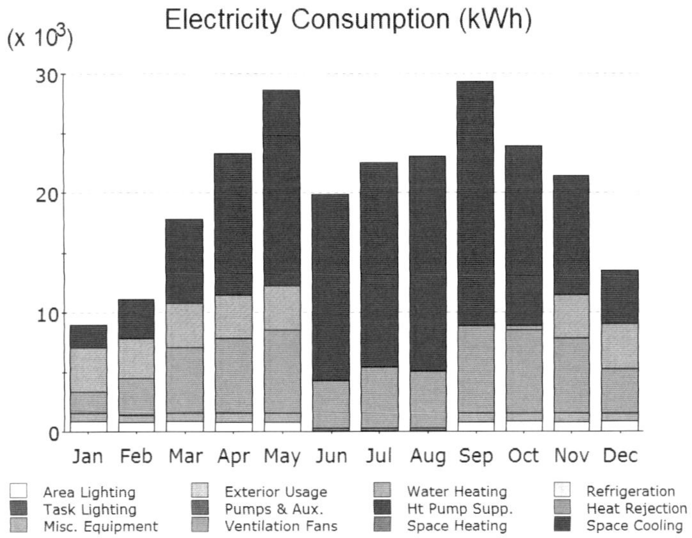

Figure 2. Simulation result of the institute's residential hall in 2016.

3.2 Prediction of Energy Usage

3.2.1 Typical Occupancy

Due to the unforeseen number of occupancy in a student residential hall, several scenarios have been attempted in order to obtain a reasonable range of energy usage. Scenario 1 simulates a typical occupancy of VTC Student Dormitory based on the recent number of residents. All other settings remain the same as the validation's case like 24 hour operated MVAC system, 50L domestic hot water for a person, etc. The total energy consumption of typical usage is 303,510 kWh in 2017. Most of the energy is used to maintain a comfortable living environment by air-conditioning and ventilation. More than a half of energy, i.e. 177,510 kWh, is consumed by space cooling and 84,420 kWh is used to operate ventilation fans. 22,480 kWh is used for water heating, while 9,190 kWh is used for miscellaneous equipment and 9,880 kWh is used for area lighting. For the trend of energy consumption, space cooling and ventilation are fluctuating due to varying weather condition. Miscellaneous equipment used and area lighting are stable for every month and the consumption is nearly a constant for both of these two items. Scenario 2 simulates a typical occupancy with an energy saving approach. MVAC system can be optimized in a more energy efficient manner after understanding the behavior of resident. A student dormitory is not equivalent to a hotel, it does not need to be maintained as perfect condition as a hotel for clients who has an option to spend their time to enjoy the facilities rather than going to the bed. It is expected that the student should go to bed or stay in their bedroom after 11 p.m. As the earliest lesson is at 9 am at the institute, they can take rest until 6 am in the morning. To reduce unnecessary energy usage at night, this scenario proposes only regular air exchange to be operated during normal sleeping hour from 11 pm to 6 am, air-conditioning will cease due to less requirement in that period. All other settings remain at the same condition as scenario 1. The total energy consumption under typical use is lowered to 278,810 kWh. Most of the energy is still used to maintain a comfortable living environment by air-conditioning and ventilation. 160,630 kWh is consumed by space cooling and 76,610 kWh is used to operate ventilation fans. Other parameters remain unchanged with the same number of resident. 22,480 kWh is used for water heating, 9,190 kWh is used for miscellaneous equipment and 9,880 kWh is used for area lighting. Compare to scenario 1, the reduction of electricity consumption is 24,700 kWh which is equivalent to a reduction rate of around 8%.

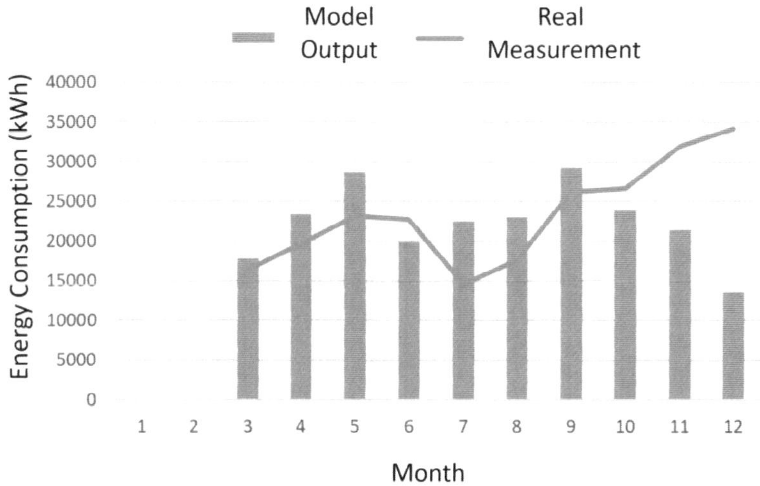

Figure 3. Data comparison between simulation and real situation

3.2.2 Peak Occupancy

Scenario 3 simulates peak occupancy of residential hall to find out the maximum energy that could be consumed. All settings remain the same as scenario 1 except number of residents. The total energy consumption under peak use is 410,360 kWh in 2017. More energy is used to maintain a comfortable living environment especially in May and September. 222,270 kWh is consumed by space cooling and 108,700 kWh is consumed to operate ventilation fans. 43,350 kWh is used for water heating, 23,410 kWh for miscellaneous equipment and 12,610 kWh for area lighting. The consumption for space cooling and ventilation increase from 261,930 kWh for typical use to 330,970 kWh for peak use. Although the quantity of resident is doubled theoretically, the energy used is not doubled. Because air-conditioning and ventilation for public place is generally fixed, the increase is mainly come from the use of the air conditioner in bedroom during hot and humid weather. On the other hand, consumption of water heating depends on occupancy rate unless students take shower in somewhere else. Thus the increasing rate is nearly double, from 22,480 kWh to 43,380 kWh. The last two parameter, miscellaneous equipment and area lighting, demonstrate a greater difference in increasing rate even though they consume at a similar level of electricity during typical occupancy. The former has a huge rise, from 9,910 kWh to 23,410 kWh, while the latter only has a slight increase. It is because the energy use factor is set larger manually due to several possibilities including higher probability of public devices usage, much more personal electrical devices usage and larger volume of cloth washing, etc. In contrast, area lighting consumption increased by 2,730 kWh to 12,610 kWh. It shows that the day light benefits both public area and bedroom as over 30% of surface area can introduce day light for illumination. Again, the consumption in public area is close to a fix value, thus the increase mainly comes from bedroom and activity room. With energy saving, i.e. scenario 4, the total energy consumption is lowered to 385,030 kWh. Most of the energy consumed is to maintain a comfortable living environment by air-conditioning and ventilation. 204,850 kWh is consumed by space cooling and 100,790 kWh is used to operate ventilation fans. Other parameters remain unchanged due to the same number of residents. 43,350 kWh is used for water heating, 23,410 kWh is used for miscellaneous equipment and 12,610 kWh for area lighting. Compared to scenario 3, the reduction of total electricity consumption is 25,330 kWh which is equivalent to a reduction rate of around 6%. The result is very close to the decline between scenario 1 and scenario 2. Since the total energy consumption is higher, the decreasing percentage is lower at a similar reduction value.

3.3 Prediction of Life Cycle Carbon Emission

PV panel and wind turbine generated 20,790 kWh and 2,800 kWh respectively per year in the residential hall. [9] Before determining the carbon emission, these energy savings are subtracted from total electricity consumption in the simulation result. No other energy source such as liquefied petroleum gas was used in the student dormitory therefore electricity is the only source of carbon footprint in the operation phase. The latest GHG emissions factors (2015) of 0.540 kg CO_2 e per kWh provided by CLP for electricity generation is adopted for the following calculation. The result will be presented in CO_2 e per m^2 and per capita for convenient comparison with other building's performance. The surface area of the residential hall from 1/F to 11/F is 15,075 m^2. The density is 440 residents for typical occupancy (50%) and 880 residents for peak occupancy (100%). The carbon emissions of each scenario are shown in Table 1.

	Typical Occ	Peak Occ
No energy saving	345.5 kg CO_2/capita	237.3 kg CO_2/capita
	10 kg CO_2/m^2	13.9 kg CO_2/m^2
Energy saving	313.2 kg CO_2/capita	221.8 kg CO_2/capita
	9.1 kg CO_2/m^2	12.9 kg CO_2/m^2

Table 1. Carbon emissions per capita and per m^2 for each scenario.

For a typical building's life span of 50 years, if energy saving mode is adopted, a total of 684 tonnes of CO_2 e can be reduced, which is equivalent to about 6.5% of carbon reduction for 50 years of operations.

3.4 Energy Conservation Measures

Currently, the following energy conservation measures have already been implemented at the residential hall:

- Use of occupancy sensor and photo sensor for lighting control.
- Introduction of massive day lighting for illumination.
- Use of T5 fluorescent and compact fluorescent light fitting.
- Use of LED televisions.
- Use of dual-flash buttons instead of press-type or handle-type water cisterns.
- Energy metering and monitoring system for energy management, etc.

More energy conservation measures are suggested for future retrofit of the building to further decrease the carbon emission, such as:

- Purchase electrical equipment with Grade 1 energy label. Energy Efficiency (Labelling of Products) Ordinance requires energy labels to be displayed on prescribed electrical products before being supplied in Hong Kong to facilitate consumers in choosing energy-efficient electrical appliances.
- Reschedule the HVAC plan. A comprehensive investigation should be conducted though questionnaire at the beginning of every semester to understand the general behavior of residents in

dormitory like when do they leave and return and how often do they stay in common area.

- Use air-conditioner in good practice. Reminders should be posted near the switches of the air-conditioner to remind the users of the best practices in using air-conditioners such as setting the air-conditioner temperature at 25.5°C.

Install tap aerators. These water saving devices control the amount of water that flows through the tap without affecting the water pressure as they mix the water with air. The aerator acts as a sieve, separating a single flow of water into many tiny streams which introduces the air into the water flow. As there is less space for the water to flow through, the water flow is reduced, resulting in water savings.

4. Conclusion

The present study involved a carbon audit for the operation phase of the student residential hall at Tsing Yi campus from 1/F to 11/F using walkthrough combined with modeling software. The scope of this study includes the energy use of MVAC system, lighting system, water heating and other electrical equipment. An existing condition was first measured and applied as a baseline study to validate the accuracy of the simulation. Four scenarios are simulated for the following year, in which two for typical occupancy and two for peak occupancy. The average carbon emission is 13.9 kg CO_2 e per m^2 per annum and 237.3 kg CO_2 e per capita per annum under peak occupancy. The carbon emission is 7,310 tonnes CO_2 eq for 50 years' operation. Scenario 4, also under peak occupancy, introduces an effective energy use pattern for air-conditioning system which only operate between 6 am to 11 pm. If this approach is to be adopted, carbon emission will be 6,831 tonnes CO_2 e and 684 tonnes of carbon emission can be reduced in the entire 50 years life span. Finally, some energy saving opportunities was suggested for this residential building. Both the material use in construction stage and selection of electrical equipment as well as the user behavior are crucial to reduce carbon emission.

Reference

1. *Hong Kong's Climate Change Strategy and Action Agenda – Consultation Document*, Environmental Bureau, HKSAR Government, September **2010**.
2. *Hong Kong Climate Change Report 2015*, Environmental Bureau, HKSAR Government, November **2015**.
3. *Hong Kong's Climate Action Plan 2030+*, Environmental Bureau, HKSAR Government, January **2017**.
4. F. Hu, X. Zheng, Procedia Engineer. *Carbon Emission of Energy Efficient Residential Building* **121**, 1096-1102 (2015).
5. S.-H. Cho, C.-U. Chae, Sustainability **8, 579** (2016).
6. China Light and Power Co. Ltd. Today's Fuel Mix. (2016).
7. Based on residence period at the institute: 1 September 2016 to 31 May 2017.
8. eQUEST only provides five options for calculation which are closed for business (0%), low occupancy (25%), typical occupancy (50%), high occupancy (75%) and peak occupancy (100%).
9. Chau, Y. C. (2017). *Life Cycle Carbon Emissions of Student Residential Hall at THEi Tsing Yi Campus* (Bachelor dissertation). Technological and Higher Education Institute of Hong Kong.

Acknowledgments

We gratefully acknowledge the financial support provided by the Research Grants Council (RGC) of the Hong Kong Special Administrative Region, China (Project No.: UGC/IDS25/15 & UGC/FDS25/E03/15) and by THEi Seed Grant (Project No.: SG1617108).

IOP Conf. Series: Earth and Environmental Science **168** (2018) 012038 doi:10.1088/1755-1315/168/1/012038

Fixed Bed Combustion of Biomass: Particulate Matter Mitigation by Optimising Temperature and Fuel Particle Size

Mesa Cicilia Kemunto, Enyu Wang, Xiang Gou*, Ali Imran Shah, George Odhiambo Adwek

School of Energy and Environmental Engineering, Hebei University of Technology, Tianjin 300401, China

e-mails: cessmesa83@gmail.com(M. C. K.); enyuwang@163.com(E.W.); *corresponding author: gouxiang@sina.com (X.G.);coaltechnology@gmail.com(A. I. S.); adwekg@gmail.com (A. O. G.)

Abstract. Experimental results are presented on the particulate matter emissions from a fixed bed combustion reactor chamber combusting two types of ambient dried biomass fuels, six different particle sizes (0.044-0.149mm) in seven different temperature zones (600-1200 °C).The fuel was ground into various particle sizes and the temperature of combustion was varied for investigation on the emission due from the various parameters. Measurements were made for particulate matter (PM) during the whole phase of combustion by a portable soot analyzer. The PM investigated here is total catch that includes solid particles of elemental carbon and fuel and condensed organic compounds due from incomplete combustion which are less than or equal to 10μm in aerodynamic diameter. The fuel was ground by a grinder and sieved into different sizes by a shake sieve with various mesh sizes. It was found that the PM emission from both biomass fuels was significantly dependent on the temperature of combustion which in turn dictated the residence time in the combustion chamber. The mean PM mass emissions from wood and peanut shell were established. The optimum combustion temperature was identified to be 900°C attracting PM mitigation of up to 63%. The least PM producing particle size for peanut shell and wood fuel was determined to be 0.149mm combusting at 12000C and 0.074mm combusting at 6000C respectively with capacity to mitigate PM production by a maximum of 30-49%. However the effect of the two parameters was established to be mutually dependent. The time taken for complete combustion reduced steadily with the increase of the temperature from 360 seconds at 600°C to 150 seconds at 1200°C.

1. Introduction

Solid biomass has been used for centuries especially in domestic and industrial energy supplies. Due to the greenhouse gases effects on the environment, there is intense interest on measures resulting in less pollution, making biomass use more attractive albeit its particulate matter (PM) challenges that have caused serious health concerns, with fine particles of below 1μm pausing the greatest risk to human health [1].Going forward the PM inventory from biomass is bound to increase. There is need for a comprehensive knowledge on biomass feedstock on both physical and chemical characteristics [2] especially to inform mitigation of PM production. While wood is the most predominant of all biomass fuel, wood combustion is associated with high emissions of polycyclic aromatic hydrocarbons (PAH) among other organics which are known to be both carcinogenic and mutagenic [1, 3, 4].

IOP Conf. Series: Earth and Environmental Science **168** (2018) 012038 doi:10.1088/1755-1315/168/1/012038

Extensive research is being undertaken on the combustion of various agricultural by-products and other unconventional fuels like maize cobs, date stones, palm stalks and grasses among others [5-9, 17]. Different systems are being used in combusting the fuel, for pulverised fuel systems the fuel particles are milled to a size between10-1000μm in an effort to optimise the combustion process that is partially controlled by residence time, turbulence and temperature [2, 10], these particle sizes are the focus of this paper. It is also well established that raw biomass has a high content of moisture and even after some time of ambient drying depending on the weather, there remains about 15-20% of moisture. When more drying is not achieved the fuel loses a lot of energy through latent heat of evaporation significantly decreasing the efficiency, alternatively, there is immense energy consumed in drying the fuel to the required moisture levels before usage and the process is energy intensive calling to focus the optimal practices[12]. There is therefore more interest on using unprocessed or semi-processed fuels.

A solid fuel is basically composed of volatile material which is usually predominant in biomass fuels and other elemental compositions like carbon and nitrogen among others. The volatile material burns as a gas and exhibits a visible flame whereas the fixed elements like carbon burns without flame as a solid form [10].Based on the relationship between chemistry and heat and mass transfer during particle conversion, fuel particle size has a major effect on the emissions [2].

There is also major interest being directed towards co-firing of biomass and coal as a way of reducing the greenhouse gases production [11, 15]. For co-firing as well the fuel particle size and temperature are principal parameters

In this paper the optimal temperature for combustion and particle size especially for pulverised systems are investigated and the relationship between particle size, temperature, residence time and particulate matter emissions is established based on experimental results. Finally the PM mass emissions per unit mass from the two fuel samples is determined. The determination of which shows that the combustion of biomass in optimal temperature and particle size can largely mitigate the production of PM.

2. Experiment Materials and Set-up

2.1. The fuel

Peanut shells are generally herbaceous biomass fuel [14]. The chemical composition of raw peanut shells is determined to be lignin (about a third of the fuel mass), cellulose (approximately one third to half of the fuel mass) and other minimal compositions of protein, ash and moisture. [21].The shells have a fibrous 'skeleton' supporting a cellulose-lignin layer. A substantial proportion of fibres in peanut shells (up to 70 wt. % [21]) significantly affects the texture of fuel particles and consequently, devolatilisation and combustion behaviour of the biomass fuel.

The peanut shells were shred into six different sizes, the largest being 0.149mm and the smallest being 0.044mm. The bigger particles took wafer like flaked dimensions while the fine particles often appeared in an agglomeration of particles that stuck together to form a circular figure especially enhanced due to the shaking process through the mesh wire.

Wood cells are composed of cellulose (40% to 50%), hemicellulose (15% to 25%) and lignin (15%-30%) [13].The wood was ground into six different particles of similar size to the peanut particle sizes. Grinding wood results into single fibrous particles that are not clustered unlike peanut shell.

Table 1 shows the ultimate and proximate analysis as well as the heating value of both fuels grown in China used in this work. It can be seen from the proximate outcome that the volatile matter from both fuels is significantly high accounting for more than two thirds of the fuel content. There is a low proportion of fixed carbon especially for wood, a rather low proportion of fuel ash for peanut shell but significantly low for wood, the moisture content for both fuels is also quite low for this ambient dried biomass fuel.

Table 1. Table Showing Proximate and Ultimate Analysis

Fuel	(M)%	(VM_{ad})%	(A_{ad})%	(FC_{ad})%	(C)%	(H)%	(S)%	(N)%	(O)%	Heat Value(kcal/kg)
Wood	11.29	80.79	2.68	5.24	38.76	4.95	0.18	1.37	40.82	4969.00
Peanut shell	9.75	73.72	3.43	13.10	37.44	4.65	0.07	1.11	43.56	5172.00

2.2. Experimental Setup

This research was to determine the effect of temperature and fuel particle size on emission of PM per unit mass of fuel combusted for the subsequent two types of biomass; common wood and peanut shell. Investigations of the combustion process were carried out in an electric fixed bed reactor with capability for a constant pre-determined temperature and air flow for a stable and complete combustion. The measurement of PM concentration per metre cubed of gas was determined using a portable Foedisch PFM 92K portable soot analyser with a stainless steel probe with a pre-amplifier. The measuring principle is dependent on the friction electric sensor with a measuring range of 0.1-1000mg/m³. Investigations on the combustion process were carried out for the parameters shown in table 2.

Table 2. Table Showing the Parameters Onsidered in the Experiment

Temperature(°C)	600,700,800,900,1000, 1100,1200
Particle size (mm)	*≥0.149;* ≥0.105; ≥0.088; ≥0.074; ≥0.053; ≤0.044
Air flow rate	*1200ml/min*
Air mixture	O_2:21%, N_2:79%

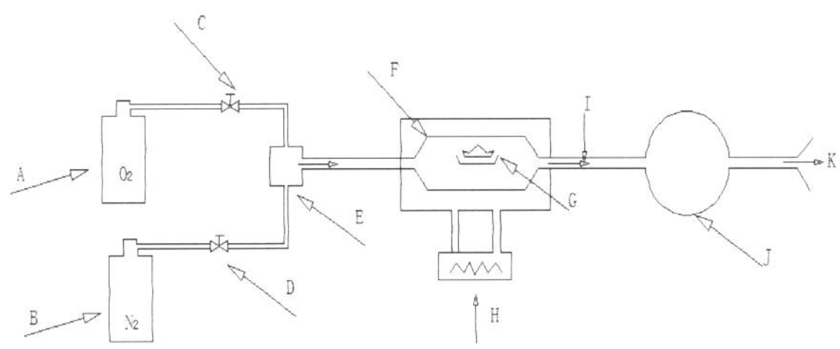

Figure 1. The experimental set.

A: Oxygen
B: Nitrogen
C: Pressure Valve for Oxygen
D: Pressure Valve for Nitrogen
E: Gas Mixing Chamber
F: Combustion Chamber
G: Ceramic Boat with fuel sample
H: Thermal Regulator
I: Flue gas
J: PM measuring device
K: Flue gas

After reaching the predetermined temperature in the combustion furnace at the set airflow, a ceramic boat with a predetermined mass of sample fuel was introduced into the combustion zone and the combustion instantly started. In that moment both readings were concurrently taken by the flue gas analyser and the portable soot analyser. After a period of time the concentration of the flue gases in the exhaust stream decreased to normal at which point the experiment was terminated. All emissions indications were recorded starting from the time when the samples were introduced into the

combustion zone until the moment where no emission indicated. The recorded concentrations were used to determine the outcome. All samples were preserved under the same conditions at room temperature. To enhance result confidence, it was necessary to repeat measurements a number of times under similar conditions. All pollutant measurements were done at least twice; a mean emission value was calculated from results obtained. Typical emission curve for PM evolution can be seen in Fig. 2.
The values (in mg per gram of combusted fuel) for Particulate Matter were derived from the following equation;

$$W_z=q.t.c_p/m_f \quad (1)$$

where W_Z is the mass of pollutant in $mg.g^{-1}$ of fuel, this measurement is taken by the portable soot analyser, q is the flowrate of gas (O_2 and N_2) supplied to the combustion chamber in $m^3.s^{-1}$ while tis the period of time considered in seconds. C_P is the PM concentration during combustion in $mg.m^{-3}$ that is was recorded once per second during the experiment. Finally, m_f is the mass of fuel sample in grams.

3. Results and Discussion

3.1. Residence Time

The mean time taken for a sample of approximately 0.030 grams to combust entirely was determined to be 360 seconds at 600ºC but steadily decreased to a mean of 150 seconds at 1200ºC. The biomass considered in this experiment had a high Volatile/ Fixed carbon ratio which dictates that the dominant form of combustion is gas phase oxidation of the volatile species. Volatility of which is influenced by both fuel particle size and combustion temperature both considered here [18]. Smaller particles in this case release more volatiles rapidly due to the increased surface area for fuel oxygen interaction and the higher temperatures of combustion considered here compel majority of the combustible fuel to be released as volatiles while the lower temperatures considered here favor the formation of char. This serves to explain the decrease in residence time; that as temperatures spiral upwards the combustion is practically almost finished at the end of the release of the volatiles which are the majority of the combustible fuel.

There is a sharp PM peak or sometimes multiple peaks at the beginning of combustion and towards the end of the process with changeable heights as the first peak could be higher and wider or lower. The multiple peaks indicates that there are more than one process taking place which sometimes overlap and sometimes subsequent one another. At the start of combustion the release of moisture and volatile evolution pursued by combustion of the volatiles can succeed one another in high temperatures hence indicating the multiple peaks. Later in the burning process the char combustion largely occurs as char oxidation which is accomplished through gas-solids reactions [18] hence multiple peaks as well can be formed due to the presence of more than one gas in the reaction.

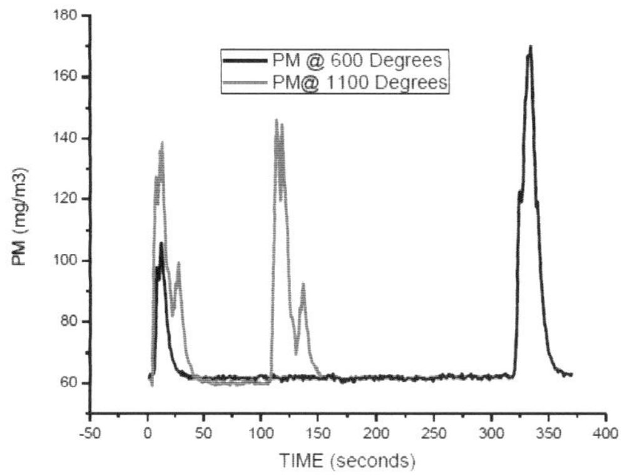

Figure 2. Figure Showing the PM Indication at 600°C And 1200°C.

There is also an observation indicating a significant gap between particles concentration points accounting for more than 70% of the fuel combustion residence time. When fuel is introduced into the combustion zone, emissions increase profoundly due to the release of primary volatile organic compounds [20], there is an ensuing charcoal forming stage that is quite low in emissions whose time can identify with the low or no particle indication observed, after which the charcoal combustion emissions are recorded towards the termination of the combustion process.

3.2. Temperature and Particle Size

1) Peanut shell

In this investigation there is a twin effect of temperature and particle size, of which temperature seems to have a domineering effect.There is a strong indication that the increase of temperature leads to less PM formation during VM combustion for peanut shell(Fig.3). PM is indicated to be more during char combustion and less during VM combustion except for combustion occurring at 700°C where PM was leading during VM combustion(Fig.5). At 900^{0}C, the lowest PM indication during both VM and char combustion was recorded, this observation has high similarity with 1200^{0}C, where the PM indication was also quite low. It has been established that combustion occurs at three major levels with adsorbed water being driven out between 57-92°C,organic matter combustion between 200 and 650°C and burning of residual char between 650 and 850°C[19].This strongly indicates that below 900°C there is a high probability of soot production due to incomplete combustion which boosts the PM formation. This supports the findings that 900°C is the optimal combustion temperature.

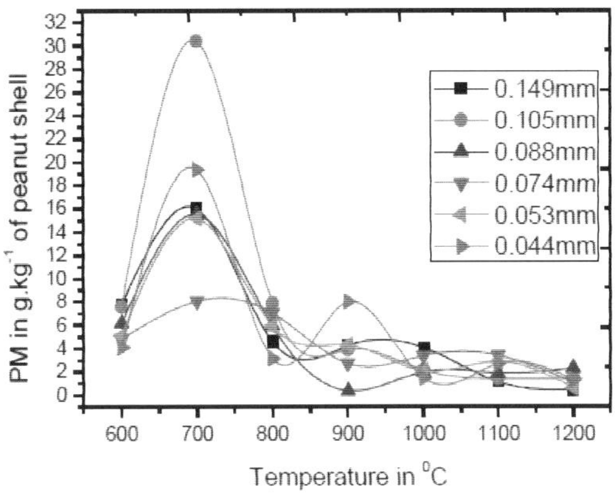

Figure 3. Showing the PM yield during VM combustion.

The mean PM yield from peanut shell combustion was 18.200±9.200 g.kg^{-1}. As shown below; the data collected at temperature zones 600 to 800°C (Fig.5) had a huge difference between the PM indicated during VM combustion and char combustion due to the effect of particle size at these temperatures which encourage volatilization of fuel and slow char combustion informing some huge standard deviation in the data, additionally, the data is from six different fuel particle sizes. There is research informing the contribution of biomass to the inventory of ambient PM for PM_{10} and $PM_{2.5}$ but there is no data related to the combustion of peanut shells. The current AP-42 document from the Environmental Protection Agency (EPA) emission factors corresponding to wild fires and fire place PM is 8.500g.kg^{-1} and 17.300 g.kg^{-1} [20],the same research findings put $PM_{2.5}$ production from below average quality rice residues at 12.950±0.300g.kg^{-1} both of which are in our neighborhood.

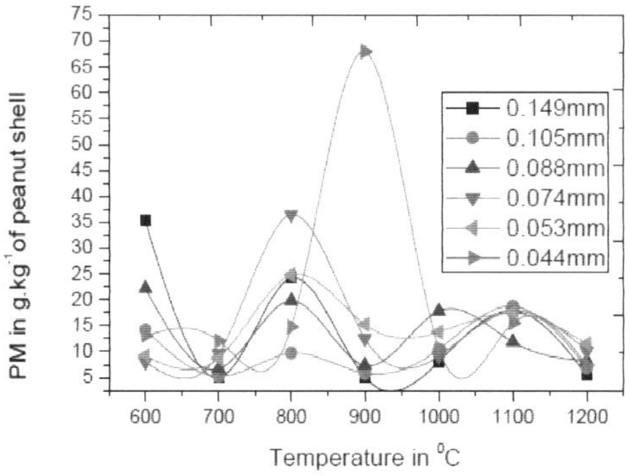

Figure 4. Showing the PM yield during char combustion.

Fuel particle size is a governing factor in PM emission, decreasing the particle size increases the formation of PM however there is need to consider the impact of the micro-particle sizes used in pulverized combustion systems [22]. From observation there is significantly less PM indication from 900°C to 1200°C for all the particles involved in this investigation (Fig.5).Considering 900^{0}C and 1200°C where there is occurrence of lowest PM and at a superimposed pattern of occurrence, 0.149mm, 0.105mm and 0.088mm are the lowest PM indicators in ascending order (Fig.5).

2) Wood

The lowest PM yield was obtained at 900°C while the highest was recorded at 1100°C with a yield increase of 200% (Fig.8).At 600°C and 700°C the PM yield was leading at VM combustion, beyond which the major PM yield is indicated during char combustion(Fig.8). The mean PM yield from wood was determined to be 13.062 ±5.840 g.kg^{-1} of wood fuel. Research findings put the $PM_{2.5}$ emission from different kinds of biomass as being from 4.000g.kg^{-1} to just above 40.000g.kg^{-1}of fuel from different kind of fires; wild fire, residential fire and agricultural fires [16]. The EPA, AP-42 document puts the emission factor for residential wood PM due from combustion at between 9.800g.kg^{-1}to 15.300 g.kg^{-1} [20]. Different firing techniques are bound to tone the PM yield differently due to the varying combustion efficiencies.

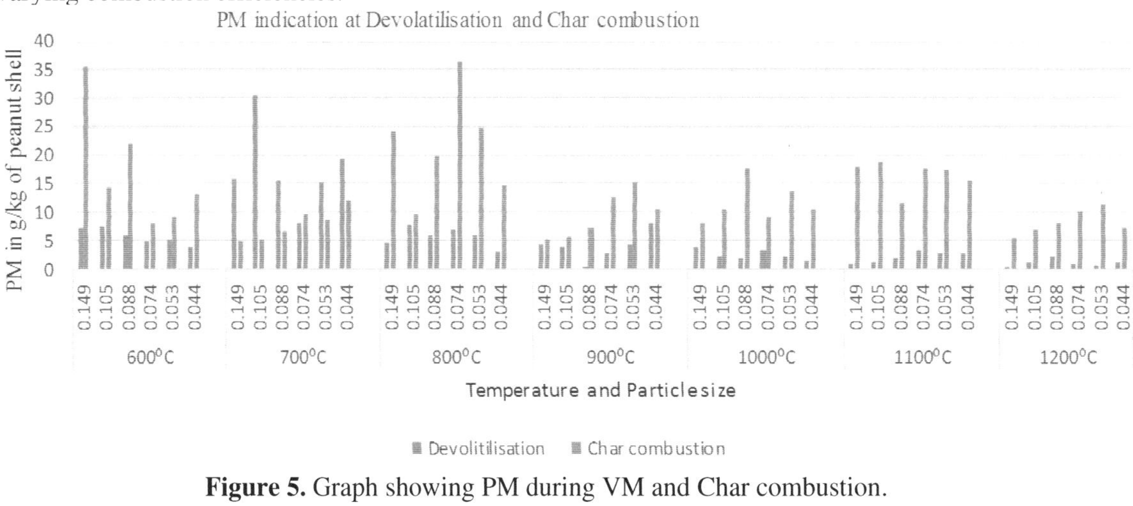

Figure 5. Graph showing PM during VM and Char combustion.

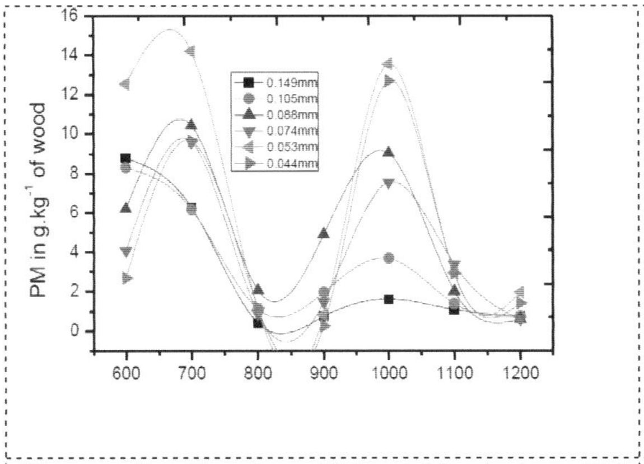

Figure 6. *Showing PM yield during VM combustion.*

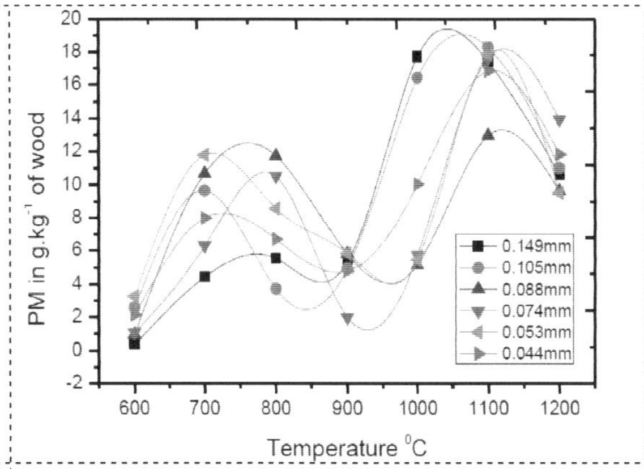

Figure 7. Showing PM yield during char combustion.

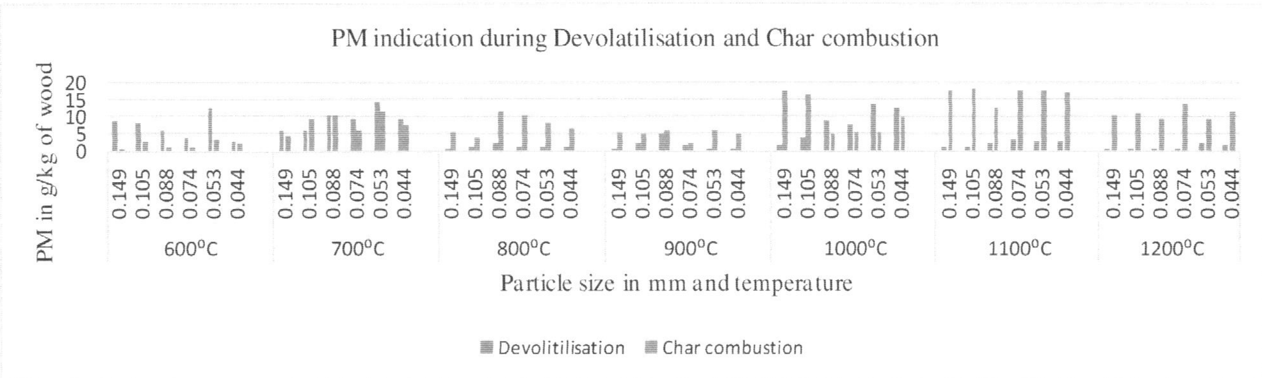

Figure 8. Graph showing PM during VM and char combustion.

Particle size seems to play a more defining role in wood fuel as opposed to peanut shell. During VM combustion the effect of particle size is well observed at 600^0C, 700^0C, 900^0C and 1000^0C (Fig.6). There is a strong pattern indicating an increase in PM as the particle size decreases (Fig.6). During char combustion at 600^0C, 900^0C and 1200^0C, the particle size impact is neutralised, nevertheless there is a weak linear increase of PM as the temperature increases during wood char combustion (Fig. 7).

4. Conclusion

Two biomass fuels; wood and peanut shell were combusted in this experiment. The total PM associated with both was significantly affected by the temperature of combustion. 900°C was identified as the optimal temperature for both wood and peanut shell combustion, at this temperature the lowest PM was indicated for both fuels respectively. The time of combustion for both fuels had an inverse relationship with the temperature of combustion. There was a strong linear decrease of PM indication during VM combustion for peanut shell and a weak one for wood. Generally, some combustion temperatures considered in this experiment serve well to neutralize the particle size impact both during VM and char combustion leading to relatively less PM yield. Due to the complex relationship and joint impact of temperature and fuel properties on PM yield, the impact of particle size on PM yield was not clearly realized. A bigger difference between the particle sizes would help magnify the link between PM and particle size, nevertheless a clear output on effect of temperature on PM yield was achieved.

5. References

[1] Kocbach Bølling, Anette, Joakim Pagels, Karl Yttri, Lars Barregard, Gerd Sallsten, Per E Schwarze, and Christoffer Boman. 2009. "Health Effects Of Residential Wood Smoke Particles: The Importance Of Combustion Conditions And Physicochemical Particle Properties". *Particle And Fibre Toxicology* 6 (1): 29. doi:10.1186/1743-8977-6-29.

[2] Williams, A., J.M. Jones, L. Ma, and M. Pourkashanian. 2012. "Pollutants From The Combustion Of Solid Biomass Fuels". *Progress In Energy And Combustion Science* 38 (2): 113-137. doi:10.1016/j.pecs.2011.10.001.

[3] Mitchell, E.J.S., A.R. Lea-Langton, J.M. Jones, A. Williams, P. Layden, and R. Johnson. 2016. "The Impact Of Fuel Properties On The Emissions From The Combustion Of Biomass And Other Solid Fuels In A Fixed Bed Domestic Stove". *Fuel Processing Technology* 142: 115-123. doi:10.1016/j.fuproc.2015.09.031

[4] Zhang, Yan, Shiro Kajitani, Masami Ashizawa, and Yuso Oki. 2010. "Tar Destruction And Coke Formation During Rapid Pyrolysis And Gasification Of Biomass In A Drop-Tube Furnace". *Fuel* 89 (2): 302-309. doi:10.1016/j.fuel.2009.08.045.

[5] Al-Omari, S.-A.B. 2006. "Experimental Investigation On Combustion And Heat Transfer Characteristics In A Furnace Fueled With Unconventional Biomass Fuels (Date Stones And Palm Stalks)". *Energy Conversion And Management* 47 (6): 778-790. doi:10.1016/j.enconman.2005.06.012.

[6] Arromdee, Porametr, and Vladimir I. Kuprianov. 2012. "Combustion Of Peanut Shells In A Cone-Shaped Bubbling Fluidized-Bed Combustor Using Alumina As The Bed Material". *Applied Energy* 97: 470-482. doi:10.1016/j.apenergy.2012.03.048.

[7] Dorez, G., L. Ferry, R. Sonnier, A. Taguet, and J.-M. Lopez-Cuesta. 2014. "Effect Of Cellulose, Hemicellulose And Lignin Contents On Pyrolysis And Combustion Of Natural Fibers". *Journal Of Analytical And Applied Pyrolysis* 107: 323-331. doi:10.1016/j.jaap.2014.03.017.

[8] Jones, Jenny M. *Pollutants Generated By The Combustion Of Solid Biomass Fuels*. 1st ed.

[9] Saldarriaga, Juan F., Roberto Aguado, Aitor Pablos, Maider Amutio, Martin Olazar, and Javier Bilbao. 2015. "Fast Characterization Of Biomass Fuels By Thermogravimetric Analysis (TGA)". Fuel 140: 744-751. doi:10.1016/j.fuel.2014.10.024

[10] Boubel, Richard Fox. *Fundamentals Of Air Pollution*. 1st ed. Elsevier/Academic Press.

[11] Baxter, Larry. 2005. "Biomass-Coal Co-Combustion: Opportunity For Affordable Renewable Energy". *Fuel* 84 (10): 1295-1302. doi:10.1016/j.fuel.2004.09.023.

[12] Bridgeman, Toby G., Jenny M. Jones, and Alan Williams. 2010. "Overview Of Solid Fuels, Characteristics And Origin". *Handbook Of Combustion*. doi:10.1002/9783527628148.hoc055

[13] "Wood". 2017. *En.Wikipedia.Org*. http://en.wikipedia.org/wiki/wood.

[14] Khan, A.A., W. de Jong, P.J. Jansens, and H. Spliethoff. 2009. "Biomass Combustion In Fluidized Bed Boilers: Potential Problems And Remedies". *Fuel Processing Technology* 90 (1): 21-50. doi:10.1016/j.fuproc.2008.07.012.

[15] Backreedy, R.I., L.M. Fletcher, J.M. Jones, L. Ma, M. Pourkashanian, and A. Williams. 2005. "Co-Firing Pulverised Coal And Biomass: A Modeling Approach". *Proceedings Of The Combustion Institute* 30 (2): 2955-2964. doi:10.1016/j.proci.2004.08.085.

[16] Hays, Michael D., Philip M. Fine, Christopher D. Geron, Michael J. Kleeman, and Brian K. Gullett. 2005. "Open Burning Of Agricultural Biomass: Physical And Chemical Properties Of Particle-Phase Emissions". *Atmospheric Environment* 39 (36): 6747-6764. doi:10.1016/j.atmosenv.2005.07.072.

[17] Limousy, L., M. Jeguirim, P. Dutournié, N. Kraiem, M. Lajili, and R. Said. 2013. "Gaseous Products And Particulate Matter Emissions Of Biomass Residential Boiler Fired With Spent Coffee Grounds Pellets". *Fuel* 107: 323-329. doi:10.1016/j.fuel.2012.10.019.

[18] Tillman, D.A. 2000. "Biomass Cofiring: The Technology, The Experience, The Combustion Consequences". *Biomass And Bioenergy* 19 (6): 365-384. doi:10.1016/s0961-9534(00)00049-0.

[19] Vassilev, Stanislav V., David Baxter, and Christina G. Vassileva. 2013. "An Overview Of The

Behaviour Of Biomass During Combustion: Part I. Phase-Mineral Transformations Of Organic And Inorganic Matter". *Fuel* 112: 391-449. doi:10.1016/j.fuel.2013.05.043.
[20] "Clearinghouse For Inventories And Emissions Factors (CHIEF) | US EPA". 2017. *Epa.Gov*. https://www.epa.gov/chief.
[21] Allen, Stephen J., Quan Gan, Ronan Matthews, and Pauline A. Johnson. 2005. "Mass Transfer Processes In The Adsorption Of Basic Dyes By Peanut Hulls". *Industrial & Engineering Chemistry Research* 44 (6): 1942-1949. doi:10.1021/ie0489507
[22] Ninomiya, Y., Zhang, L., Sato, A., & Dong, Z. (2004). Influence of coal particle size on particulate matter emission and its chemical species produced during coal combustion. *Fuel Processing Technology*, *85*(8-10), 1065-1088. http://dx.doi.org/10.1016/j.fuproc.2003.10.012

Acknowledgments
I sincerely thank my professors; Enyu Wang and Gou Xiang for their brilliant guidance, my fellow students Ali Imran Shah, George Adwek and Shen Yun Xiang for their continuous support and contribution. This work was funded and undertaken in Hebei University of Technology at the Energy and Environmental Engineering department.

IOP Conf. Series: Earth and Environmental Science **168** (2018) 012039 doi:10.1088/1755-1315/168/1/012039

Chemical and fractional composition of bio-oil obtained from Arthrospira platensis by hydrothermal liquefaction

Mikhail Vlaskin[1], Anatoly Grigorenko[1], Grayr Ambaryan[1,*], Nadezhda Chernova[1,2], Sophia Kiseleva[1,2], Yury Kostyukevich[1,3], Gleb Vladimirov[1,3] and Eugeny Nikolaev[3]

[1]Joint Institute for High Temperatures of the Russian Academy of Sciences, Izhorskaya 13 Bldg 2, Moscow 125412, Russia
[2]Lomonosov Moscow State University, Leninskiye Gory, 1, Moscow 119991, Russia
[3]Skolkovo Institute of Science and Technology Novaya St., 100, Skolkovo 143025, Russia

*E-mail: ambaryan1991@gmail.com

Abstract. This work is devoted to the study of chemical and fractional composition of the bio-oil obtained from Arthrospira platensis by hydrothermal liquefaction in the temperature range of 240-330 °C. For bio-oil analysis standard elemental analysis, ultrahigh resolution Fourier transform ion cyclotron resonance mass spectrometry (FT ICR MS) and thermogravimetric analysis were used. It was shown that the yield of bio-oil was increased from 12.4% at 240 °C to 37.2% at 330 °C. With temperature increasing the spectrum of compounds found in bio-oil using FT ICR MS narrowes and moves to the region of low-molecular compounds. It was found that among the components containing nitrogen and oxygen, compounds containing 1 and 2 nitrogen atoms, as well as ON and O_2N_3 classes, dominate in the bio-oil. With the increase in the temperature of the liquefaction process, classes of compounds containing several oxygen atoms ON, O_2N_3, O_3N_2, ON_2, transforms into the classes N and N_2. By thermogravimetric analysis it was shown that the gasoline fraction (with boiling temperatures up to 200 °C) in bio-oil samples obtained from Arthrospira platensis at 240-330 °C was in the range of 20-22%. The fraction of light-boiling components in bio-oil was slightly decreased when the temperature of hydrothermal process was increased.

1. Introduction

In recent decades, the studies devoted to biofuel production from cyanobacteria/microalgae were focused mainly on cultivation methods. In this context, the main tasks were first of all the increasing of productivity and lipid content in biomass by strains selection [1] and cultivation conditions optimization [2]. At the same time, not enough attention was paid to the problem of biomass to biofuels conversion. The rational solution of this problem is one of the key ways to improve the energy efficiency of the process of biofuel production. The traditional conversion method usually includes drying, solvent extraction of lipids and transesterification with the production of fatty acid methyl esters composing the biodiesel fuel. Obvious disadvantages of biodiesel production method are high energy costs and the use of dangerous organic solvents (such as methanol). Moreover, in case of biodiesel production only lipid part of biomass is processed, while great part of biomass including proteins and carbohydrates is not involved in the production of liquid biofuel. At the same time, it is known that lipid-rich strains have relatively low biomass productivity [3].

The problem of biofuel production is connected primarily with high moisture content when leaving the cultivation stage (80-90 % by weight). For the processing of wet biomass into biofuels the hydrothermal process seems to be more favorable (in general, a "hydrothermal process" means any process that takes place in the presence of water or steam above 100 °C [4-6]). Hydrothermal liquefaction (HTL) with the production of crude bio-oil (biocrude oil) as end product is of most interest [6]. One of the main advantages of HTL is that feedstock pre-drying is unnecessary in this case. Biomass raw can be fed into HTL reactor in humid state, e.g. in the form of an aqueous suspension. Another advantage is that by HTL the resulting bio-oil is composed not only from lipids but also from carbohydrates and proteins that increases the overall yield of the product. Additional advantages of HTL are relatively low temperatures of the process and the possibility of organizing the one-step continuous process.

During HTL the biomass is thermally treated under humid conditions at temperatures up to 370 °C and pressures up to 25 MPa. Due to this treatment, the biomass components are passed through hydrolysis and pyrolysis reactions forming a number of liquid hydrocarbons, both soluble and insoluble in water, as well as gaseous and solid reaction products. HTL products are bio-oil, aqueous solution, solid residue and gas product. The target product of HTL is the so-called "crude bio-oil" (biocrude) - liquid hydrocarbons separated from the solid phase and an aqueous solution.

After HTL the bio-oil, aqueous solution and solid residue are extracted from the reactor usually together as common condensed product. Solid residue is often separated by filtration or sedimentation. To separate bio-oil from aqueous solution in laboratory conditions the organic solvents are used. The yield and properties of bio-oil obtained from microalgae by hydrothermal liquefaction depends on many parameters including the strain of microalgae, temperature and duration of treatment, solvent etc.

Arthrospira platensis is well-known culture that is used in the production of many commercial products including healthy food supplements, food for animals, cosmetics and pharmaceutical products. It is grown usually in an open manner in a large scale. A. platensis is classified as cyanobacteria by International Code of Bacterial Nomenclature (ICNB), although according to the International Code of Nomenclature for Algae, Fungi and Plants (Melbourne Codex 2012) it is classified as blue-green microalgae.

Cyanobacteria/microalgae A. platensis is one of the most promising sources of renewable biofuel . When it is used as biomass raw for biofuel production it is often considered as microalgae. Advantages of A. platensis are relatively simple and cheap technologies for its cultivation and harvesting and lability of biochemical composition, i.e. predictable response to changes in the external environment (illumination, temperature, limitation and starvation of the main biogenic elements) [7]. The productivity of A. platensis and its oil (lipids) content may be ten times higher than that of terrestrial biomass [8].

Hydrothermal liquefaction of A. platensis has been studied early [9-10]. In [9] by the comparison of two methods – hydrothermal liquefaction and pyrolysis, it was established that the hydrothermal liquefaction method is more suitable for the conversion of biomass to liquid biofuel. In [10] the optimal conditions of hydrothermal liquefaction were established: 350 °C, 1 hour holding time and 20% biomass concentration in the initial aqueous suspension. In another work [11], in the study of hydrothermal liquefaction of S. platensis, the optimal temperature was determined as 300 ° C (holding time at a maximum temperature of 30 min), which provides a 30% yield of bio-oil. In [10-11], in addition to standard elemental analysis, FT-IR spectroscopy and gas chromatography-mass spectrometry were used.

In this work, a study of chemical composition of the bio-oil obtained from Arthrospira platensis by hydrothermal liquefaction is presented. The study was conducted in the temperature range of 240-330 ° C, with the use of standard elemental analysis and ultrahigh resolution Fourier transform ion cyclotron resonance mass spectrometry (FT ICR MS). The latter method allows in one spectrum to detect tens of thousands of compounds (without fractionation) and determine the gross formulas of the identified molecules. Recently, the FT ICR MS was used in [12], where the bio-oil produced by hydrothermal liquefaction of Nannochloropsis salina at a temperature of 350 °C was investigated. It was shown that a large number of heteroatomic nitrogen-containing compounds compose the bio-oil, and the most common are pyrroles, pyridines, pyrazines, imidazoles and their derivatives, which are the products of protein decomposition. To study the bio-oil obtained from the biomass of S. platensis, FT ICR MS was previously used in [13], however, the study depending on temperature was not carried out yet. For fractional composition analysis in present study the thermogravimetric method was used.

2. Experimental

2.1. Initial biomass

HTL experiments were carried out with cianobacterium Arthrospira platensis. We used Arthrospira platensis rsemsu 1/02-P strain with straight trichomes formed due to natural morphological variability during prolonged cultivation for 20 years under laboratory conditions.

Biomass was obtained by semi-continuous cultivation in open pond with volume of 500 l, illumination of 55 ± 5 μE/(m2 × sec), steady-lighting conditions; T=21 °C. Pond is equipped by near-surface mixing. A. platensis was cultivated using Zarrouk's medium: $NaHCO_3$ – 16,8 g/l, KNO_3 – 3.0 g/l, $K_2HPO{\cdot}3H_2O$ – 0.66 g/l, K_2SO_4 – 1.0 g/l, $MgSO_4{\cdot}7H2O$ – 0.2 g/l, NaCl – 1.0 g/l, $CaCl_2$ – 0.04 g/l, $FeSO_4{\cdot}7H_2O$ – 0.018 g/l, EDTA – 0.08 g/l, Zarrouk's trace metal solution – 1.0 ml/l, (per 1 l of distilled water). Molecular genetics identification and phylogenetic analysis with a high bootstrap support level based on the nucleotide sequence of the 16S rRNA and ITS gene showed that this culture has a 100% similarity with A. platensis strains from the NCBI strain bank, i.e. belongs to the A. platensis species.

The elemental composition of A. platensis was determined using a VARIO EL III Elementar Analysensysteme GmbH elemental analyzer. The content of ash in A. platensis was determined by pyrolysis at 800 °C and mass measurements using an analytical balance Sartorius Cubis MSA324S. The results of elemental analysis (wt.% daf basis) and ash content determination for A. platensis are presented in Table 1. Biochemical composition of A. platensis is presented in Table 2.

Table 1. Elemental composition (wt.% daf basis) and ash content of A. platensis. Oxigen content is determined by difference.

Sample	A. platensis
C	49.86
H	7.29
O	27.56
N	11.3
S	3.99
Ash	6,0

Table 2. Biochemical composition of Arthrospira platensis, wt.%.

Sample	Proteins	Lipids	Carbohydrates
A. platensis	60,7	12,1	7,1

2.2. Hydrothermal treatment

Hydrothermal treatment of A. platensis was carried out on a laboratory plant, the scheme of which is shown in Fig. 1. The reactor represents an autoclave with volume of 500 cm3, the maximum exploitation temperature and pressure for the reactor are 400 °C and 25 MPa correspondingly. Reactor heating is external ohmic. The heating process (heating rate, maximum temperature and exposure time) is controlled by a PC operator using an automated control system (supplied by National Instruments).

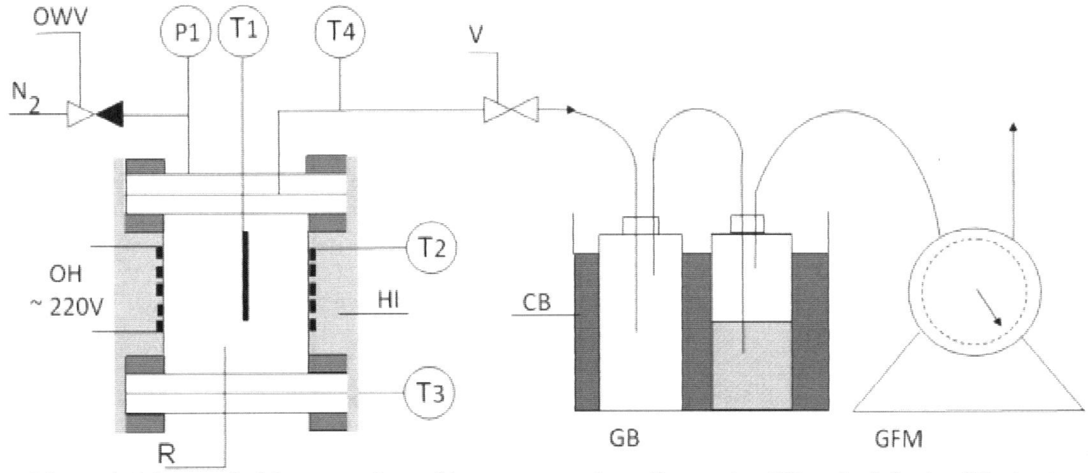

Figure 1. Scheme of a laboratory plant with reactor-autoclave. R – reactor, OH – ohmic heater, HI – heat insulation, CB – cooling bath, GB – gas bubblers, GFM – gas flow meter, OWV – one way valve, V – valve, T and P – temperature and pressure sensors.

Before hydrothermal treatment microalgae was centrifuged and dried at temperature of 80 °C. The reactor was loaded with 150 grams of distilled water and 30 grams of dried A. platensis. Then, the reactor was sealed and purged with nitrogen. Then, the reactor was heated up to a temperature of 240, 280 or 330 °C. The duration of the heating process to a maximum temperature was about 120 minutes, the exposure at maximum temperature was 60 minutes. In the reactor, a pressure close to the pressure of saturated water vapor corresponding to maximum temperature was established and maintained with the help of valve. In the end of exposure the ohmic heater was turned off and the reactor cooled to room temperature during about 5 hours.

Further, the reactor was opened and a solvent (dichloromethane) was poured into the reactor in an amount of 200 grams. After a day, about 100 ml of a solvent were removed from the bottom of the reactor with organic substances dissolved in it. The solvent was placed in an oven at 40 °C until the weight change of the residue ceased. The mass was measured using an analytical balance Sartorius Cubis MSA324S. The residue after evaporation of the solvent was a viscous liquid of a dark brown color. This product is reffered to as bio-oil.

Bio-oil samples obtained at different temperatures process were studied on the Thermo Scientific Flash 2000 HT elemental analyzer and LTQFTUltra high-resolution mass spectrometer (ThermoElectronCorp., Bremen, Germany) equipped with a superconducting magnet (7 Tesla). The bio-oil was dissolved in the MeOH to the concentration 1g/l. Ions were generated by an IonMax Electrospray ion source (Thermo Electron Corp., Bremen, Germany) in positive and negative ESI mode. The temperature of the desolvating capillary was set to 300 oC. The infusion rate of the sample was 1 μl /min and the needle voltage was 3000 V. The archived resolving power was 400 000, each spectrum was the averaging of 100 scans. Prior to the analysis the LTQ FT was calibrated using the standard Thermo calibration mixture. The analysis of the obtained data was carried out using the previously described approach, which makes it possible to detect homologous series using the weighted mass defect histogram of Kendrick [13].

To assess the fractional composition, a thermogravimetric analysis of the samples of bio-oil obtained at different temperatures was carried out. The thermogravimetric analysis was carried out on a thermal analyzer STA PT1600 (Linseis Messgeraete GmbH). A sample of bio-oil in an amount of about 50 mg was placed in a corundum crucible. Heating of the crucible was carried out in argon. The heating rate to 500 °C was 2 °C/min, and then the heating rate to 800 °C was 5 °C/min.

IOP Conf. Series: Earth and Environmental Science **168** (2018) 012039 doi:10.1088/1755-1315/168/1/012039

3. Results and discussion

Table 3 shows the yield of bio-oil as a function of temperature. With temperature increasing, the yield of bio-oil increases: from 12.4% at 240 °C to 37.2% at 330 °C. Basing on the results of elemental analysis it can be supposed that at a temperature of 240 ° C, mainly lipids contribute to the yield of bio-oil, whereas at temperatures of 280-330 °C, proteins and carbohydrates are subjected to hydrothermal liquefaction. The values of yield obtained in this work are in good agreement with the results of previous studies [10-11]. In [10], the yield of the light fraction of bio-oil at temperatures of 200 and 250 °C was about 18 and 25%, respectively. At the same time, the yield of the heavy fraction of bio-oil at these temperatures was insignificant. The maximum yield of bio-oil was reached at a temperature of 350 °C, at which the yields of light and heavy fractions amounted to about 14 and 25% respectively.

Table 3. The yield of bio-oil obtained from hydrothermal liquefaction of A. platensis, and its elemental composition (in the dry ash-free state), depending on the liquefaction temperature.

Parameter	Temperature, °C		
	240	280	330
Yield of bio-oil, %	12.4	26.5	37.2
C, %	65.5	69.1	73.7
H, %	6.1	6.0	6.0
O, %	22.3	18.0	11.1
N, %	3.5	4.1	2.8
S, %	2.6	2.8	3.0

Table 3 also shows the elemental composition of the bio-oil obtained from A. platensis by hydrothermal liquefaction. Compared with the initial biomass, higher carbon content and lower oxygen and nitrogen contents are observed in bio-oil samples. The hydrogen and sulfur content of the bio-oil is slightly lower than their concentration in the initial biomass. As the temperature increases, the carbon content increases. In bio-oil produced at 330 °C, the carbon content (for the dry ash-free state) is 73.7%. The oxygen content decreases with temperature increasing, but it remains still high (11.1% at 330 °C). The nitrogen content increases with temperature increasing, that agrees with the results of previous studies in this field [10].

Fig. 2 shows the mass spectra of the obtained samples of bio-oil. It is seen that if temperature increases, the spectrum becomes narrower and moves to the region of low-molecular compounds. With temperature increasing, the spectrum simplifies, the molecular-mass distribution takes the form of a Gaussian distribution, which resembles the mass spectrum of traditional oil [14].

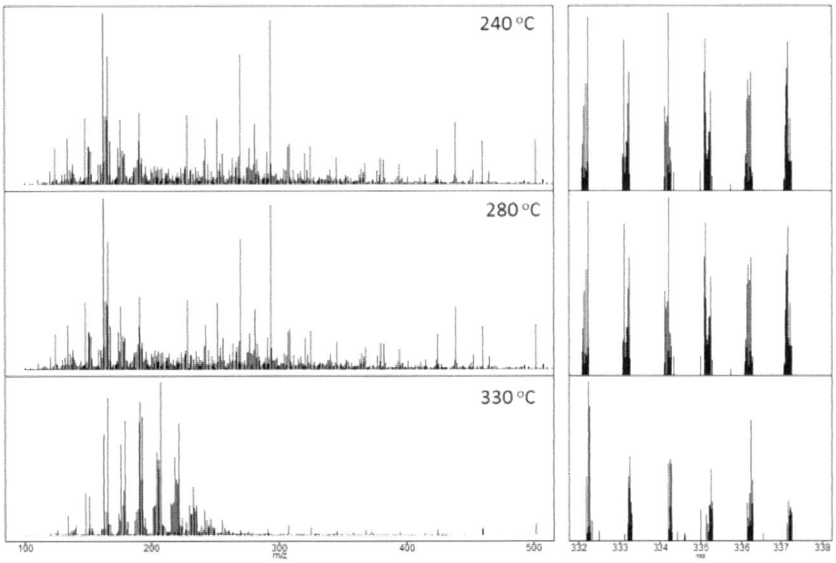

Figure 2. Mass-spectra of bio-oil samples obtained at different temperatures of the process of hydrothermal liquefaction of Arthrospira platensis.

For each of the samples, a Kendrick mass defect diagram, a weighted histogram of Kendrick mass defects were obtained, and the molecular formulas of the substances in the final products were determined. The results are shown in Fig. 3. The number of reliably determined molecular formulas after filtration and exclusion of isotopes was 1336, 1357 and 933 for temperatures of 240, 280 and 330 ° C, respectively. It can be seen that compounds containing 1 and 2 nitrogen atoms dominate among all the nitrogen-containing and oxygen-containing components obtained in all the samples (at all temperatures), and also classes ON and O_2N_3 are present. It is seen that, at low process temperatures, the diversity of the classes of compounds obtained is much broader, intensive classes of compounds containing one or several oxygen atoms are observed: ON, O_2N_3, O_3N_2, ON_2 and others. With an increase in the process temperature to 330 °C, it can be observed that the intensity of these classes decreases significantly and only N and N_2 classes dominate in the spectrum, which, in general, agrees with the results of elemental analysis. It can be seen that the problem of a high nitrogen content in the bio-oil with a temperature change in the range 240-280 ° C does not go away, it undergoes qualitative and quantitative changes, information on which can be useful in choosing the optimal method for further bio-oil processing.

Fig. 4 presents the results of the thermogravimetric analysis of the samples of bio-oil obtained at temperatures of 240 and 330 °C. It can be seen that the curves in the figure go close to each other. The curve for bio-oil obtained at 280 °C lies between the curves for 240 and 330 °C and it isn't shown in fig. 4. The mass of samples after the heating to 800 °C decreases by approximately 80%. The decrease in the mass of the samples at temperatures of 300 °C and 400 °C is also almost the same and amounts to about 43 and 63% (for both shown curves), respectively. The main differences are in the temperature range up to 300 °C. The gasoline fraction (weight change to 200 °C) in the sample of biooil obtained at a lower temperature was about 22%, which is 2% higher than the sample of bio-oil obtained at a temperature of 330 °C. The fraction of light-boiling components (change in mass to 200 °C) is also higher in a sample of bio-oil obtained at a lower temperature. Thus, with the increase in the temperature of hydrothermal liquefaction, the contribution of the diesel fraction to the increase in the yield of bio-oil is greater than the contribution of the low-boiling fractions. This result agrees with the results of previous work [10], where it was shown that in the temperature range 250-350 °C the yield of the heavy fraction of bio-oil increases and the yield of the light fraction decreases.

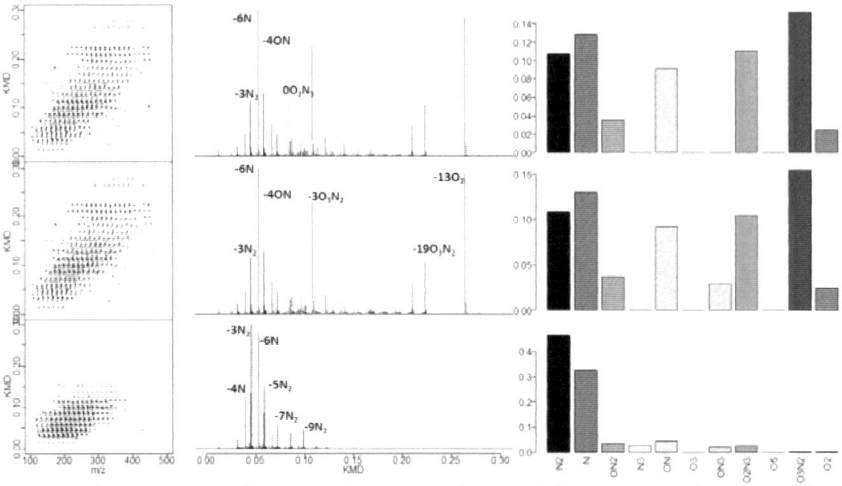

Figure 3. The visual analysis of the samples: Kendrick mass defect plot, weighted Kendrick mass defect histogram and portions of compound classes. Each class of compounds is labeled with unique color. The upper row of figures corresponds to the temperature of hydrothermal liquefaction of 240 °C, the intermediate row – to 280 °C, and the lower row – to 330 °C.

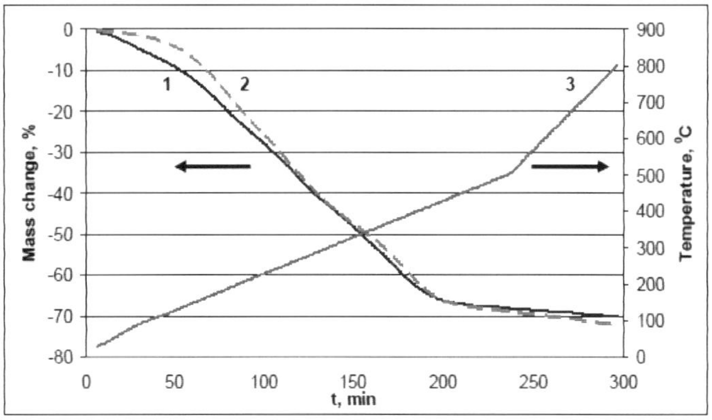

Figure 4. The change in the mass of the samples of bio-oil during the heating in argon medium: 1 - sample of bio-oil obtained at a temperature of 240 °C; 2 - sample of bio-oil obtained at a temperature of 330 °C; 3 - temperature versus time in the experiment.

4. Conclusion

With the change in the temperature of hydrothermal liquefaction of Arthrospira platensis, the yield increases and the chemical composition of the resulting bio-oil is also changed. An increase in temperature leads to an increase in the yield of bio-oil, a decrease in the oxygen content in it, and an increase in the content of carbon and nitrogen. The yield of bio-oil was increased from 12.4% at 240°C to 37.2% at 330°C. With temperature increasing the spectrum of compounds found in bio-oil using FT ICR MS narrowes and moves to the region of low-molecular compounds. From the performed analysis of mass spectra, it was found that among the components containing nitrogen and oxygen, compounds containing 1 and 2 nitrogen atoms, as well as classes ON, O_2N_3, dominate in the bio-oil. With the increase in the temperature of the liquefaction process, classes of compounds containing several oxygen atoms ON, O_2N_3, O_3N_2, ON_2, transforms into the classes N and N_2. By thermogravimetric analysis it was shown that the gasoline fraction (with boiling temperatures up to 200 °C) in bio-oil samples obtained from Arthrospira platensis at 240-330 °C was in the range of 20-22%. The fraction of light-boiling components in bio-oil was slightly decreased when the temperature of hydrothermal process was increased. The fractions corresponded to boiling temperatures of 300°C and 400°C were almost the same for all bio-oil samples obtained (in the temperature range of 240-330°C) and equaled to about 43 and 63%, respectively.

References

[1] P. Neofotis, A. Huang, K. Sury, W. Chang, F. Joseph, A. Gabr, S. Twary, Algal Research 15 (2016)
[2] Y. Zheng, Z. Chi, B. Lucker, S. Chen, Bioresource Technology 103, 1 (2012)
[3] D. López Barreiro, C. Zamalloa, N. Boon, W. Vyverman, F. Ronsse, W. Brilman, W. Prins, Bioresource Technology 146 (2013)
[4] M.S. Vlaskin, A.V. Grigorenko, A.Z. Zhuk, A.V. Lisitsyn, A.E. Sheindlin, E.I. Shkol'nikov, High Temperature 54, 3 (2016)
[5] A.V. Lisitsyn, L.A. Dombrovsky, V.Y. Mendeleyev, A.V. Grigorenko, M.S. Vlaskin, A.Z. Zhuk, Infrared Physics & Technology 77 (2016)
[6] M.S. Vlaskin, N.I. Chernova, S.V. Kiseleva, O.S. Popel', A.Z. Zhuk, Thermal Engineering 64, 9 (2017)
[7] N.I. Chernova, S.V. Kiseleva, M.S. Vlaskin, Y.Y. Rafikova, MATEC Web Conf. 112 (2017)
[8] N.I. Chernova, S.V. Kiseleva, O.S. Popel, Thermal Engineering 61, 6 (2014)
[9] U. Jena, K.C. Das, Energy & Fuels 25, 11 (2011)
[10] U. Jena, K.C. Das, J.R. Kastner, Bioresource Technology 102, 10 (2011)
[11] S.S. Toor, H. Reddy, S. Deng, J. Hoffmann, D. Spangsmark, L.B. Madsen, J.B. Holm-Nielsen, et al, Bioresource Technology 131 (2013)
[12] N. Sudasinghe, B. Dungan, P. Lammers, K. Albrecht, D. Elliott, R. Hallen, T. Schaub, Fuel 119 (2014)
[13] Y. Kostyukevich, M. Vlaskin, G. Vladimirov, A. Zherebker, A. Kononikhin, I. Popov, E. Nikolaev, European Journal of Mass Spectrometry 23, 2 (2017).
[14] Y. Kostyukevich, L. Borisova, A. Kononikhin, I. Popov, E. Kukaev, E. Nikolaev, European Journal of Mass Spectrometry 22, 6 (2016)

Acknowledgements
The study was supported through a grant of the Russian President No. MK-6302.2018.8.

IOP Conf. Series: Earth and Environmental Science **168** (2018) 012040 doi:10.1088/1755-1315/168/1/012040

Contribution in the Study and Numerical Investigation of the Flow Characteristics in a Solar Chimney

[1]**Khaled Mahdi**, [2]**Nadir Bellel**

[1]Hydrometeorological Institute for Training and Research Oran, Algeria
[1,2]Energy Physics Laboratory, University of Mentouri, Constantine 1, Algeria

e-mail: khaled_mahdi@yahoo.fr; bellelnadir@yahoo.fr

Abstract. In this present work, the power generation performance in a solar chimney system (SCS) was investigated using computational fluid dynamics (CFD). The main objective of this study is to examine the impact of the chimney shape in the power output of a SCS. The mathematical model mentioned below was solved by the commercial CFD FUENT 6.3. The model is validated by experimental results from the literature review and used as predictive tool. The influence of the absorber temperature, the geometrical shapes of the solar chimney system on the power output of the SCS. These results lead to the better understanding of the impact on these parameters on the thermal updraft wind SCS operating system.

1. Introduction

The solar chimney system (SCS) converts solar energy into mechanical energy and then using a wind turbine /generator, this mechanical energy converted into electrical energy. The system uses the greenhouse effect in the collector to generate updraft in the chimney, which can drive wind turbines for electricity production. SCS has advantages in its low cost of construction and function, and the supply of renewable energy to communities without aggravating environmental pollution and intensifying climate change. SCS is considered to have high potential for application in developing countries with extensive available field and abundant solar isolation. The solar chimney power plant concept was originally proposed in 1903 by Isidoro Cabanyes [1]. In 1931, a description of a solar chimney power plant was presented by Günther [2].

The basic study on the solar chimney concept was performed by Schlaich in the 1970, and in 1981 he began the construction of a 50kW pilot solar chimney power plant in Manzanares, Spain [2], [3]. Studying the impact of different components in a solar chimney is a primary task for designing a solar chimney power plant. Haaf *et al.* [3], [4] indicated that the critical roles of chimney height and air temperature in determining the system power output. Pasumarthi *et al.* [4], [5] established a mathematical model for estimating the heat transferring process and the airflow in a SCS.

2. Experimental Setup

The basic dimension of the SCS was built in Wind Engineering Section at Kyushu University research institute for applied mechanics (RIAM) Japan [6] are as follows:

Chimney height for cylindrical and conic shape, 2 m; chimney radius, 0.16 m; collector radius, 1.5 m; height between the inlet of the collector and the center, 0.2 m (as shown in Figure 1).

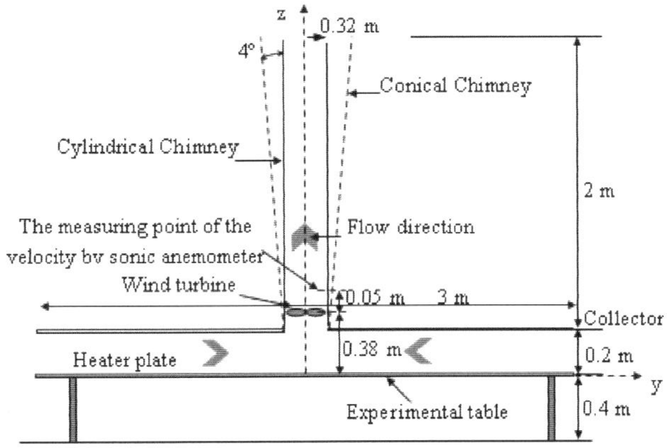

Figure 1. Schematic diagram of a SCS.

3. CFD Simulation Methods

The solar chimney geometry was created using GAMBIT (Version 2.4.6), the pre-processing module of the FLUENT (Version 6.3.26) code. The schematic appearance of the different shape of the chimney is shown in Figure 2. The following tow assumptions have been made:

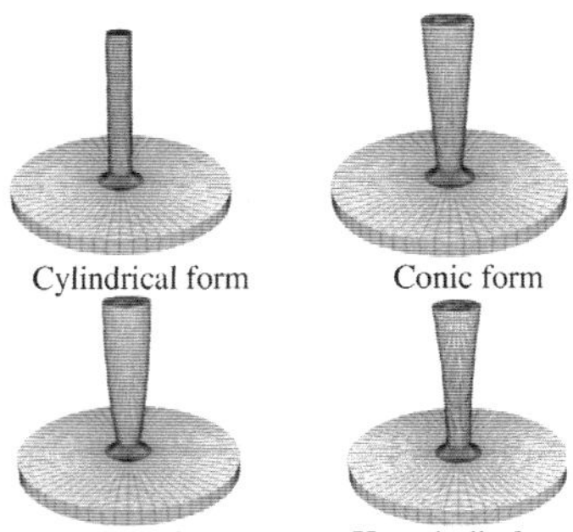

Figure 2. Three-dimensional view of the different shapes solar chimney.

1) The absorber surface of the collector is assumed to be uniform and smooth.
2) The temperature of the absorber surface of the collector is assumed the same.

In actual condition of the absorber temperature of the collector, a temperature difference exits between the middle and the edge of the absorber. The air temperature variation exists of the edge at the center of the absorber caused by the surrounding environment of the experiment. This is because of the natural convection heat loss from the chimney aperture to the ambient. Therefore, in the present numerical analysis a temperature difference (ΔT) of 20°C to 40°C has been considered between absorber surface and ambient temperature of the laboratory. The total numbers of cells has been fixed approximately 860725 tetrahedral cells. The grid generation for computational domain is shown in Figure 1.

3.1. Numerical Procedure

Four different chimney shapes: cylindrical, conic, parabolic and hyperbolic are investigated in this study. Each form is modelled as show in Figure 2 using the Gambit tool of Fluent CFD software package [7]. The geometry and mesh were created using GAMBIT 3.and duplicating the actual geometry of the different chimney form. For modelling a solar chimney, The 3D modelling and grid generation was carried out in the GAMBIT 2.0.4 package.

3.2. Numerical Solution

IOP Conf. Series: Earth and Environmental Science **168** (2018) 012040 doi:10.1088/1755-1315/168/1/012040

The basic equations including the models discussed till now were numerically solved with the help of the commercial simulation program FLUENT (version 6.23). Fluent uses the finite volume method for the solution of the differential equations. The basic equations were solved numerically with the help of the SIMPLE algorithm (Semi-Implicit Method for Pressure-Linked-Equations). The accuracy of all approximations was second order. The standard relaxation parameters were adapted for this. Many cells in the radial direction and relatively few in the x direction are used in the collector of the upwind power plant. The reverse is valid in the chimney. All numerical calculations had to be carried out with the solver with double precision. A convergence criterion of 10^{-3} was imposed on the residuals of the continuity equation momentum equation. The convergence criterion 10^{-6} was given on the residual of energy equation. A typical numerical simulation process is described in Figure 3.

TABLE I. OUNDARY CONDITIONS AND MODEL PARAMETERS RADIATION PROPRIETIES OF WALL

materials	α	ρ	τ
Glass	0.06	0.95	0.84
Absorber plat	0.95	0.95	0

3.3. Mathematical Model and Boundaries Conditions

Since the purpose of this numerical study is to understand the flow characteristics and the effect of the geometry on the solar chimney performance that was built in RIAM, not the assessment of turbulence models, it is decided that a standard k–ε model [8] would serve the purpose which is also adopted by Motoyama *et al.* and Sangi *et al.* [9], [10]. The conservation equations and the standard k–ε turbulent model used in this study is shown below:

The continuity equation:

$$\frac{\partial u_i}{\partial x_i} = 0 \tag{1}$$

Conservation of momentum is described by:

$$\frac{\partial u_i}{\partial t} + u_i \frac{\partial u_i}{\partial x_i} = -\frac{1}{\rho}\frac{\partial p}{\partial x_i} + v\frac{\partial^2 u_i}{\partial x_j \partial x_j} + \rho f_i \tag{2}$$

where p is the static pressure, $v\frac{\partial^2 u_i}{\partial x_j \partial x_j}$ is the stress tensor (described below), and ρf_i is the gravitational body force and external body forces.

The energy equation in the following form:

$$\frac{\partial(\rho c_p T)}{\partial t} + \frac{\partial(\rho u_j c_p T)}{\partial x_j} = \frac{\partial}{\partial x_j}\left(\lambda \frac{\partial T}{\partial x_j}\right) + T\beta \frac{\partial p}{\partial t} + S_h \tag{3}$$

For the standard k–ε models, the constants have the following values [10]:

$$C_{1\varepsilon}=1.44,\ C_{2\varepsilon}=1.92,\ C_{\mu}=0.09,\ \sigma_k=1.0,\ \sigma_\varepsilon= 1.3. \tag{4}$$

3.4. Boundary Conditions

The boundary conditions such as absorber collector and ambient temperature are defined from the laboratory conditions [11], [12], [13]. The relative static pressure of the collector inlet is zero, and the temperature is the ambient temperature:

$$P_{int} = 0, \quad \text{and} \quad T_{int} = T_\infty \tag{5}$$

$$\text{by:} \quad \frac{\partial T}{\partial x} = 0, \quad u = 0, \quad v = 0 \tag{6}$$

The above relations express an adiabatic wall and the non-slip conditions respectively.

Data for an acrylic sheet and absorber plate in terms of optical factor is illustrated in Table II.

TABLE II. PHYSICAL PROPERTIES OF WALLS OF THE SOLAR CHIMNEY PROTOTYPE

materials	α	ρ	τ	ρ(kg/m³)	C_p(J/kg/K)	λ(W/m/K)
acrylic sheet	0.06	0.95	0.84	1190	1465	0.19
Absorber plate	0.95	0.95	0	7830	491	70

3.5. Numerical Method Validation
In order to validate our numerical method, we performed a numerical study using the Japanese prototype data for the experience at laboratory used by the validation reference work (Haaf, 1984). Transient numerical results for the fluid motion established in the Japanese prototype are presented, including the required validation with appropriate experimental results. The numerical results for the air temperature at the inlet section of the wind turbine, during the whole day, follows the same trend as the experimental results of Japanese [5], with an average deviation of 12% (Figure 3). The results obtained for the air velocity in the same section are shown in Figure 4. On the same figure, values measured just prior 8:00 and around 20:00 are discarded (the respective rapid increase and decrease reached in the stack effect can be due to external atmospheric agents). The average deviation is, at most, equal to 12%. It can be observed, on figure 10, the expected power from the present simulation and the one given by Haaf (1984). We notice the same trend and an average deviation of around 9%.

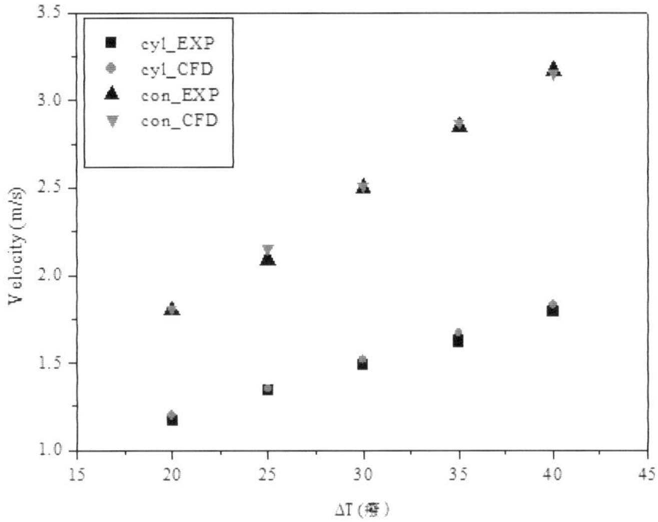

Figure 3. Comparison of velocity between CFD model and experimental data at y= 0.38 m.

TABLE III. CURVE FITTING IN THE CYLINDRICAL AND CONICAL CHIMNEY AND THE PREDICTION OF OPTIMAL GEOMETRY

Curve equation	$u = A + B\times\Delta T$			
	cyl_EXP	Cyl_CFD	con_EXP	Con_CFD
A	0.570	0.566	0.382	0.458
B	0.030	0.031	0.070	0.068
R^2	0.99905	0.99984	0.99883	0.99915

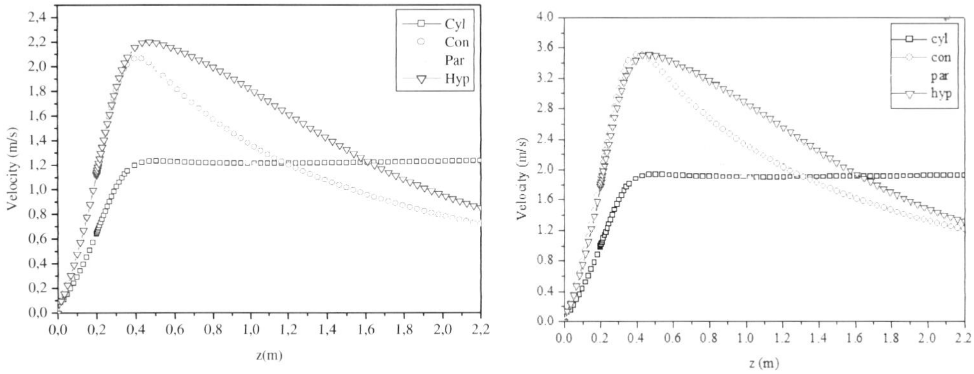

Figure 4. Effect of the chimney shapes on the updraft velocity, , ΔT=20°C and 40°C.

Figure 4 and Figure 5 shows the velocity updraft, through the height of the chimney. We notice that velocities in the three shapes (cylindrical, conic, parabolic and hyperbolic) increase regularly with increasing temperature difference between absorber plate and ambient temperature.

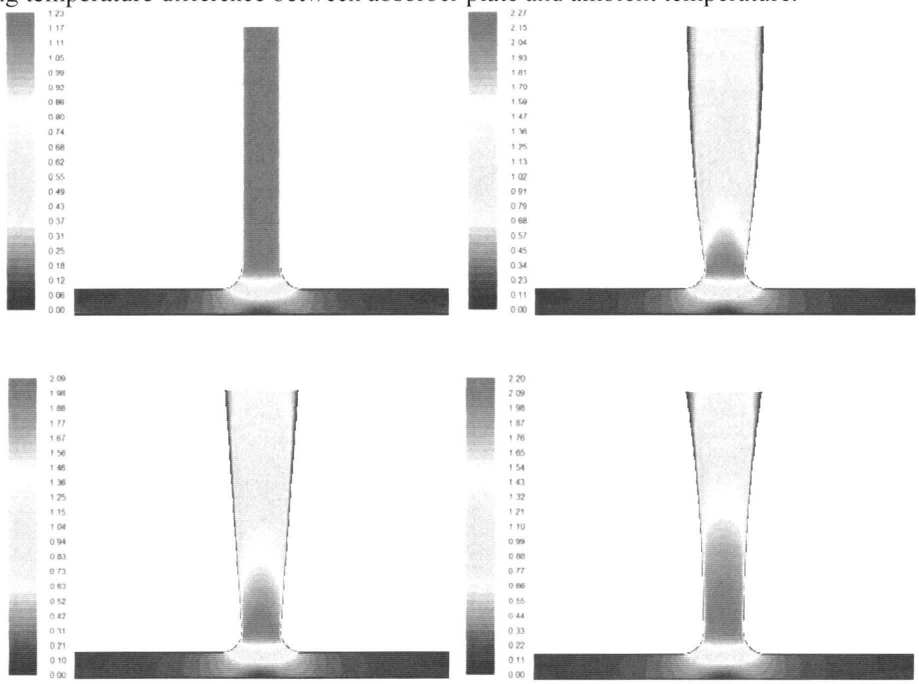

Figure 5. Profiles of air velocity through the chimney for different shapes of chimney, ΔT=20°C.

We see in Figure 6 and 7 the temperature is decreased because airflow is rising in the chimney as show in Table III. The temperature rapidly falls because of the velocity updraft of three forms (conic, parabolic and hyperbolic) compared from cylindrical shape, because as hot air rises with important velocity updraft, air temperature rapidly falls; outer portion of the chimney is insulated from four shapes. Air velocity in collector was slow due to larger height (0.2 m) compared of the chimney radius. In addition, at the point of z = 0.38 m and it rises rapidly at the point z = 0.5 m due to the sudden contraction in chimney sections.

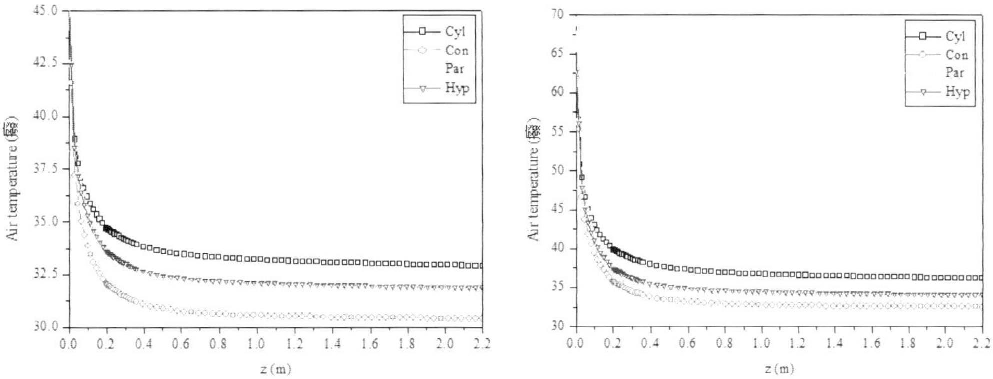

Figure 6. Effect of the chimney shapes on the updraft velocity, ΔT=20°C and 40°C

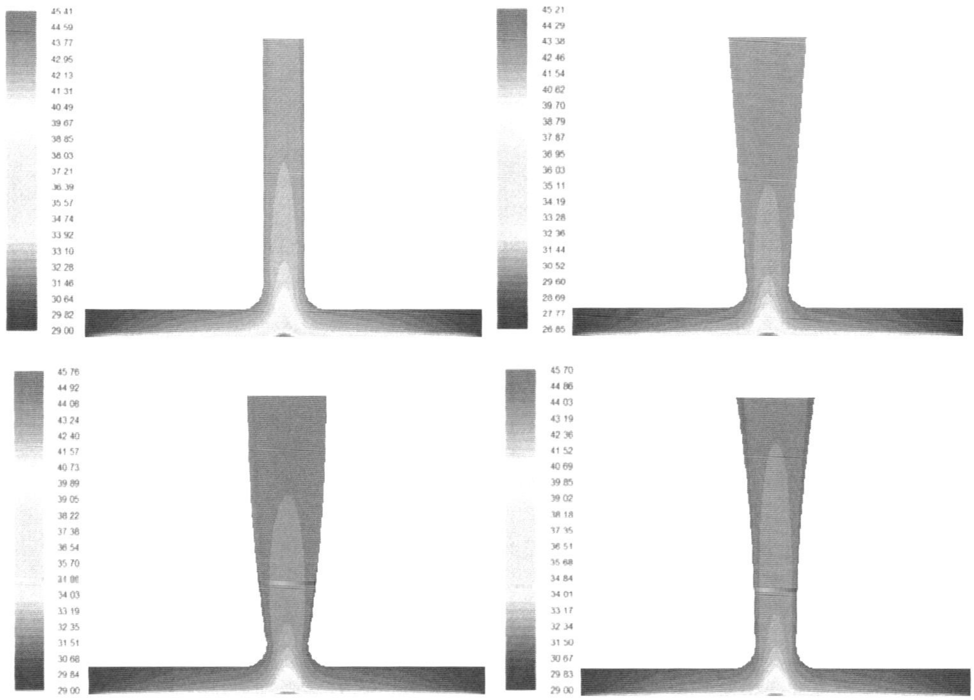

Figure 7. Profiles of air temperature through the chimney for different shapes of chimney, ΔT=20°C.

Table IV shows that the maximum velocity is occurring inside the chimney at different levels, same for Power out and comparative obtained by CFD simulation up to 0.7m of th height.

TABLE IV. COMPARISON BETWEEN THE AIR VELOCITIES OBTAINED BY CFD SIMULATION

chimney	z (m)	u_{max} (m/s)	$\overline{u}$ (m/s)	P_{max} (mW)	$\dot{m}$ (kg/s)	ratio
cyl	0.50	1.87	1.82	22.68	0.1651441	1
con	0.39	3.44	2.43	141.26	0.3182465	4.74
par	0.36	3.38	2.18	134.72	0.3182036	4.12
hyp	0.43	3.39	2.70	135.42	0.3019360	5.63

4. Conclusion

The investigation of the velocity updraft from solar chimney shapes was undertaken numerically and was validated using the published experimental results for two different chimneys geometries (cylindrical and conical shape). A good agreement between experimental and numerical results was obtained, it is determined that due to the shape of chimney is affected, performance of the system increases positively. In the conclusion, one can say that the shape of the conical, parabolic and hyperbolic chimney gives almost the same maximum available power compared with cylindrical shape of the SCS.

We have obtained:

- The available power of the SCS increases with increases of the absorber plate temperature differences.
- The main power available of the hyperbolic SCS is five times greater than the cylindrical SCS.

Nomenclature

Symbol	Nom,	unité
A	area,	m^2
k	kinetic energy of turbulence,	m^2/s^2
T	temperature,	K
u	velocity ,	m/s
$\overline{u}$	mean velocity,	m/s
z	chimney height,	m
$\dot{m}$	mass rate,	kg/s
Cp	power coefficient	
p	pressure,	pa
R	correlation coefficient	
Greek letters		
α	absorptivity	
ρ	Reflectivity	
τ	transmittivity	
ε	rate of dissipation,	m^2/s^2
β	thermal expansion coefficient	1/K
δij	kronicer symbol	
∞	*ambient*	
Acronym		
cyl	cylindric	
con	conic	
par	parbolic	
hyp	hyperbolic	
EXP	experimental	
CFD	computational fluid dynamics	
RIAM	research institute for applied mechanics	

5. References

[1] J. Lorenzo, Las Chimneas solares: de una propuesta espanola en 1903 a de Mansanares, http://www.fotovoltaica.com/chimenea.pdf.

[2] H. Günther, Hundred Years-Future Energy Supply of the World, Franckhsche Verlagshandlung, Stuttgart, Germany, 1931.

[3] W. Haaf, "Solar chimneys, part II: preliminary test results from the Manzanares pilot plant," International Journal of Solar Energy, vol. 2, May.1984, pp. 141-161,doi:10.1080/01425918408909921

[4] W. Haaf, K. Friedrich, G.Mayr, and J. Schlaich, "Solar chimneys, part I: principle and construction of the pilot plant in Manzanares," International Journal of Solar Energy, vol 2, Apr. 1983, pp. 3-20,doi: 10.1016/B978-0-12-384988-5.00032-2

[5] S. W.Yuan, "Foundations of Fluid Mechanics", 3rd ed., Prentice Hall International, London, pp. 104-113, 1988.

[6] Fluent 6.2 user guide, Fluent, Inc., Lebanon, NH, 2004.

[7] M. Motoyama, K. Sugitani, Y. Ohya, T. Karasudani, T. Nagai and S. Okada, "Improving the power generation performance of a solar tower using thermal updraft wind," Energy Power

Eng, vol. 6, Sep. 2014, pp. 362-370, doi: 10.1145/566654.566623

[8] M. Motoyama, K. Sugitani, Y. Ohya, T. Karasudani, T. Nagai And S.Okada, "Improvement of power generation on solar tower by thermal updraft wind", 23rd Conference Wind Engineering , 2014, doi: 10.14887/kazekosymp.23.0_109

[9] R. Sangi, M. Amidpour and B. Hosseinizadeh, "Modelling and numerical simulation of solar chimney power plants," Sol Energy, vol.85, 2011, pp. 829-838, doi: 10.1145/566654.566623

[10] T. Ming, W. Liu and G. Xu , "Analytical and numerical investigation of the solar chimney power plant systems," Int. J. Energy Res, vol. 30, Apr. 2006, pp. 861–873, doi: 10.1002/er.1191

[11] B.E. Launder and D.B. Spalding, "The numerical computation of turbulent flows," Comput Methods Appl. Mech Eng, vol. 3, March. 1974, pp. 269–289, doi: 10.1016/0045-7825(74)90029-2

[12] M.O. Hamdan and O. Rabbata, "Experimental Solar Chimney Data with Analytical Model Prediction," in Proceedings of the Solar Conference, vol. 1, pp. 327-332. World Renewable Energy Forum, WREF2012, ISBN:9781622760923, Denver, Colorado, 13–18 May 2012.

[13] Fluent CFD software package, version 6. 3. 26, 2006.

IOP Conf. Series: Earth and Environmental Science **168** (2018) 012041 doi:10.1088/1755-1315/168/1/012041

A Practical Classification Evaluation Approach for Comprehensive Competitiveness of Distributed Energy Supply Systems

Xiang Huang, Sen Ouyang, Shaohua Ma

School of Electric Power, South China University of Technology, Guangzhou, China

e-mail: huangx1226@foxmail.com, ouyangs@scut.edu.cn, 454427183@qq.com

Abstract. The distributed energy supply system(DESS) is widely used in power grids. This paper proposes a practical classification evaluation approach (PCEA) based on analytic hierarchy process-entropy weight method (AHP-EWM), with the goal of the quantitative evaluation values of DESS. Firstly, a practical evaluation index system (PEIS) considering soft competitiveness and hard competitiveness is constructed and scenes of DESS are classified into types under the consideration of developmental characteristics. Then, to break the hierarchies between intertypes and improve the fairness of evaluation results, an index-contributing coefficient (ICC) based on EWM is introduced to estimate the influence degree of different indexes on classification evaluation between different types of scenes. Next, with the input valuables of practical evaluation indexes from the PEIS, the classification evaluation results of DESSs are obtained by PCEA. Finally, the DESSs in southern part of China are taken for an example for analysis and further results verify the adaptability of PCEA.

1. Introduction

DESSs can achieve energy cascade utilization by adopting combined supply mode [1-2]. Due to the characteristics of small scale, small capacity and advantages of high energy efficiency and environmental protection, DESSs are widely implemented in hospitals, hotels, supermarkets and etc.

Currently, remarkable works providing evaluation models of DESSs are available and reviewed [3-5]. Paper [5] puts forward to optimize the operation based on primary energy consumption and carbon dioxide emissions for different conditions. It is apparent that each evaluation work mentioned above has certain features. Although contents of them are different, the basic idea is roughly consistent. Improvement of this idea as well as innovation of PCEA can be summed up as follows.

Existing evaluation indexes lack adequate attention of the holistic system. Most works tend to focus only on one or several aspects, such as energy cost [6], environmental protection [7-8] and other operating characteristics [9]. Practical indexes closely correspond to different aspects of comprehensive competitiveness and are chosen on the principle of higher contribution rate, easier accessibility and quantifiable characteristic.

As characteristics of different scenes vary with types of the application field[10], in this paper, DESSs are mainly classified into three types: building sites, regional sites and island sites.

Analytic hierarchy process (AHP) is an assignment method to apportion weight of indexes by combining qualitative method with quantitative method[11]. However, some indexes of one type of DESSs have larger (or smaller) values comparing with that of other types. This will cause the problem that evaluation results by AHP are excessively concentrated by the type. The PCEA is designed in this paper to deal with the problem by introducing ICC, which is solved out by EWM. As ICC can be used to estimate the influence degree on different types of DESSs by the same index, this method can break the hierarches between them.

Purpose of existing evaluations for DESS can be summarized as follows: to compare different configuration schemes or operating strategies in order to make valuable and optimal decision-making[7], to conduct assessment of various benefits and improve the renewable energy system based on lower cost or less emission[8]. However, evaluations of comprehensive competitiveness from a macro perspective to provide reference for investment and construction planning are non-existent at present. Therefore, PCEA is proposed to deal with this issue. With regard to the soft competitiveness, analysis of operational status and weak links are given to provide experience for operation and maintenance management of similar scenes. As to the hard one, conditions for development of DESSs are gathered to analyze economic, environmental benefits.

This paper is organized as follows. In Section II, practical evaluation index system is established. Section III describes the classification framework. The procedures of PCEA are elaborated in Section IV. Section V presents case studies. Finally, this paper concludes in Section VI.

2. Practical Evaluation Index System

2.1. Influencing Factors of Classification Evaluation

Comprehensive competitiveness of DESSs contains two aspects: soft and hard competitiveness. Its specific components are shown in Fig.1.

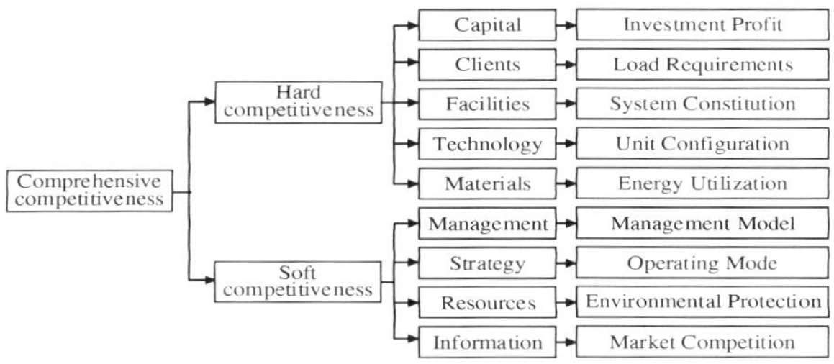

Figure 1. Specific components of comprehensive competitiveness.

According to the theoretical analysis, factors affecting the evaluation of comprehensive competitiveness of DESSs are refined into five aspects, which are described below:

(1) Economic Benefit

As reflected in capital, management and information, economic benefit plays a significant role in popularizing prospect of DESSs. Economic benefit is the primary influencing factor.

(2) Characteristic of Combined Supply

In different sites, there are mainly four kinds of supply requirements: power, cooling-power, heating-power and cooling-heating-power. Characteristic of combined supply, which reflected specifically in clients and technology, is used to measure the hard competitiveness of DESSs.

(3) Energy Utilization

Energy utilization mainly focuses on the energy reserves. It evaluates the abilities to effective mining local diverse energy sources and realizing the cascade utilization of DESSs.

(4) Operating Mode

The main research contents of operating mode are aspects related to operation performance when DESSs are connected to the power grid.

(5) Environmental Benefit

Environmental benefit is to reflect the energy-saving and emission-reduction benefit provided by DESSs. While renewable energy and fuel cell replace the traditional fossil fuel power, pollution emissions will be significantly reduced.

2.2. Practical Evaluation Index System

According to the principle of higher contribution rate, easier accessibility and quantifiable characteristic, a practical evaluation index system is established, which contains nine practical indexes.

(1) Investment Per kW: x_1(yuan/kW)

This index reflects the investment feasibility of DESSs from side and directly measures the raw material cost, moving cost, etc.

(2) Profit Rate of Investment: x_2 (%)

If the profit rate of investment is not less than the standard investment profit rate, the project is considered to be acceptable, otherwise unfeasible.

(3) Payback Period of Investment: x_3 (year)

Payback period of investment represents the duration lasting from the date when the DESS project starts working to the day on which the original investment is totally paid off by using the net income of this project.

(4) Capacity of Power Supply: x_4 (MW/hour)

This index gives a direct reflection of load capacity and scale of the system. It can be utilized to provide valuable experience for operation and maintenance management of DESSs.

(5) Energy Utilization Ratio: x_5 (%)

Energy utilization ratio means the rate of consumption amount of the energy accounting for the total supplied to the system. The utilization efficiency and degree of energy of DESS is better reflected by this index.

(6) Power Generation Ratio of New Energy: x_6 (%)

Power generation ratio of new energy equals to the proportion that new energy power generation accounts for the total power supply per year. The technology level of new energy development can be revealed by this index.

(7) Annual Utilization Hours of Power: x_7 (hour)

Annual utilization hours of power are counted by hours and means annual operational time of DESSs. Ability of power supply is well reflected by this index.

(8) Total Emission Reduction: x_8 (thousand tons)

When comparing with the traditional centralized fossil fuel power under circumstance of providing the same load supply, total emission reduction refers to the total -emission reduction of DESSs.

(9) Total Emission Reduction Rate: x_9 (%)

Total emission reduction rate means the ratio that amount of total emission reduction by DESSs accounts for that by traditional fossil fuel power per year while providing the same load supply. This index is more obvious to reflect the environmental value of DESSs.

3. Classification Framework

In the classification framework, DESSs are mainly divided into three types based on the scale and application field. These three types are defined as building sites, regional sites and island sites.

3.1. Building Sites

Building sites consist of hotels, hospitals, commercial centers, public buildings, etc. Their economic relationships are simple, which can result in flexible application. In addition to small scale, less varieties of terminal loads and lower energy efficiency, building sites have the characteristic of long-time operation per year as well. Hence, they will be competitive only under certain conditions. Meanwhile, load requirements in this type can roughly coincide with the supply ratio of cooling and power or heating and power of gas engine. Besides, grid-connected operation as well as maximum energy cascade utilization is achievable for building sites [10].

3.2. Regional Sites

Regional sites are mainly composed of industrial parks, university towns, shopping malls, integrated parks, large residential parks, etc. Such large-scale type of sites has various terminal loads, such as

IOP Conf. Series: Earth and Environmental Science **168** (2018) 012041 doi:10.1088/1755-1315/168/1/012041

production areas, tourist centralized service areas, large commercial facilities and so on. Load types may be summed up as follows: civil heat load, industrial heat load, etc. Annual operating time of regional sites can be flexibly adjusted according to actual needs. They are highly competitive on the economic front which can be proved by the significantly lower investment Per kW and high profit rate. While meeting the power supply, these sites can realize complementary operation with large power grid by means of peak load shifting. DESS of Guangzhou University Town is currently the largest distributed energy projects in China, which is a regional site.

3.3. Island Sites

The island type of scenes, whose main developing form is smart micro-grid, generally refers to the independent island with high permeability of distributed generations. The entire system of this type is "single-unit and single-grid" with fluctuating loads and isolated operation. To satisfy the internal rate of return, power price of island end-users will reduce and economic efficiency of power will increase. Areas around the island are rich in renewable energy. Effective development of island energy can greatly ease the power shortage. In the planning of island sites, it is necessary to forecast the output of distributed energy, optimize the energy storage and build a comprehensive energy network.

4. Practical Classification Evaluation Approach

4.1. Designing Idea of PCEA

In the longitudinal dimension, considering five influencing factors: economic benefit, characteristic of combined supply, energy utilization, operating mode and environmental benefit, the practical evaluation index system is constructed. Meanwhile, in the horizontal dimension, DESSs are divided into three types based on the scale and application field.

4.2. The Overall Procedure of AHP-EWM

The AHP-EWM mainly focuses on solving two basic issues. The first issue is how to make full use of such many evaluation indexes to evaluate DESSs. Under the consideration of expert opinion, AHP can be utilized to deal with the first issue. The other one is the problem that evaluation results are excessively concentrated by the type. Based on the information theory, the information entropy presented by the same index of different types of DESSs is different. Given to this point, ICC based on EWM is introduced to break the hierarchies between the evaluation results of different types of DESSs.

Procedures of the approach are listed as follows.

(1)Data pre-processing:

The number of evaluated DESSs is defined as: $m=m_L+m_Q+m_H$, where m_L、m_Q、m_H represents the number of building sites, regional sites and island sites. The number of practical evaluation indexes is defined as n. Hence, raw $m\times n$ matrix is written as follows:

$$X_{m\times n}=\begin{bmatrix} x_{11} & x_{12} & \cdots & x_{1n} \\ x_{21} & x_{22} & \cdots & x_{1n} \\ \cdots & \cdots & \cdots & \cdots \\ x_{m1} & x_{m2} & \cdots & x_{mn} \end{bmatrix} \tag{1}$$

Z-score standardized transformation is used to improve the representation of sample data considering the universality of the normal distribution. The standardized equation is written as follows:

$$x_{ij}^{'}=\frac{x_{ij}-x_{jmean}}{x_{jstd}} \tag{2}$$

Where $i \in \{1,2,\ldots,m\}$, $j \in \{1,2,\ldots,n\}$. x_{jmean} and x_{jstd} are respectively the mean value and standard deviation value of the column data of the raw matrix. Then the $x'_{m\times n}$ is solved out by Z-score standardized transformation.

Numerical deviations of different indexes are very large. Indexes with larger values play a strengthen role in comprehensive analysis while indexes with smaller values play the opposite role. To deal with this problem, 0-1 standardized transformation is adopted to eliminate the influence.

$$r_{ij} = \frac{x'_{ij} - x'_{ij\min}}{x'_{ij\max} - x'_{ij\min}} \tag{3}$$

Where x'_{ijmax} and $x'_{ijmin}$are the minimum value and maximum value of the matrix$x'_{m\times n}$. Through the two-step transformation, the raw evaluation matrix is transformed into the standardized matrix$R_{m\times n}$.

(2)Calculation of index-contributing coefficient (ICC):

Evaluation values of a certain type of DESSs will be collectively too large (or too small), which results from that some indexes of this type have larger (or smaller) values comparing with these indexes of other types. This causes the problem that evaluation results are excessively concentrated by the type. Hence, EWM based ICC is introduced to measure the influencing degree of the same index from different types of DESSs.

Detailed steps are listed as follows:

Firstly, information entropy of a index from different DESSs can be solved by the following equation:

$$E_{\alpha j} = -\ln(m)^{-1} \sum_{i=1}^{m_\alpha} p_{ij} \ln p_{ij} \tag{4}$$

$$p_{ij} = \frac{r_{ij}}{\sum_{i=1}^{m_\alpha} r_{ij}} \text{, } (p_{ij} \neq 0) \tag{5}$$

$$\lim_{p_{ij}\to 0} p_{ij} \ln p_{ij} = 0 \text{, } (p_{ij}=0) \tag{6}$$

Then the ICC of a practical evaluation index from different types of DESSs is defined as $C_{\alpha j}$ and can be solved by the following equation:

$$c_{\alpha j} = \frac{1 - E_{\alpha j}}{n - \sum_{j=1}^{n} E_{\alpha j}} \tag{7}$$

ICC matrix: $C_{3\times n}=[C_L\ C_Q\ C_H]^{\mathrm{T}}$ is composed of ICC from all types of DESSs. Influences made by practical evaluation indexes on each type are different.

(3)Evaluation results:

AHP is adopted to calculate weights of practical evaluation indexes for DESSs. This method avoids the limitations as a result of simply relying on the data.

Firstly, according to the importance between each pair of indexes combined by the accuracy requirements of evaluation results, the judgment matrix $D_{n\times n}$ is obtained.

Secondly, to calculate the maximum eigenvalue λ_{max} and the corresponding eigenvector $W_{AHP}=\{W_{1(AHP)},\ W_{2(AHP)},\ \cdots,\ W_{n(AHP)}\}$of the judgment matrix $D_{n\times n}$.

Finally, due to the estimation error, which will destroy consistency of the judgment matrix $D_{n\times n}$, the consistency checking is indispensable. If the inconsistency caused by deviation is acceptable, the normalized eigenvector W_{AHP} will be the weight vector, otherwise the judgment matrix should be re-constructed. Procedures of the consistency checking are presented as the reference [11].

Equation for evaluation results of the comprehensive competitiveness without ICC is written as follows:

$$M_i=\sum_{j=1}^{n} r_{ij}\omega_{j(AHP)}=\begin{bmatrix} r_{11} & r_{12} & \dots & r_{1n} \end{bmatrix}\begin{bmatrix} \omega_{1(AHP)} \\ \omega_{2(AHP)} \\ \dots \\ \omega_{n(AHP)} \end{bmatrix} \tag{8}$$

Where M_i is the evaluation result of DESS i. $w_{j(AHP)}$ represents the weight of index j.
Equation for evaluation results of the comprehensive competitiveness with ICC is written as follows:

$$W_{\alpha j}=\frac{c_{\alpha j}\omega_{j(AHP)}}{\sum_{j=1}^{n} c_{\alpha j}\omega_{j(AHP)}} \tag{9}$$

$$M_{m_\alpha}=\begin{bmatrix} M_{\alpha 1} & M_{\alpha 2} & \dots & M_{\alpha m_\alpha} \end{bmatrix} \tag{10}$$

$$M_{\alpha i}=\sum_{j=1}^{n}\left(r_{ij}W_{\alpha j}\right) \tag{11}$$

Where $W_{\alpha j}$ refers to the weight of index j for some type of DESSs based on ICC. M_{m_α} is made up of evaluation results of all evaluation objects.$M_{\alpha i}$ represents evaluation result of DESS i.
Through the practical evaluation index system and weights of indexes based on AHP combined with ICC, PCEA transforms the comprehensive competitiveness into quantitative evaluation results. By comparison of evaluation results of comprehensive competitiveness, competitive level of DESSs can be obtained. If evaluation result is larger, the DESS will be more competitive in accomplishing high effective utilization of local diversified energy, giving full play to potential economic value and providing valuable reference.

5. Case Studies

Statistical data of practical evaluation indexes from 21 DESSs in southern part of China were collected for case studies. In order to facilitate the comparison of evaluation results, numbers of three sites are 7. The raw statistical data are shown in Tab.1, in which X1~X9 represent practical evaluation indexes, L1~L7 represent building sites, Q1~Q7 represent regional sites, H1~H7 represent island sites.

Table 1. Raw Statistical Data

No.	X1	X2	X3	X4	X5	X6	X7	X8	X9
L1	1.140	7.720	11.000	28.014	72.300	0.243	3.378	0.500	8760
L2	1.345	6.099	14.520	40.060	61.455	0.287	2.669	0.660	8760
L3	1.630	7.257	9.350	33.056	95.436	0.348	3.175	0.425	8000
L4	1.345	6.099	14.520	40.060	61.455	0.287	2.669	0.660	8000
L5	1.630	7.257	9.350	33.056	95.436	0.348	3.175	0.425	8760
L6	1.345	6.099	14.520	40.060	61.455	0.287	2.669	0.660	8760
L7	1.630	7.257	9.350	33.056	95.436	0.348	3.175	0.425	8500
Aver.	1.438	6.827	11.801	35.337	77.568	0.307	2.987	0.536	8506
No.	X1	X2	X3	X4	X5	X6	X7	X8	X9
Q1	0.533	7.880	12.210	108.207	73.630	0.468	7.030	0.700	8000
Q2	0.608	6.068	16.606	160.147	38.288	0.533	5.413	0.952	8400
Q3	0.789	7.013	6.349	123.356	100.137	0.692	6.256	0.364	7980
Q4	0.608	6.068	16.606	160.147	38.288	0.533	5.413	0.952	7000
Q1	0.789	7.013	6.349	123.356	100.137	0.692	6.256	0.364	7900
Q2	0.608	6.068	16.606	160.147	38.288	0.533	5.413	0.952	6500
Q3	0.789	7.013	6.349	123.356	100.137	0.692	6.256	0.364	8000
Aver.	0.675	6.732	11.582	136.959	69.843	0.592	6.005	0.664	7683
No.	X1	X2	X3	X4	X5	X6	X7	X8	X9
H1	1.141	16.550	19.240	26.360	81.000	0.488	1.041	0.650	2500
H2	1.403	13.075	21.741	37.695	68.850	0.600	0.822	0.735	2600

H3	1.631	15.392	16.354	32.423	91.530	0.698	0.968	0.553	2000
H4	1.403	13.075	21.741	37.695	68.850	0.600	0.822	0.735	3000
H5	1.631	15.392	16.354	32.423	91.530	0.698	0.968	0.553	4000
H6	1.403	13.075	21.741	37.695	68.850	0.600	0.822	0.735	4000
H7	1.631	15.392	16.354	32.423	91.530	0.698	0.968	0.553	4500
Aver.	1.463	14.564	19.075	33.816	80.306	0.626	0.916	0.644	3229

At first, raw statistical data of practical evaluation indexes from case DESSs were normalized by the data pre-processing according to equations (2)~(3).
Secondly, according to the AHP method, the measurements of importance degree between each pair of indexes were set as: 1, 1.2, 1.4, 1.6, 1.8, 2.0, 2.2, 2.4, 2.6, where 1 represents the same importance degree between two indexes, 2.6 represents the largest disparity between two indexes. The disparity of importance degree varies from small to large corresponding with figures from small to large. Weights of practical indexes by using AHP are shown in Tab.2.
Thirdly, index-contributing coefficient matrix $C_{3\times n}$ for practical evaluation indexes was carried out by equations (4)~(7). Results are shown in Tab.3.
Finally, according to equation (10), evaluation results and corresponding radar chart of comprehensive competitiveness without ICC are shown in Tab.4 and Fig. 2(a). And, evaluation results by equations (9)~(11) are shown in Tab.5 and Fig.2(b).
Comparing Tab.1 and Tab.2, on the one hand, the practical evaluation index X2 has the largest weight of 0.18. Values of index X2 from island sites are far bigger than that from building sites and regional sites. Moreover, weights of indexes X5 and X8 are relatively higher. They are 0.16 and 0.14 respectively. Average values of X5 and X8 from island sites are respectively not less than or only slightly smaller than that from the other two types of DESSs. On the other hand, in terms of the building sites and the regional sites, average value of X4 with a relatively higher weight from regional sites is even up to more than four times of that from building sites.
As shown in Tab.4 and Fig.2(a), evaluation results of island sites are always larger than that of regional sites, which are always greater than that of building sites. This may lead impractical results that DESS with lower comprehensive competitiveness from island sites may have larger evaluation result than that of the DESS with higher comprehensive competitiveness from other types.
To avoid the above problems and improve the evaluation rationality of comprehensive competitiveness from each type of DESSs, EWM was used to introduce the ICC. Results of ICC are shown in Tab.3. Through the horizontal comparison of ICC of each index, X1 and X4 have higher contribution degree to comprehensive competitiveness. Furthermore, objective contribution coefficient of index X2 and index X5 are smaller than that of index X1 and index X4. This helps solve the problem that evaluation results from island sites are always higher than that from the other two types. Thus, objective ICC based on the EWM reduces the impact of pure subjective weights calculated by AHP. Evaluation results of comprehensive competitiveness are restrained and corrected by objective ICC, which improves the objective rationality of evaluation results.

Table 2. Weights of Indexes Based on AHP

No.	X1	X2	X3	X4	X5	X6	X7	X8	X9
Weight	0.0905	0.1899	0.0787	0.1044	0.1684	0.1203	0.0688	0.1484	0.0607

Table 3. The index-contributing coefficient(ICC)

No.	X1	X2	X3	X4	X5	X6	X7	X8	X9
L	0.1233	0.1040	0.0969	0.1233	0.0969	0.1233	0.1040	0.0969	0.1316
Q	0.1063	0.1018	0.1081	0.1063	0.1081	0.1063	0.1018	0.1081	0.1532
H	0.1356	0.0993	0.0987	0.1356	0.0987	0.1356	0.0993	0.0987	0.0985

Table 4. The Evaluation Results of Comprehensive Competitiveness without ICC

No.	1	2	3	4	5	6	7
L	0.3116	0.3417	0.3997	0.3349	0.4066	0.3417	0.4042
Q	0.4766	0.4864	0.5042	0.4739	0.5034	0.4694	0.5043
H	0.5442	0.5448	0.6090	0.5484	0.6270	0.5574	0.6315

Table 5. The Evaluation Results of Comprehensive Competitiveness with ICC

No.	1	2	3	4	5	6	7
L	0.4414	0.5029	0.6131	0.4476	0.6684	0.5029	0.6495
Q	0.5367	0.4852	0.5786	0.4331	0.5756	0.4144	0.5794
H	0.4365	0.5054	0.6267	0.5127	0.6634	0.5311	0.6726

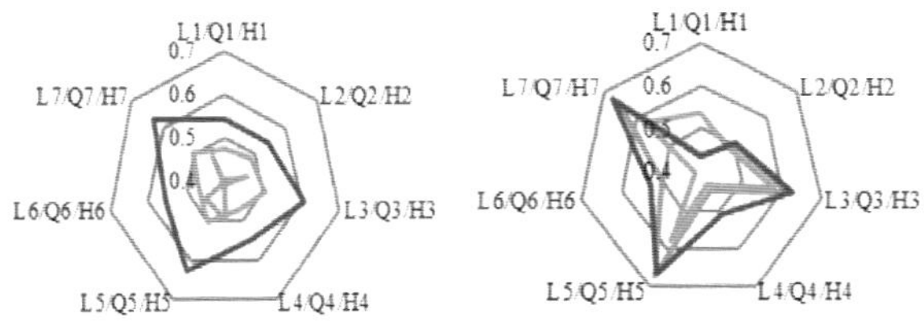

(a)without ICC (b) with ICC

Figure 2. Radar chart of evaluation results.

Eventually, as in Tab.5 and Fig.2(b), brief summaries are listed as follows:

(1) From the microscopic classification perspective:

1). Evaluation results of comprehensive competitiveness from building sites {L5, L7}, regional sites {Q3, Q5, Q7} and island sites {H5, H7} are larger in their respective types. These DESSs have higher reference values for operation and maintenance management of respective type of DESSs.

2). Evaluation results of comprehensive competitiveness from building sites {L1, L4}, regional sites {Q4, Q6} and island sites {H1} are smaller in their respective types. Appropriate measures should be taken to rectify the management of operation and maintenance to improve the competitiveness in time.

(2) From the macroscopic overall perspective:

Introduction of the ICC solves the problem that evaluation results are excessively concentrated by the type. Taking the Q1 and H1 as an example, before introducing the ICC, evaluation results are $M_{H1} > M_{Q1}$; but after introducing the ICC, evaluation results are just the opposite $M_{H1} < M_{Q1}$.

6. Conclusion

The PCEA is intended to address the quantification problem in investment and construction planning of DESSs. Innovation points and main research contents can be concluded as follows:

Practical evaluation index system closely corresponds to different aspects of comprehensive competitiveness. Classification framework is established under the consideration of developing features. Indexes and classification results can be applied to classification evaluation of DESSs AHP efficiently applies evaluation indexes to comprehensive evaluation of DESSs. By introducing ICC based on EWM, the issue that evaluation results are excessively concentrated by the type can be solved. Hence, approach proposed in this paper will provide valuable evaluation results of DESSs.

In summary, on the one hand, PCEA takes full consideration of differences among different types of DESSs, on the other hand, evaluation results can not only provide competitive DESS model for planning, but also discover low competitive DESSs timely. Rationality and potential promotional value of PCEA are verified by case results of DESSs in southern part of China.

7. Acknowledgment
This work was supported by the Natural Science Foundation of Guangdong Province (2016A030313476).

8. References
[1] RD Mcwhinney, N Rastogi, "A Novel Optimal Operational Strategy for the CCHP System Based on Two Operating Modes,"Power System, vol. 27, no. 2, pp. 1032-1041, May. 2012.

[2] M Liu, Y Shi, "Load Forecasting and Operation Strategy Design for CCHP Systems Using Forecasted Loads,"IEEE Trans on Control Sys Tech, vol. 23, pp. 1672-1684, Sep.2015.

[3] Y. T. Ge, S. A. Tassou and I. Chaer, "Performance evaluation of a tri-generation system with simulation and experiment," Appl. Energy, vol. 86, no. 11, pp. 2317–2326, Nov. 2009.

[4] H. B. Ren and W. J. Gao, "A MILP model for integrated plan and evaluation of distributed energy systems," Appl. Energy, vol. 87, no. 3, pp. 1001–1014, Mar. 2010.

[5] Heejin. Cho, "Evaluation of CCHP systems performance based on operational cost, primary energy consumption, and carbon dioxide emission by utilizing an optimal operation scheme," Appl. Energy, vol. 86, no. 12, pp. 2540–2549, Dec. 2009.

[6] Lahdelma R, Hakonen H, "An efficient linear programming algorithm for combined heat and power production," Eur J Oper Res,vol. 148, no. 1, pp. 141-151, Jul. 2003.

[7] Damiana Chinese, Antonella Meneghetti . "Optimisation models for decision support in the development of biomass-based industrial district-heating networks in Italy," Appl Energy, vol. 82, no. 3, pp. 228-254, Nov. 2005.

[8] Mancarella P, "Global and local emission impact assessment of distributed cogeneration systems with partial–load models," Appl Energy, vol. 86, no. 10, pp. 2096-2106, Oct. 2009.

[9] Chen H, Ooka R, Iwamura K, Huang H, and Yoshizawa N, " Study on sustainable redevelopment of a densely built-up area in Tokyo by introducing a distributed local energy supply system," Energy Build, vol. 40. no. 5, pp. 782-792, 2008.

[10] Ruan Y, Liu Q and Zhou W, "Optimal option of distributed generation technologies for various commercial buildings," Appl Energy, vol. 86, no. 9, pp. 1641-1653, Sep. 2009.

[11] P Dehghanian, M Fotuhi-Firuzabad, and S Bagheri-Shouraki, "Critical Component Identification in Reliability Centered Asset Management of Power Distribution Systems Via Fuzzy AHP,"IEEE Syst Jour, vol. 6, no. 4, pp. 593-602, Nov 201

Evaluation of Ancillary Service Market Mechanism to Facilitate High Penetration Renewable Integration in China

Wang Caixia[1], Lei Xuejiao[1], Li Qionghui[1], Sun Yong[2], Cao Zheng[2], Li Baoju[2]

[1]Department of Renewable Energy and Statistics, State Grid Energy Research Institute, Beijing, China
[2]Department of Power Dispatch and Control, State Grid Jilin Electric Power Supply Company, Changchun, China

e-mail: wangcaixia@sgeri.sgcc.com.cn, caozheng@jl.sgcc.com.cn

Abstract. Renewable energy curtailment is increasingly a critical issue in China. Mechanism barriers are becoming a key constraint for renewable energy integration, in which downward regulation mechanism that incentivizes conventional generators to be dispatched down is playing a key role in some regions, such as Northeast China. Pilots of such market mechanism, i.e. downward regulation ancillary service mechanism, have already been initiated in Northeast China. This paper evaluates the impact of downward regulation ancillary service mechanism in power systems with high penetrations of renewable energy. Firstly, the current situation of renewable energy development in China is analyzed. Secondly, the current practice of downward regulation ancillary service market mechanism in China are summarized and analyzed. Thirdly, the evaluation methodology as well as the case study system are described. Fourthly, simulation results are presented and analyzed. Finally, implications and suggestions are proposed.

1. Introduction

By the end of 2016, the cumulative installed capacity of wind and solar generation in China reached 226.06 GW, which accounts for 14% of the total generation capacity. The total amount of energy generated from wind and solar in 2016 was 307.2 TWh, which accounts for 5% of the total energy being generated [1]. Wind and solar generation has turned to the second largest source of electricity generation in 16 provinces[2]. The proportions of the wind and solar generation capacity in Gansu, Ningxia, Xinjiang and Qinghai are over 30%. However, renewable energy integration has been a critical issue in recent years . In 2016, the amount of wind curtailment in China hit a record height, reaching 49.7 TWh [3]. The wind curtailment rate in Gansu, Xinjiang, Jilin and Inner Mongolia are over 20%. Solar curtailment amounted to 7.04 TWh. Almost all the solar curtailment are concentrated in Northwest China, i.e. Gansu, Xinjiang, Qinghai, and Ningxia.

Mechanism barriers such as a lack of compensation mechanism for conventional generators that provide downward regulation services to wind and solar, inflexible prices, etc. are increasingly becoming key bottlenecks for renewable energy integration in China. Among them, downward regulation services are playing a key role. In this paper, the downward regulation services mean the services provided by the generators when they are dispatched down to make room for renewable energy. Due to a lack of compensation mechanism for conventional generators that provide downward

regulation services to wind and solar, conventional generators do not have any incentives to be dispatched down to help integrate more renewables, and thus partially resulted in the serious renewable energy curtailment issue.
On March 2015，CPC Central Committee and the State Council released *Opinions Regarding the Deepening of the Power Sector's Reform* (Paper #9), marking the official launch of the new-round power sector reform. One of the basic principles of the new electricity reform is to promote the increase of renewable generation and distributed energy systems in electricity supply. Proper market mechanism design is critical for alleviating renewable energy curtailment problem. In some parts of China, pilots of downward regulation ancillary service market have already been initiated [4]. Evaluation of the effectiveness of the market is of significance for its improvement and future establishment of a mature electricity market that facilitate renewable energy integration.
This paper is organized as follows. Part II analyzes the current practice of downward regulation ancillary service market in China. Part III describes the evaluation methodology as well as the case study system. Part IV presents the evaluation results. Part V concludes the paper with implications and suggestions.

2. Current Status of Downward Regulation Ancillary Service Market in China

It is worth to note that downward regulation service is a unique concept of power system dispatching in China. In power systems in an electricity market context, these services are provided by the spot markets, and are not in the arena of ancillary services. Since there is no mature electricity market, especially there is not a spot market in China, the downward regulation services are classified as one type of ancillary services and incentives for these services commonly seen in the documents of "grid-connected power plant ancillary services management "in different provinces and regions [5][6]. In these documents, downward regulation services, named compensated deep downward regulation, are classified as one type of ancillary services. More specifically, in these documents, there is a benchmark downward regulation service, which differs region by region and generally is 50% of the maximum generation capacity of a generator. If the generation is called upon to provide downward regulation services with the output of the generator being above the benchmark, these services are free. If the output of the generator is below the benchmark while providing these services, the amount of energy below the benchmark will be compensated.
With high penetration of renewable energy in the system, more frequent and deeper downward regulation services are needed from conventional generators, as in many wind intensive areas in China, wind blows hard during the mid-night when there is low electricity demand in the system. The downward regulation compensation specified for a non-renewable energy or low renewable energy power system is not enough to incentivize conventional generators. For example, in Northwest China, the amount of energy below the benchmark regulation services are paid 0.30RMB/kWh, while in North China, they are paid only 0.05 RMB/kWh. This greatly limits the system's downward regulation capability.
With the advancement of China's new round of power market reform, some areas are actively carrying out market explorations to unlock power system flexibility, which help renewable energy integration. One of the most notable trial is the ancillary service market in Northeast China, which started in October 2014 and improved in 2016. In Northeast China, a market mechanism is introduced in procuring downward regulation ancillary services from thermal generators. Instead of setting up a fixed compensation rate by the government beforehand, thermal generators can bid for their capabilities to be dispatched down and they will be compensated based on the clearing price. The more they can be dispatched down, the higher compensation rate they will get. The bidding cap for downward regulation and cycling are shown in Table I and Table II.

Table 1. Bidding Cap and Floor for Downward Reguation

Period	Type	Loading rate	Bid Floor (RMB/kWh)	Bid Cap (RMB/kWh)

Non-heating	Non-CHP	40%<R<=50%	0	0.4
	CHP	40%<R<=48%		
	All	R<40%	0.4	1
Heating	Non-CHP	40%<R<=48%	0	0.4
	CHP	40%<R<=50%		
	All	R<40%	0.4	1

Table 2. Bidding Cap for Cycling

Capacity (MW)	Bidding cap per cycling (10^3 RMB)
100	500
200	800
300	1200
500～600	2000
800～1000	3000

Besides Northeast China, other regions in China are also in the process of initiating downward regulation ancillary service market pilots. For example, Shandong Province in May 2017 issued the "Shandong Electricity Ancillary Service Market Operating Rules (draft)" [7]. The main idea is also introducing a market mechanism to incentivize downward regulation services from thermal generators, similar to that of Northeast China. In Shandong, the trans-regional interconnections are considered as a provider of downward regulation services. The details on bidding intervals and cost allocation are different from Northeast China.

3. Evaluation Methodology and Case Study System

3.1. Evaluation Methodology

In this paper, an annual production simulation is used to evaluate the performance of downward regulation ancillary service market to facilitate renewable energy integration. The production simulation model is built upon least cost unit commitment in the paper. The model includes a long-range unit-commitment for the scheduling of the generating units and a economic dispatch model to calculate system performance indices, such as annually system operation cost, wind curtailment rate, generation cost per kWh, etc.

In the long-range unit-commitment, the generators are scheduled for a period of 7 days (the optimization horizon), after which the scheduling period rolls forward by 5 days (the rolling step). There are 2 days of look ahead, that is, there is an overlap of 2 days between each consecutive optimization horizon. This overlap gives the model some extra foresight to account for the reduction in demand that typically occurs at weekends. Also, if there was no overlap, an excessive number of units would be shut down towards the end of each optimization horizon. With respect to system operation in practice, long-range scheduling such as this facilitates fuel purchasing and contracts, as well as generator maintenance scheduling.

(1) Objective function

The objective function is the sum of generation cost $C_{Gen}(t)$ and load curtailment penalty cost $C_{LC}(t)$, for all the time steps in the scheduling horizon.

$$f_{obj}=\sum_{t\in T} C_{Gen}(t)+\sum_{t\in T} C_{LC}(t) \tag{1}$$

The generation costs $C_{Gen}(t)$, for each time step is given in Eq. (2). It includes three parts, i.e. startup cost, no load cost and incremental cost. For no load cost and incremental cost, both fuel cost and CO_2 emission cost are included. U_I is the group set of all generators, which are grouped according to rated capacities, and u_I is index for members of U_I. The decision variables $V_{start}(u_I,t)$, $V_{online}(u_I,t)$, and $V_{gen}(u_I,t)$, correspond, respectively, to whether the unit started, whether it was online, and its level of production for that hour.

$$C_{Gen}(t)=\sum_{u_I\in U_I}\begin{pmatrix} c_{start}(u_I)\cdot V_{start}(u_I,t)+ \\ c_{noload}(u_I)\cdot V_{online}(u_I,t) \\ +c_{incr}(u_I)\cdot V_{gen}(u_I,t)\end{pmatrix} \tag{2}$$

The load curtailment penalty cost $C_{LC}(t)$, for each time step is given in Eq. (3). It is calculated with value of lost load c_{voll} and the amount of load curtailment $V_{loadcurt}(t)$, which is a decision variable.

$$C_{LC}(t)=c_{voll}\cdot V_{loadcurt}(t) \tag{3}$$

(2) Main constraints

Power balance:

$$\sum_{u_I\in U_I}V_{gen}(u_I,t)+P_{wind}(t)-V_{windcurt}(t)+V_{loadcurt}(t)=P_{dem}(t),\forall t \tag{4}$$

The decision variables $V_{gen}(u_I,t)$, $V_{windcurt}(t)$, and $V_{loadcurt}(t)$ correspond respectively to the level of production, wind curtailment and load curtailment at time step *t*. Wind curtailment and load curtailment are used as slack variables in the model for some instances when they are necessary to satisfy the constraint. $P_{wind}(t)$ is wind power forecast at time step *t*, and $P_{dem}(t)$ is load at time step *t*.

Reserve target:

$$P_{res\_up}(t)\le\sum_{u_I\in U_I}\left(V_{online}(u_I,t)\cdot P_{\max}(u_I)\right)-\sum_{u_I\in U_I}\left(V_{gen}(u_I,t)\right)+V_{windcurt}(t),\forall t \tag{5}$$

$$P_{res\_dn}(t)\le\sum_{u_I\in U_I}\left(V_{gen}(u_I,t)\right)-\sum_{u_I\in U_I}\left(V_{online}(u_I,t)\cdot P_{\min}(u_I)\right)+P_{wind}(t),\forall t \tag{6}$$

Equations (5) and (6) are upward and downward reserve respectively. $P_{res\_up}(t)$ and $P_{res\_dn}(t)$ are respectively upward and downward reserve target. Wind power is allowed to provide reserve in both the reserve constraints ($V_{windcurt}(t)$ and $P_{wind}(t)$ in Equations (5) and (6)). At high wind, for example, wind can be used as downward reserve, or having been curtailed, can be used as positive reserve.

Minimum and maximum output:

$$\sum_{u_I\in U_I}\left(V_{gen}(u_I,t)\right)\le P_{\max}(u_I)\cdot N(u_I),\forall t \tag{7}$$

$$\sum_{u_I\in U_I}\left(V_{gen}(u_I,t)\right)\ge P_{\min}(u_I)\cdot N(u_I),\forall t \tag{8}$$

$$V_{gen}(u_I,t)\ge V_{online}(u_I,t)\cdot P_{\min}(u_I),\forall t,u_I \tag{9}$$

$$V_{gen}(u_I,t)\le V_{online}(u_I,t)\cdot P_{\max}(u_I),\forall t,u_I \tag{10}$$

$P_{\max}(u_I)$ and $P_{\min}(u_I)$ are respectively the maximum and minimum production of generator group u_I. And $N(u_I)$ is the generator number in group u_I.

Unit start identification:

$$V_{start}(u_I,t) \geq V_{online}(u_I,t)\text{-}V_{online}(u_I,t\text{-}1), \forall t,u_I \quad (11)$$

3.2. Case Study System

(1) Case system description

A test system with generators and electric demand from a provincial power system in Northeast China has been used. For load data, measured load data for the provincial power system in Northeast China is used. The total installed capacity of coal-fired generation is 12.9 GW. For the generator data, the case region's generator parameters are shown in Table III. In Northeast China, the large proportion of CHP units is a critical limiting factor for wind power integration. Due to the heat demand, the minimum electrical outputs of the generators are higher. As this paper mainly discuss the performance of downward regulation ancillary service markets, the minimum output of all the generators are assumed to be 60% throughout the year for simplicity. Future studies on unlocking the flexibility of heat generation will be conducted.

Table 3. Information on Coal-Fired Generators in the System

Type	Cap(MW)	Number
Coal-XL	600	5
Coal-M1	300	5
Coal-M2	300	10
Coal-M3	300	4
Coal-S1	200	6
Coal-S2	200	11
Coal-XS	100	8

(2) Simulation scenarios

Three set of scenarios are simulated in the paper. The first set of scenarios (Scenario Set I) assumes no downward regulation ancillary service market in the system, and varies wind capacity in the system, to show the system performances with different wind power penetrations in the system, as shown in Table IV.

The second (Scenario Set II) and third (Scenario Set III) set of scenarios assumes downward regulation ancillary service market to incentivize deeper downward regulation in the system. In Scenario Set II, different amount of 600MW generators in the system are considered to be flexible, i.e. bid to provide deeper downward regulation. It is assumed that these flexible generators can reduce their minimum output to as low as 30% of their maximum output. In Scenario Set III, different amount of 300MW generators in the system are considered to be flexible, and their minimum outputs are assumed to be 30% as well.

The heat rate curves of the generators are given in the simulation, which shows higher heat rate, thus higher fuel consumption and generation cost per kWh of electricity generated, with the decrease of the output level of the generators.

System performances of these scenarios are calculated with the production simulation model. They are compared and analyzed in Section IV. System performance indices include annual system operation cost, generation cost per MWh, wind curtailment rate, downward regulation cost, etc.

4. Simulation Results

4.1. System Performance with No Downward Regulation Ancillary Service Market

The variation of annually system operation cost, generation cost per kWh, wind curtailment rate, with different penetrations of wind energy are shown respectively in Fig.1 and Fig.2.

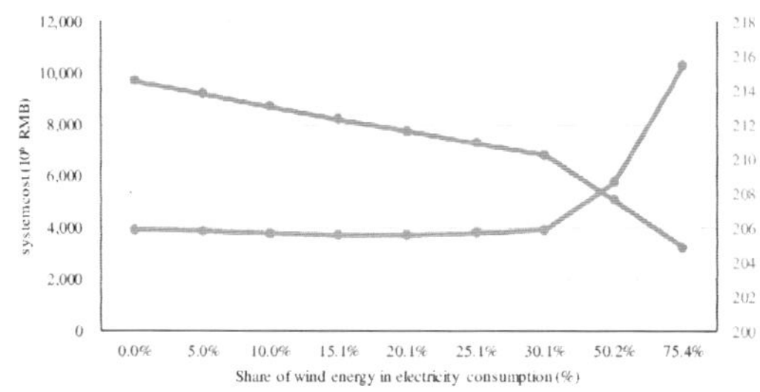

Figure 1. Annually system operation cost and generation cost per MWh with different penetrations of wind energy.

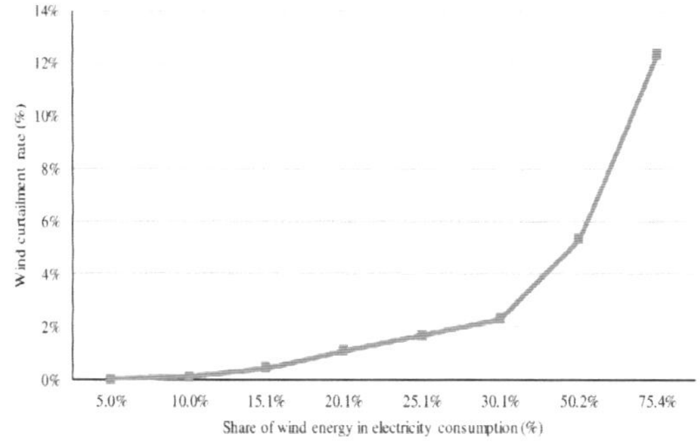

Figure 2. Wind curtailment rate with different penetrations of wind energy.

Results show that with increasingly penetration of wind energy, the annual system operation cost, which consists only the conventional generation cost decreases. However, the conventional generation cost per MWh decreases first, and then increases sharply at high wind penetrations. This is due to the increased fuel consumption rate of coal-fired generators at reduced output level. As the system regulation capability is limited, wind curtailment rate increases with increasing wind energy penetrations.

4.2. System Performance with Downward Regulation Ancillary Service Market

(1) When the 600MW generators in the system are providing downward regulation services (Scenarios Set II))

Fig.3 and Fig.4 shows the system cost and wind curtailment rate with increased number of flexible 600MW generators. System cost includes conventional generation cost and the downward regulation services cost induced in the system. Results show that when there are more flexible 600MW generators in the system, wind curtailment rate decreases, which indicates the effectiveness of downward regulation ancillary service markets in facilitating wind integration. This however, would lead to increased system cost, as the downward regulation services provided by the flexible 600MW generators need to be paid. This can be seen as the price paid by the system to integrate more renewable energy.

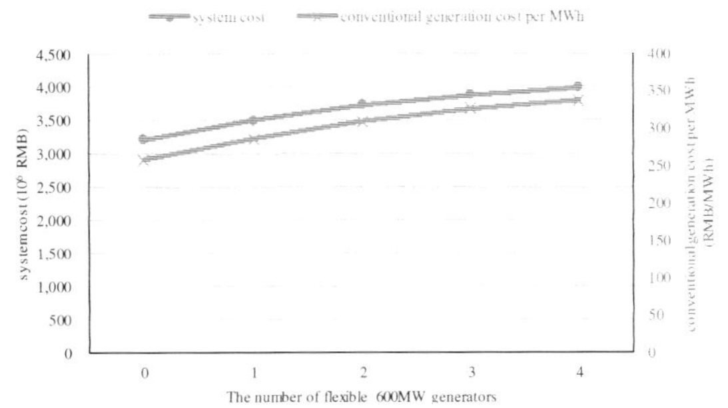

Figure 3. System cost with increased number of flexible 600MW generators.

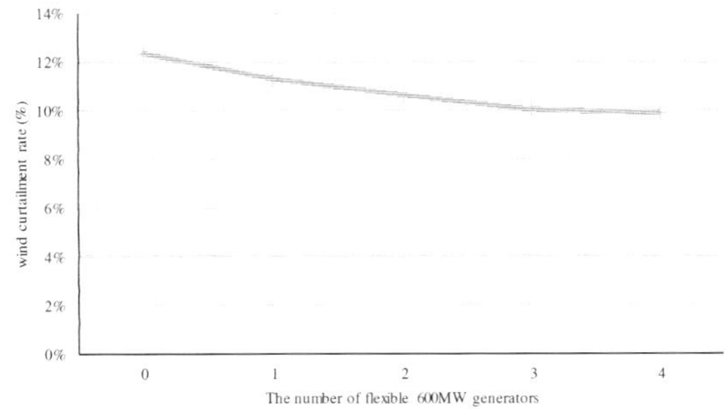

Figure 4. Wind curtailment rate with increased number of flexible 600MW generators.

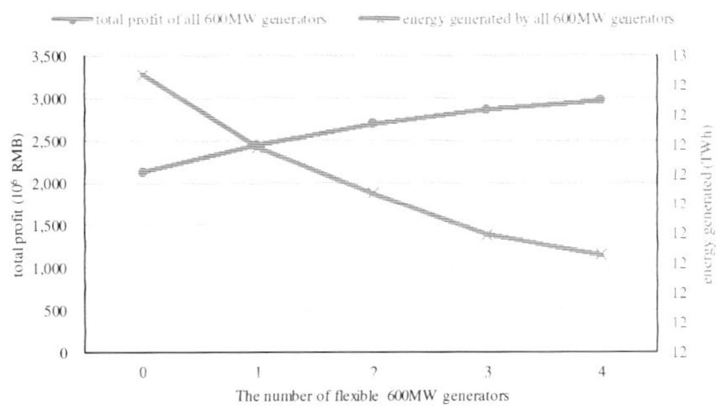

Figure 5. Total profit and energy generated by all 600MW generators.

Fig.5 shows the energy generated by all the 600MW generators and their profit. As shown, when the 600MW generators, which are the least cost generators in the system become more flexible, their income from electricity generation will decrease but they will have extra income from providing the flexible services, which finally leads to increased profit.

(2) When the 300MW generators in the system are providing downward regulation services (Scenarios Set III))

Although system cost and wind curtailment rate present similar patterns when the 300MW generators are providing flexible services, but the impact of such flexibility on 300MW generators are different, as both their profit and energy generation increased. as shown in Fig. 6. The increased generation is due to their flexibility even though their generation cost is high compared to 600MW generators.

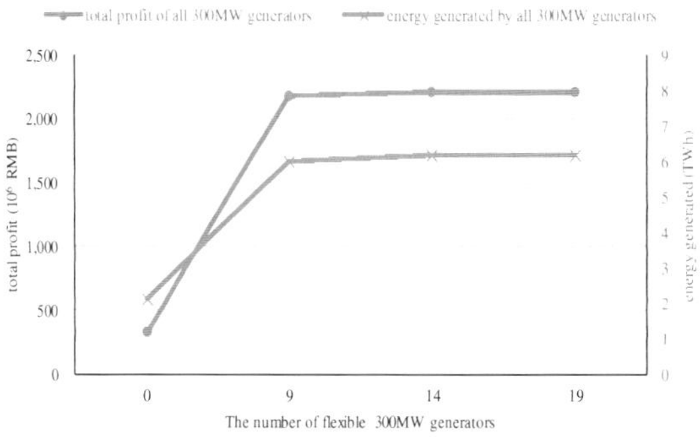

Figure 6. Total profit and energy generated by all 300MW generators

5. Conclusions

Mechanism barriers are increasing becoming a constraint in renewable energy integration. The new round of power sector reform provides opportunities for solving this issue. Pilots of downward regulation ancillary service market have already been initiated in some parts of China. Evaluation of the effectiveness of such a market is of significance for future electricity market development. This paper conducts studies on the performance of the current downward regulation ancillary service market. Results show that downward regulation ancillary service market that incentivizes deeper downward regulation of conventional generators help renewable energy integration. The system cost, however, will increase as the regulation services provided by the conventional generators need to be paid. For conventional generators themselves, increasing the flexibility will make them more profitable. The change of profit differs for different generators. For generators with low fuel rate, which are considered more efficient generators, increasing their flexibility reduces the energy they generate, but their total profit will still increase due to the extra profit brought by the ancillary service market. For generators with high fuel rate, increasing their flexibility will increase their generated energy as well as their profit. This indicates that the value of flexibility are rewarded in the ancillary service market. Judging from the standard that a good market design should reflect system needs as well as reward those providing these services, the downward regulation ancillary service market is playing a positive role in facilitating renewable energy integration.

It is worth to note that the downward regulation ancillary service market is a specific product of the current market transition period in China, i.e. transitioning from a non-market system to an electricity market. In the future, spot markets should be established to facilitate renewable energy integration.

6. Acknowledgment

The work is supported by the research project-Research and Development of Renewable Energy Priority Exchange Technology in the Context of Electricity Market (SGLNSY00FZJS1601337) of State Grid Corporation of China, and the research project-Study on the Scheme of High Penetration Renewable Energy Integration Pilot Projects funded by Energy Foundation (G-1512-24143). The authors thanks State Grid Corporation of China, State Grid Shenyang Electric Power Supply Company, and the Energy Foundation for funding the project.

7. References

[1]. China Electricity Council. National power sector statistics 2016, 2016.

[2]. State Grid Corporation of China. "White paper on facilitating renewable energy development and integration". 2016.

[3]. http://www.nea.gov.cn/2017-01/26/c_136014615.htm. The status of wind operation in 2016. January 2017.

[4]. Pilot scheme of ancillary service market in Northeast China. Northeast China Energy Regulatory Bureau. November 2016.

[5]. Grid-connected power plant ancillary services management. Northwest China Energy Regulatory

IOP Conf. Series: Earth and Environmental Science **168** (2018) 012042 doi:10.1088/1755-1315/168/1/012042

Bureau. September 2015.
[6]. Grid-connected power plant ancillary services management. North China Energy Regulatory Bureau. September 2015.
[7]. Pilot scheme of ancillary service market in Shandong Province (draft). Shandong Energy Regulatory Bureau. May 2017.

Blockchain Technology Hyperledger Framework in the Internet of Energy

Dongpo Zhang[1], Zhenyuan Zhang[1], Luan Chen[1], Shu Li[1], Qi Huang[1], Yihong Liu[2]

[1]School of Energy Science and Engineering, University of Electronic Science and Technology of China, Chengdu, China
[2]China Academy of Space Technology, Beijing, China
e-mail: 751583778@qq.com

Abstract. Internet of Energy, a combination of power grid and Internet, contains many energy forms. Different form of energy is distributed in Internet of energy, which means that it is closely related to distributed technology. Different forms of energy needs to interconvert, so it is necessary to make power trading in this distributed system, the current power trading model is not adaptive to distributed power trading, so a new distributed trading model is important to Internet of energy. Blockchain technology with distribution and decentralization features is widely used in electronic currency transactions, Hyperledger-Fabric is the one of typical framework of blockchain, which could provide a platform for distributed ledger, it has distributed blocks and ledgers to make decentralized trading by using blockchain technology for Internet of energy.

1. Introduction

Energy is the foundation of social operation, the reform of Energy is driving social development [1]. Internet of energy (IOE) is a significant energy reform combining energy with Internet in the information age, which is a new energy system that deeply integrates the energy system and the Internet, where openness is the key idea and deep integration of the Internet ideas and technologies is the key feature [2]. IOE can increase the development of new energy because it is totally based on varieties of energy integration. In addition, it adapts to new energy access and improves energy efficiency [3]. IOE is different to traditional power grid in user part, the traditional power grid users in the IOE are not simple consumers, and they can also be producers when new energy and power storages are widely used [4], so IOE can provide intelligent demand response to the user side. But these above features greatly increase the complexity of the IOE, it is a distributed system with various energy forms, different form of energy need to interconvert, it also means users need make distributed trading to constitute the energy network [3], a central model can not meet the diverse needs of the IOE, Moreover, it faces challenges because existing trading model lack the method of distributed power trading, it urgently requires a operation model [4].

The blockchain is a decentralized distributed accounting technique, which is gradually becoming popular with the boom of Bitcoin, which is a special currency that relies on encryption algorithms [6]. In essence, blockchain can be understood as an open general ledger based on a computer program, it records all transactions which is occurred on blockchain system [7]. Each node in the system can update its recorded data to the network. Each of participants can copy a complete data from blockchain container, which constitutes a decentralized distributed data to achieve the distributed trading and interaction [8].

This emerging technology can fundamentally change the trading network links, it can solve bottleneck and reduce costs and risks compared with the traditional center system .As a new-type decentralized trading technology, the blockchain is widely used in the financial field [9], and its decentralized transaction model is consistent with the IOE, the decentralized trust mechanism provided by the blockchain satisfies no third-party transaction in the IOE model, the information in the IOE will be shared for each users, this way of trading could reduce transaction costs and improves power trading efficiency and precision, The information in blockchain can not be tampered, it can ensure the accuracy of power measurement and reduce the power metering mistake[10]. However, Power trading is more complicated than financial transactions, the customers in financial field only make financial transactions by information interaction, but power trading in IOE includes information interaction, energy form conversion, power dispatching. Although blockchain technology is compatible to the distributed trading, a framework of blockchain which is compatible to the power trading is needed.

This paper discusses a blockchain framework in power trading for the Internet of energy, analyzes the advantages of decentralized trading, moreover, this paper explains the process of hyperledger which is used to power trading for the Internet of Energy.

2. Centralized and Decentralized Power Trading

2.1. Centralized Power Trading Model

The traditional power trading model be shown in Fig. 1, power trading system as the center of the transaction, it has a high authority in the transaction structure, each unit must trust the power trading system and all transactions need to pass the center of the certification, user is a node of the center in this framework. There are some disadvantages in centralized power trading model, firstly, every user must trust this center, and this system will crash if the power trading system gets error. In addition, users are isolated in this centralized model, the transaction line must pass the power trading system if one user want make power trading which could cause wasting of time and resource .

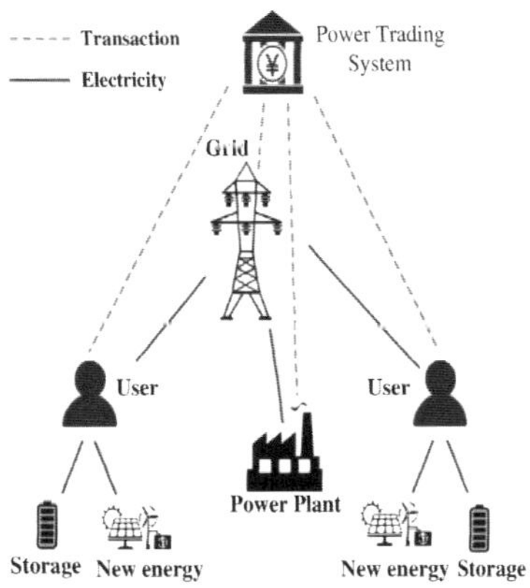

Figure 1. Example of a centralized power trading model.

2.2. Decentralized Power Trading Model

The decentralized power trading model of the structure is shown in Fig. 2, it shows that a blockchain system with decentralized construction. Users are important part of the blockchain system rather than node of center, all trading are maintained by each user to ensure the blockchain system has the characteristics of no repudiation and decentralization. Therefore, the benefits of every user can be protected. And in this model, Transaction line can connect user to user, it is a peer to peer model to be compatible with the power trading in the internet of energy.

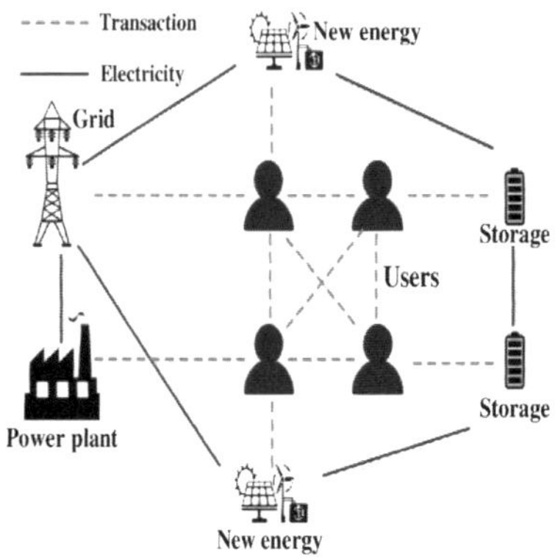

Figure 2. Example of decentralized power trading model.

3. Hyperledger-Fabric and Power Trading.

3.1. Blockchain Framework in Power Trading.

Centralized power trading requires accreditation of central authority, but in blockchain system, there is not a central authority. Every node of this blockchain system will preserve all of trading information. The information will write in block, once the written process is done, the block information will be the unique record and cannot be rewrite anymore. The block will synchronize with blockchain container to ensure the transactions are reliable. Power dispatch system will execute dispatching after trading success and sync success. The blockchain framework in Power Trading is shown in Fig. 3.

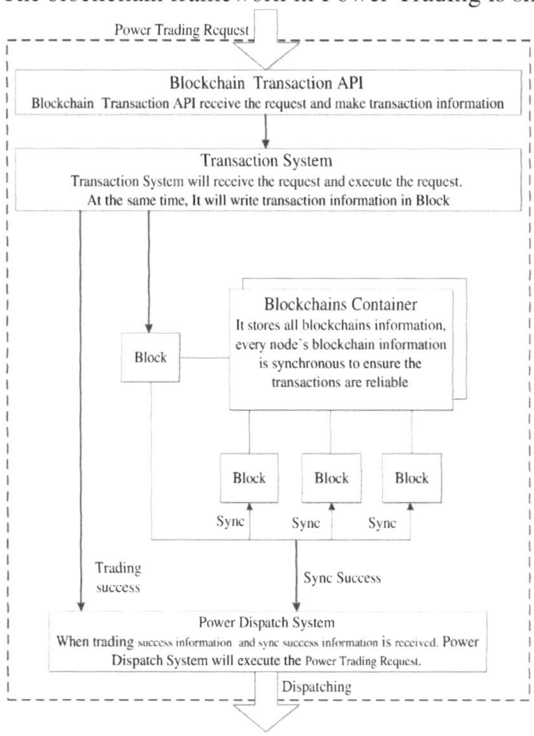

Figure 3. Blockchain framework in power trading.

When power trading request is called, the blockchain transaction API will receive this request and process it, if the request is available, transaction information will be sent to Transaction System, Transaction System will receive the request and execute the request, at the same time, it will write transaction information in Block. When one block is writing, the writing information will sync other block of blockchain container, when the process of sync is complete, power dispatch system will receive the status and execute the power trading request.

3.2. Hyperledger Framework in the Internet of Energy.

Hyperledger-Fabric is a collaborative project under the management of the Linux Foundation. The Hyperledger project is an IBM open source blockchain project with a good community and ecosystem. Unlike some traditional blockchain technology such as Ethereum, the Hyperledger-Fabric does not set any tokens in itself. The most famous token in blockchain is the Bitcoin which is the first generation of blockchain applications. The feature of Bitcoin is using the Proof of Work (POW) mechanism to ensure that the entire blockchain system balance and stability. The users through the computer terminal compute equations to get the token of the special reward. The special solution of this system is limited; it means that the total amount of tokens is certain. The difficulty of the equation will increase with the reduction of the remaining token. Generating a token on each meter requires other users to agree. Unlike most currencies, Bitcoin does not rely on a specific currency institution, which is generated by a large number of calculations based on a particular algorithm. The bit currency economy uses a distributed database of many nodes in the entire P2P network to confirm and record all transactions, and the use of cryptographic design to ensure that all aspects of currency circulation security.

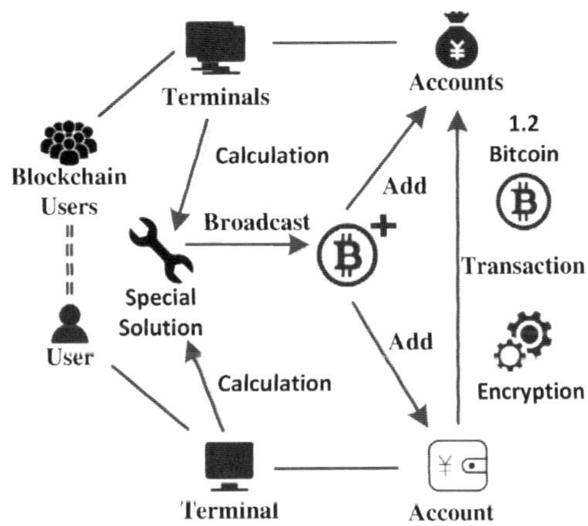

Figure 4. Transaction model of bitcoin.

Fig. 4 is a transaction model of Bitcoin; the user calculates the equations and gets some tokens. These tokens can be traded by encryption methods, the generation and transaction of each token requires the members of the system to consensus agreement. Each user has an account to store tokens. There is not exist one user has more computing power than all other users. Therefore, the system is maintained by everyone in the system. The stability of the encryption algorithm ensures the reliability of the transaction and the user's consensus on all transactions ensures that the transaction record can not be changed.

But energy can not be well stored such as electricity, and the energy generation is not dependent on the solution equation. However, the blockchain can meet the precise measurement of energy Internet and transaction traceability and other characteristics, so the energy of the blockchain application does not require setting tokens. So a distributed ledger technology based on the blockchain enables the energy Internet to be combined well. Hyperledger is a distributed ledger system which is a framework

IOP Conf. Series: Earth and Environmental Science **168** (2018) 012043 doi:10.1088/1755-1315/168/1/012043

of blockchain technology; it uses Kafka as its consensus mechanism to create a transparent and non-tampering distributed ledger. The core module of the Hyperledger-fabric runs in the Docker container, Docker is an open platform for developers to build and run distributed applications, it also is an open source application that allows developers to package their applications to a portable container. Docker is published to any popular Linux machine, it can also achieve virtualization. The container is completely using the sandbox mechanism, and there is no contact between them.

This paper designs a Hyperledger-based power trading architecture, NodeSDK unit, a tool to operate Hyperledger, is used to interact with the terminal and other units, Dispatch unit for the implementation of physical transactions, chain-code is used to interaction with Blocks and Ledgers, CouchDB, a distributed database, is used to a database query for Blocks value.

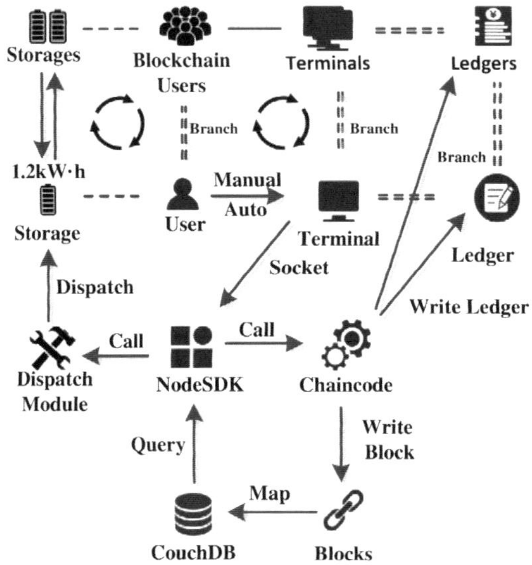

Figure 5. Power trading model with hyperledger.

The process of energy transaction with Hyperledger is shown in Fig. 5, when one of the users in the blockchain system want to make a trading, this user can log in through the blockchain system terminal, manually select or through the Chiancode automatically select appropriate transaction. when the transaction has submitted, the transaction information will be sent to NodeSDK unit. And then NodeSDK will analyse transaction data submitting to the chain-code and initiate the transaction, after the peers and orderer complete the proccess of calculation, kafka consensus, and broadcast, chain-code will write block and ledger. The information in Blocks is mapped to CouchDB in the form of key and value for user query. When the block write is complete, the NodeSDK unit will call the Dispatch unit, it will carry out electrical control to complete the dispatching, to achieve power trading. At the same time, the transaction information will be recorded by all ledgers. In this blockchain system, each user can trade between users, so there is not an authoritative, centralized trading platform. The trust mechanism is maintained by every user within the system, blockchain as a network of information between users of the network, constitute a network of mutual trust and communication. Chain-code can make the transaction automation and intelligence. Thus, it can make the IOE intellectualized through use the smart chain-code. Public ledgers can provide a non-repudiation and non-tampering of the trust mechanism to ensure that the benefits of users within the Blockchain system.

3.3. Hyperledger-Fabric Transaction Process

In the transactions of energy, the IOE is a distributed system, it require a consensus trading model. Hypeledger provides a decentralized trading framework,in this framework, all transactions logs will be recorded. Every peer will join transactions. It can make the transactions secure and stable.

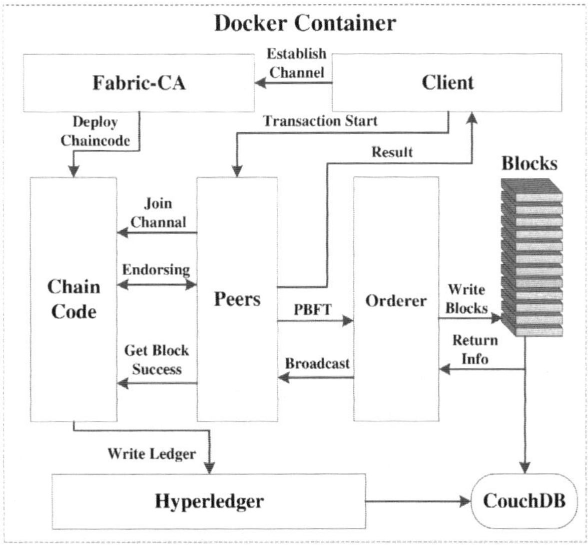

Figure 6. Trading process of hyperledger framework.

Transaction process of Hyperledger framework is shown in Fig. 6, the core module of the Hyperledger-fabric runs in the Docker container, there are some Docker images of Hyperledger including fabric certificate components Fabric-CA, Client is used to be a access gateway for users, chain-code is smart contract for trading, Peers are endorsement node, and Orderer is consensus node, Blocks is a block container of Hyperledger system, CouchDB is distributed databases of Hyperledger. When the system run, all peers will join the channel and deploy the chain-code to the Docker and initialize the chain-code. At this time, Docker images is up, And the Genesis block will be generated, genesis block is the first block of this system. In the subsequent transaction, the client initiates the transaction and sends the transaction request to each Peer. Peers receive the request and then perform the transaction separately through the chain-code, then peers return the result to the client, and submit the result to the Orderer, the Order processes the consensus and writes the result to Blocks. When the block is written successfully, Blocks will return the results to Orderer, and then, Order broadcast the results to each peer to complete the endorsement. After each peer verifies the transaction, Hyperledger will complete this transaction.

Hyperledger information is stored in CouchDB, the form of information is 'key and value', it provides users query API. When the transaction is carried out several times, once a new peer is added, the channel will be reestablished by the peers and the new peer will be deployed. The new image will generate new blocks, and the previous container will be saved to ensure the continuation of the transaction and the integrity of the previous transaction records.

Orderer mainly provides two Grpc interfaces: Broadcast and Deliver, two interfaces are used to client communication. The client sends a message to the Orderer via the broadcast, and the Orderer retrieves the client's message via Broadcast and Orderer will save this message. The client obtains data from the Order through the deliver interface.

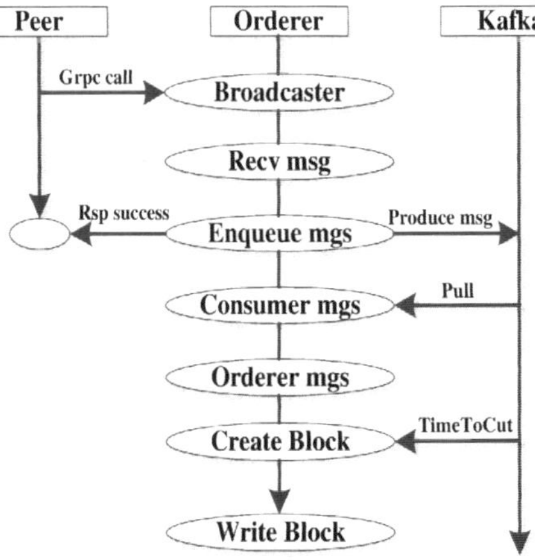

Figure 7. Broadcast process of hyperledger.

Broadcast Process is shown in Fig. 7. Peer initiates communication via GRPC, after the connection to Orderer is successful; it can send a message to Orderer. When Orderer receives the message sent by the Peer via the receive interface (Recv msg), Orderer will push the message to Kafka, When Kafka finish the consensus, it will pull the success message to Orderer, and Orderer will use chain-code to create or write blocks. In this framework, distributed transaction will be well processed. Distributed transaction is an important factor of distributed system. Therefore, it will suit the architecture of the Internet of Energy well.

4. Conclusion

Distributed power trading is an important part of Internet of Energy, and blockchain is a good method to solve the problem in the power trading. Hyperledger is a blockchain technology framework, Hyperledger use no-tokens model compared with Bitcoin so that it is more compatible for power trading, furthermore, Hyperledger has provided a decentralized and trustworthy trading model that uses consensus tools to make certification. This paper has introduced blockchain technology and Hyperledger, designs a framework in the Internet of Energy by using Hyperledger.

This trade model is accordant with the concept of energy interconnection and information interconnection. Therefore, it is a feasible method for the development of power trading in the Internet of Energy.

5. References

[1] Power Grid Technology 2015, 11 (39), 3005-3013 DOI: 10.13335 / j.1000-3673.pst.2015.11.001.

[2] Sun Hongbin, Guo Qinglai, Pan Zhaoguang. Energy Internet: concept, architecture and forward prospects [J]. Power System Automation, 2015,39 (19): 1-8. [2017-08-25].Yanping Zhang. SDN-enabled Platform for Energy-efficient Relay Protection Optical Port in IOE [A].Proceedings of 2016 6th International Conference on Information Engineering for Mechanics and Materials(ICIMM 2016)[C].

[3] Research on Framework and Typical Application in Energy Internet [J]. Proceeding of the CSEE, 2016, (15): 4011-4023. ZHANG Jian, WANG Yi, KANG Chongqing, CHENG Gengnan, HE Dawei.

[4] Cao Yin. Energy + block chain, a better future or theme speculation [J]. Electric Age, 2016, (09): 28-31.

[5] Chen Hao. Economic Analysis of Bitcoin [D]. Zhejiang University, 2015.

[6] Cheng Hua, Yang Yunzhi. Development trend of block chain and coping strategies of

commercial banks [J] .Journal of Financial Supervision, 2016, (06): 73-91.
[7] (School of Economics and Management, Nanjing University of Science and Technology, Nanjing 210009, China); Research on the Development of Financial Industry in Block Chain Driving [J]. Journal of Nanjing University of Posts and Telecommunications (Natural Science Edition), 2016, (05): 1-9.
[8] ZHOU Li-qun, LI Zhi-hua.Application of Block Chain in Supply Chain Finance [J]. Information Systems Engineering, 2016, (07): 49-51.
[9] Huang Jiehua, ultra high spirit, Xu Yuzhuang Bai Xiaomin, hu. The intelligent raise block chain contract design [J]. Journal of information security research, 2017, (03) : 211-219.
[10] Jun WANG, Peng WU, Xiangyu WANG, Wenchi SHOU. The outlook of blockchain technology for construction engineering management [J]. Frontiers of Engineering Management, 2017, (01):67-75.

IOP Conf. Series: Earth and Environmental Science **168** (2018) 012044 doi:10.1088/1755-1315/168/1/012044

Multi-level Fuzzy Statistics and Maximum Entropy Rule Based Comprehensive Evaluation Strategy for Development Level of Global Energy Interconnection

Xi Ye[1], Quan Tang[1],Weiting Xu[1], Xi Zhang[2], Xiaoling Jin[3], Xianzong Dai[3]

[1] State Grid Sichuan Economic Research Institute,Chengdu,Sichuan, China
[2]State Grid Chengdu Power Supply Corporation,Chengdu,Sichuan,China
[3]Energy Research Institute, State Grid, Beijing, China
Email: 35174586@qq.com, rocky_135@qq.com

Abstract. Global energy interconnection is the next development stage of future power grid. In order to evaluate its development level and identify the weakness spot in the developing process, a comprehensive evaluation strategy for development level of global energy interconnection is proposed in this paper. This strategy is proposed based on the fuzzy statistics method and maximum entropy rule. The effectiveness of this evaluation strategy is illustrated through calculation examples. The calculation results show that the development degree of global energy interconnection in China, United States and Germany has reached a very good level.

1. Introduction

With the rapid increase of renewable energy installed capacity, it is necessary to build a world's platform to achieve the unified development, allocation and accommodation of renewable energy. Therefore, the concept of global energy interconnection [1] is put forward. Global energy interconnection is a global energy allocation platform on the base of ultra-high voltage power gird and smart grid. The main task of global energy interconnection is to transmit renewable energy on a global scale. Global energy interconnection is the next development stage of future power grid, which conforms to the development trend of world energy and power grid.

The basic theoretical system of global energy interconnection has been established, including concepts, layout frame, development stage and main features. However, in order to identify the weakness spot in the construction process of global energy interconnection, the comprehensive evaluation method of its development level still needs further study.

Lots of work has been done focusing on the development level evaluation of smart grid. A maturity model for smart grid has been proposed by IBM [2]. The United States department of energy has developed a smart gird development evaluation index system, which summarizes six characteristics of smart grid's development level [3]. A revenue assessment system has been put forward by European Union, which summarizes the driving factors of smart gird development into three aspects: market, security and power quality, environment. The above research results have provided good foundation for the development level evaluation for global energy interconnection. However, global energy interconnection is a grand goal which involves complex and numerous influence factors. Therefore, current research results for smart gird cannot be used directly.

A comprehensive evaluation strategy for development level of global energy interconnection is proposed in this paper. Firstly, the hierarchical evaluation structure of index system is proposed. Then, a qualitative index's quantified method based on fuzzy statistics and expert evaluation is put forward. Afterwards, a weighting method for indexes based on maximum entropy rule is analyzed. Finally, the effectiveness of this comprehensive evaluation strategy is illustrated through calculation examples.

2. Technological Process Chart of Comprehensive Evaluation Strategy

The technological process chart of comprehensive evaluation strategy is proposed first in fig.1, in which the core methods are multilevel fuzzy statistics and maximum entropy criterion based weight design.

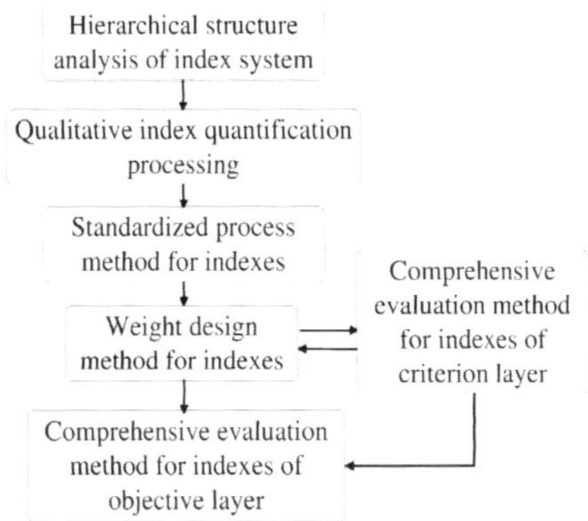

Figure 1. Technological process chart of comprehensive evaluation strategy.

Firstly, the hierarchical structure of index system is analyzed. Based on this, the value of comprehensive evaluation index is obtained using analytic hierarchy process. In this work, the index system is divided into three layers: objective layer, criterion layer and index layer. Secondly, the qualitative indexes are quantified using fuzzy statistics and expert evaluation method. The indexes are normalized using range transformation method. Afterwards, the weights of indexes are obtained using maximum entropy criterion based optimal design method. Finally, the values of comprehensive evaluation index for objective layer is calculated.

3. Hierarchical Structure Analysis of Index System

There are lots of influence factor and indexes of development level for global energy interconnection. It will be very complicated if all the indexes are calculated directly. Therefore the hierarchical structure for indexes is proposed in table1.

There are three layers in this hierarchical index system. Objective layer is the ultimate goal of evaluation and it is development level for global energy interconnection in this paper. Criterion layer is the detail description and extension of objective layer, which is divided into nine classes according the main influence factor. Index layer is the further detail decomposition for each influence factor in Criterion layer. The subordinate relation of each influence factor is determined based on the hierarchical structure. The essence of this development level evaluation is a multilevel and multi-factor comprehensive evaluation problem considering the influence of uncertainty.

Table 1. Hierarchical Structure of Indexes System

Objective layer	Criterion layer	Index layer	Index attribute
Development level for global energy interconnection	Concept promotion	Energy structure adjustment target	qualitative
		Electrification target	qualitative
		Emission reduction target	qualitative
		Interconnectivity target	qualitative

		Education degree	qualitative
		Domestic political stability	qualitative
	Technical level	Voltage level of main power gird	quantitative
		Technology level of ultra-high voltage(UHV)	qualitative
		Technology level of power forecasting for renewable energy	qualitative
		Technology level of renewable integration	qualitative
	Engineering construction	Number of interregional transmission lines	quantitative
		Capacity of interregional transmission lines	quantitative
		Presence of UHV engineering construction	quantitative
		Per capita transmission line length	quantitative
		Per capita installed capacity	quantitative
		Development level of renewable energy	qualitative
	Market operation	Power demand growth level	quantitative
		Opening degree of international market	qualitative
		Investment risk for electric power project	qualitative
	Extensive interconnection	Power gird covering range	quantitative
		Complementary capability	qualitative
		Import and export activity	quantitative
	Reliability and security	N-1 pass rate of transmission rate	quantitative
		Regulation capability	quantitative
		Power outage time	quantitative
	Economy and efficiency	Line loss rate	quantitative
		Wind power curtailment rate	quantitative
		Photovoltaic power curtailment rate	quantitative
		Generation cost for renewable energy	quantitative
	Emission reduce	Development plan for renewable energy	quantitative
		Investment for renewable energy	quantitative
		Per capita renewable energy installed capacity	quantitative
		Generation hours for renewable energy	quantitative
		Carbon emission for unit power generation	quantitative
		Per capita electric vehicle ownership	quantitative
		Per capita electric vehicle charging infrastructure	quantitative
	Smartness and interaction	Flexible pricing mechanism	quantitative
		Investment for smart grid	quantitative
		Demand response participation	quantitative
		Installed rate for smart meters	quantitative
		Interactive technology level	quantitative

4. Qualitative Index Quantification Method Based on Fuzzy Statistics

It is difficult to quantify the qualitative indexes in table1, which brings great difficulty to the comprehensive evaluation method. Therefore, a fuzzy statistical [4-6] based method is proposed here to quantify the qualitative indexes, considering the fuzzy linear transformation and subordinate degree synthesis principle.

The membership degree for each element is obtained using fuzzy test in fuzzy statistical method. There are four basic elements for fuzzy test: (1) Theory of domain U, which is the scope of researched problem. (2) A certain element u in U. (3) A fuzzy set A^* related to a common set of random motion A. (4) Condition S, which is linked to the objective or psychological factors of segmentation process for fuzzy concept.

There are three main steps for fuzzy statistical method. Firstly, the theory of domain U and influence factor set should be determined. Secondly, the test participants should vote for whether a certain point in U belongs to each element of influence factor set. Finally, analyzing the vote results, the membership frequency can be calculated using equation (1). As the increase of n, the membership frequency will become stable, which converge to the membership degree of u belonging to A_*.

$$f_n(u) = \frac{\theta(u)}{n} \tag{1}$$

Where $f_n(u)$ is the membership frequency of u belonging to A_*, $\theta(u)$ is the number of A_* covering u, n is the number of total test.

The score set can be divided into five grades first, which are "very good", "good","common","poor" and "very poor", respectively. Then, the membership degree of a qualitative index belongs to each score grade is obtained using fuzzy statistical method. The membership degree matrix of index system can also be obtained, shown in equation (2).

$$\begin{bmatrix} R_1 \\ R_2 \\ \cdots \\ R_g \end{bmatrix} = \begin{bmatrix} r_{11} & r_{12} & \cdots & r_{15} \\ r_{21} & r_{22} & \cdots & r_{25} \\ \vdots & \vdots & \cdots & \vdots \\ r_{g1} & r_{g2} & \cdots & r_{g5} \end{bmatrix} \tag{2}$$

Where R_g is the g-th qualitative index, r_{gm} is the membership degree of index g belongs to score grade m.

The membership degree matrix should be quantified to obtain the comprehensive score corresponding to each index. Firstly, each score grade is given a certain score value. Then, the comprehensive score is obtained through the multiplex operation between score vector and membership degree matrix, as in shown in equation (3).

$$\begin{bmatrix} F_1 \\ F_2 \\ \cdots \\ F_g \end{bmatrix} = \begin{bmatrix} r_{11} & r_{12} & \cdots & r_{15} \\ r_{21} & r_{22} & \cdots & r_{25} \\ \vdots & \vdots & \cdots & \vdots \\ r_{g1} & r_{g2} & \cdots & r_{g5} \end{bmatrix} \bullet \begin{bmatrix} c_1 \\ c_2 \\ \vdots \\ c_5 \end{bmatrix} = \begin{bmatrix} \sum_{i=1}^{5} r_{1i}c_i \\ \sum_{i=1}^{5} r_{2i}c_i \\ \vdots \\ \sum_{i=1}^{5} r_{5i}c_i \end{bmatrix} \tag{3}$$

Where c_1~ c_5 is the score value of each score grade, respectively. For the small value prefer indexes, such as investment risk, it is considered that a smaller index value corresponds to a higher score value. For the larger value prefer index, such as economic benefit, it is considered that a larger index value corresponds to a higher score value.

5. Index Normalization Processing Method Based on Range Transformation Method

In order to calculate and analyze the indexes with different dimension, all the indexes should be normalized. A range transformation method [7] is introduced here to normalize the indexes. For a positive index r_{ij} prefers high value, it can be normalized using equation (4). For a negative index r_{ij} prefers low value, it can be normalized using equation (5).

$$y_{ij} = \frac{x_{ij} - \min\limits_{1\le t\le n} x_{it}}{\max\limits_{1\le t\le n} x_{it} - \min\limits_{1\le t\le n} x_{it}} \left(1 \le i \le m, 1 \le j \le n\right) \tag{4}$$

$$y_{ij} = \frac{x_{ij} - \min\limits_{1\le t\le n} x_{it}}{\max\limits_{1\le t\le n} x_{it} - \min\limits_{1\le t\le n} x_{it}} \left(1 \le i \le m, 1 \le j \le n\right) \tag{5}$$

All the indexes normalized by this method satisfy the range of $0 \le y_{ij} \le 1$. In addition, no matter negative or positive index, the optimal value is 1and the worst value is 0.

6. Optimal Design of Index Weights Based on Maximum Entropy Criterion

The main factor affects the rationality the multiple indexes' comprehensive evaluation is the weight of each index. Index weight is the quantitative distribution of importance for evaluation object from different aspects. Through the reasonable setting of index weight, the real contribution degree of each index for evaluation object can be obtained. Comparing to the subjective weighting method, the

IOP Conf. Series: Earth and Environmental Science **168** (2018) 012044 doi:10.1088/1755-1315/168/1/012044

weighting method based on entropy has higher reliability. In addition, the computational complexity of weighting method based on entropy is lower comparing to most of objective weighting method.

A maximum entropy rule [8-9] based index weighting method is used here to determine the weights of indexes in index layer relative to the corresponding indexes in criterion layer, and the weights of indexes in criterion layer relative to the objective index. Based on the maximum entropy mathematical model and optimization method, the optimal weights can be obtained in equation (6).

$$w_j = \frac{\exp\left(-\left[1+\delta\sum_{i=1}^{n}\left(1-r_{ij}\right)^2/\left(1-\delta\right)\right]\right)}{\sum_{j=1}^{m}\exp\left(-\left[1+\delta\sum_{i=1}^{n}\left(1-r_{ij}\right)^2/\left(1-\delta\right)\right]\right)}, j=1,2,\text{L},m \quad (6)$$

Where n is the number of calculation sample, r_{ij} is the normalized index value of i-th index in calculation sample j. δ is a parameter that can be assigned of any value in the range $0 \le \delta \le 1$.

7. Calculation Examples

10 countries are considered as research objective in the calculation examples here, which are United States, Australia, Thailand, China, Russia, Brazil, Pakistan, Germany, Ethiopia and Saudi Arabia, respectively. The current development level of global energy interconnection construction in 2015 of each country is calculated using the method proposed above. The values of quantitative index are obtained using public data from internet.

7.1. Weights Analysis of Index Layer Relative to Criterion Layer

The weights of indexes in the index layer relative to corresponding index in the criterion layer are analyzed here using maximum entropy rule based method. Taking technical level and engineering construction for example, the weighting results are shown in fig.2 and fig. 3.

From fig.2, it is shown that the technology level of renewable integration has the most obvious influence on technical level index, while the voltage level of main power gird has the minimum influence. From fig. 3, it can be seen that the development level of renewable energy has the most obvious influence on engineering construction index, while the capacity of interregional transmission lines has the minimum influence.

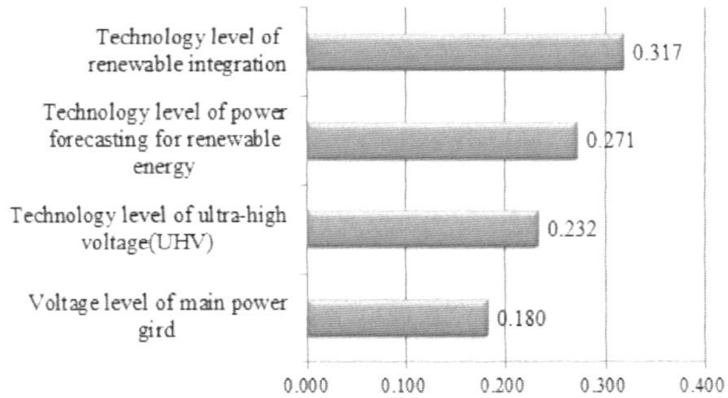

Figure 2. Weights of indexes relative to technical level.

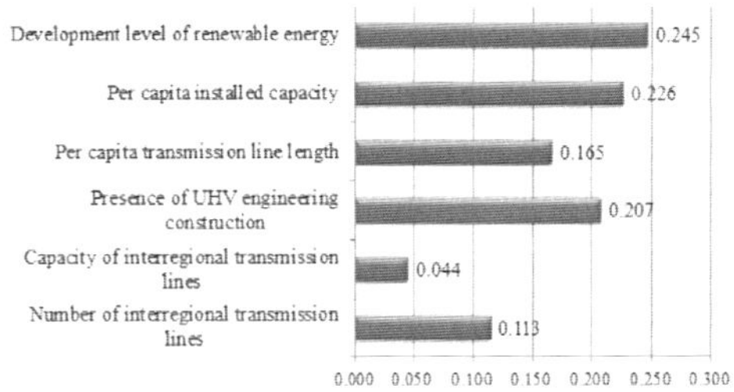

Figure 3. Weights of indexes relative to engineering construction.

7.2. Comprehensive Evaluation Value of Indexes in Criterion Layer

The comprehensive evaluation values of indexes in criterion layer for each country are calculated here. The results of United States, China, Ethiopia are given in fig. 4~5.From fig.4 and fig. 5, it can be seen that the nine criterion indexes of United States and China have reached a relatively high level, which reflects a relatively good development level of global energy interconnection in this country. However, from fig. 6, we can see that the nine criterion indexes of Ethiopia is very low, which reflects a very poor development level of global energy interconnection limited by poor domestic political and economic environment.

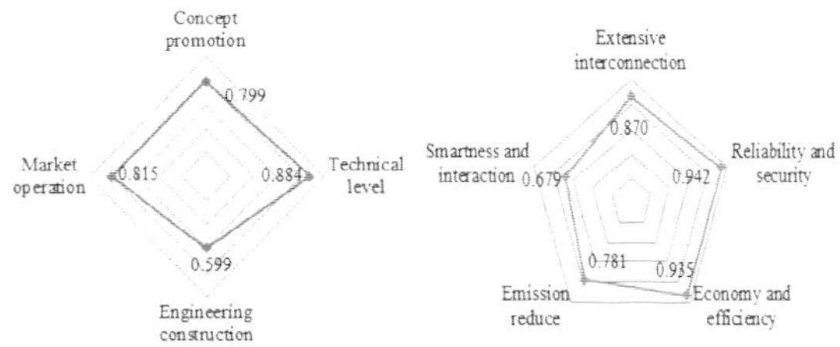

Effective indexes Executive indexes

Figure 4. Value of indexes in criterion layer for United States.

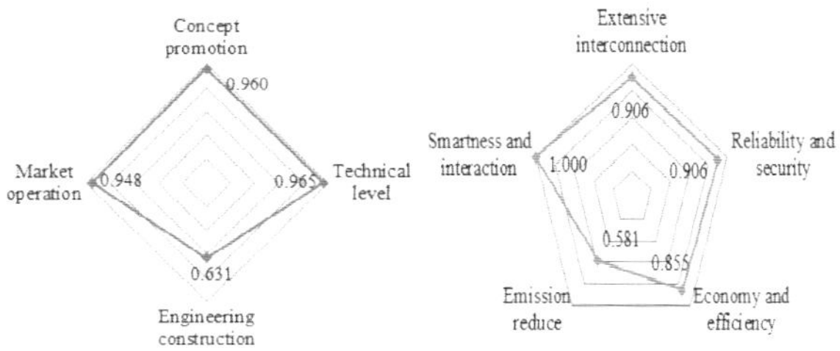

Effective indexes Executive indexes

Figure 5. Value of indexes in criterion layer for China.

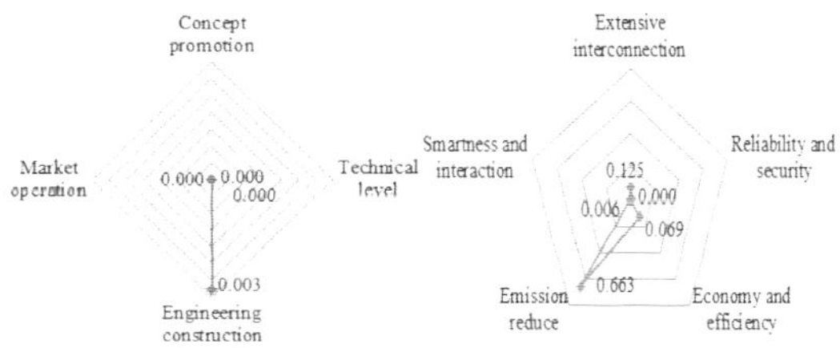

Effective indexes Executive indexes

Figure 6. Value of indexes in criterion layer for Ethiopia.

7.3. Analysis of Effective Index and Executive Index

The comprehensive evaluation values of effective index and executive index for each country are shown in fig.7, of which effective indexes including concept promotion, technical level, engineering construction and market operation, while executive indexes including extensive interconnection, reliability and security, economy and efficiency, emission reduce, smartness and interaction.

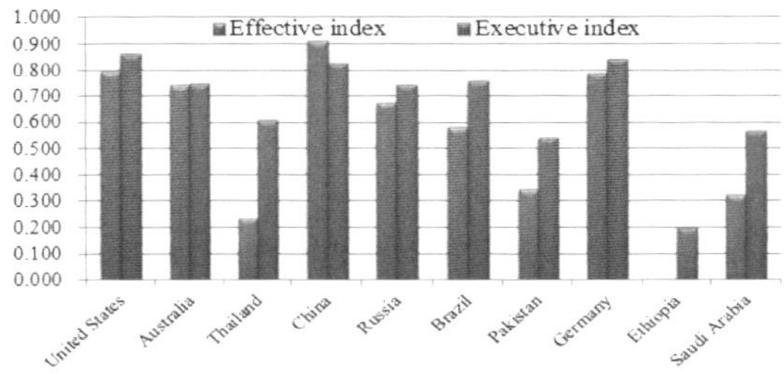

Figure 7. Effective and executive indexes for each country.

From the results in fig.7, it can be seen that both of the effective and executive indexes for United States, China and Germany have reached a relatively high level, which reflects that these countries have invested a lot in the construction process of global energy interconnection and these investments have achieved good effect.

The values of executive indexes for Thailand, Pakistan, Saudi Arabia and Brazil are higher than the effective indexes obviously, which shows that some investment of these countries in the construction process of global energy interconnection has not fully achieved benefit yet.

7.4. Developmet Level Evaluation Value of Global Energy Interconnection

The comprehensive development level evaluation values of global energy interconnection for ten countries are shown in fig.8. From the results we can see that the development level of global energy interconnection in China lies in the first place while Ethiopic lies in the last place. The development of these ten countries can divide into four classes: China, United States and Germany are in the first class with a very good development level. Australia, Russia and Brazil are in the second class with a good development level. Saudi Arabia, Thailand and Pakistan are divided in the third class with a common development level. While the development level of Ethiopia is very poor.

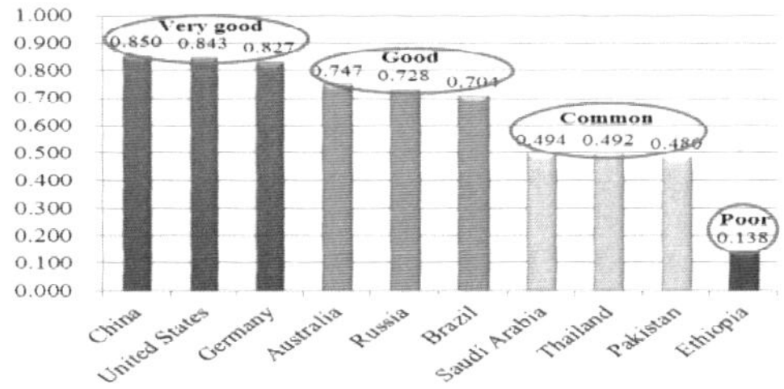

Figure 8. Developmet level evaluation value of global energy interconnection for each country.

8. Conclusion

Global energy interconnection is the next development stage of future power gird on the base of smart grid. The construction of global energy interconnection conforms to the development trend of world energy and power grid.

In order to identify the weakness in the development process of different countries, the development level global energy interconnection should be quantified. Therefore, a multi-level fuzzy statistics and maximum entropy rule based comprehensive evaluation strategy for development Level of global energy interconnection is proposed in this paper. The calculation examples for development level evaluation of ten typical countries illustrate the effectiveness of this strategy. From the results, the main factors that influence development level in terms of different aspects are identified through the weights analysis of index layer. The final evaluation values show that the development degree of global energy interconnection in China, United States and Germany has reached a very good level.

9. References

[1] Zhenya Liu. Gloval Energy Interconnection, Elsevier Press, 2015.
[2] IBM. Smart grid maturity model, presented in NWPPA E&O Conference, April, 2009.
[3] Electricity Advisory Committee. 2014 US Smart Grid System Report, 2014.
[4] Yanwei Xu,Chaohuan Hou and Shefeng Yan, "Fuzzy statistical normalization CFAR detector for non-rayleig data," IEEE transactions on aerospace and electronic systems, vol.51, Iss.1, pp. 383–396, 2015
[5] Y. Y. Chen. "Fuzzy analysis of statistical evidence," IEEE transactions on fuzzy systems, vol.8, Iss.6, pp. 796–799, 2000.
[6] O Hryniewicz. "Possibilistic decision and fuzzy statistical tests", Fuzzy sets and systems, vol. 157, Iss. 19, pp.2665-2673, 2006.
[7] Zhang Lijun,Yuan Nengwen. "Comparison and selection of index standardization metho in linear comprehensive evaluation medel(in Chinese)", Statistics and information forum, Vol.25, No.8, pp.10-15, Aug.2010.
[8] Ding Yuanming, Qiu Lei and Jia Liangchen."Combination weight model of evaluation index based on the maximum entropy theory(in Chinese)", Mathematics in practice and theory, Vol.44, No.4, pp.16-21, Feb.2014.
[9] Yuannian Li, Sifeng Liu and Zhigeng Fang. " Study on the weight of evalustion index based on the rule of maximum entropy", 2007 IEEE International Conference on Grey Systems and Intelligent Services, pp.1347-1351, 2007.

Acknowledgment
This work is supported by the technology project of State Grid Corporation of China 5230HQ160009.